PATHOGENESIS-RELATED PROTEINS in PLANTS

PATHOGENESIS-RELATED PROTEINS in PLANTS

edited by

Swapan K. Datta, Ph.D.
Plant Breeding, Genetics and Biochemistry Division
The International Rice Research Institute
Manila, Philippines

Subbaratnam Muthukrishnan, Ph.D.
Department of Biochemistry
Kansas State University
Manhattan, Kansas

CRC Press
Boca Raton London New York Washington, D.C.

Library of Congress Cataloging-in-Publication Data

Datta, S. K.
Pathogenesis-related proteins in plants / edited by Swapan K. Datta, Subbaratnam Muthukrishnan,
p. cm.
Includes bibliographical references and index.
ISBN 0-8493-0697-3 (alk. paper)
1. Plant proteins. 2. Plants – Diseases and pest resistance. 3. Plant diseases – Molecular aspects.
I. Datta, S. K. II. Muthukrishnan, Subbaratnam. III. Title.
QK898.P8D38 1999
572′. 62—dc21 99-12438
CIP

International Standard Book Number 0-8493-0697-3
Library of Congress Card Number 99-12438
Printed in the United States of America 2 3 4 5 6 7 8 9 0
Printed on acid-free paper

The Editors

Swapan K. Datta, Ph.D., pioneered genetic engineering in indica rice in 1990 while working at ETH-Zurich, Switzerland, in association with Professor Ingo Potrykus. Dr. Datta received his B.Sc. (Hons.) from Presidency College (1972), and M.Sc. and Ph.D. degrees from Calcutta University, India (1980). He joined RKMVC College in 1976 as a lecturer in botany and served as a lecturer and reader in botany at Visva-Bharati University, Santiniketan (1979–1985). He received a DAAD Fellowship (1985–1986) and worked in Germany on resistance genetics in barley and wheat, with Professor G. Wenzel. He was awarded the Friedrich Miescher Institüt (FMI) fellowship (1987) in Basel, Switzerland. He took the senior scientist position at ETH-Zurich, Switzerland (1987–1993) and worked on gene technology for crop plants and established genetically engineered rice with a number of genes including PR-genes in collaboration with Professor S. Muthukrishnan. He spent six months (1990) as a Visiting Associate Professor at University of California-Davis, USA. In 1993, he joined the International Rice Research Institute, Philippines, as a Senior Scientist–Tissue Culture Specialist.

He is associated with several professional societies, including the International Society for Plant Molecular Biology, International Association of Plant Tissue Culture. He has mentored 13 Ph.D. students, 7 postdoctoral associates, and has published/contributed about 100 research papers on various aspects of plant biotechnology dealing with haploids, chemodifferentiation, and genetic engineering of plants with abiotic and biotic stress-related protection genes, gene expression, function, and application for crop improvement.

Subbaratnam Muthukrishnan, Ph.D., received his B.Sc. (Chemistry) and M.Sc. (Biochemistry) degrees from Madras University, India, in 1963 and 1965, respectively, and his Ph.D. (Biochemistry) from the Indian Institute of Science in 1970. He carried out postdoctoral research at the University of Chicago (1971–1973), the Roche Institute of Molecular Biology (1973–1976), and the National Institutes of Health (1976–1980). He joined Kansas State University in 1980, where he is currently a Professor of Biochemistry. He has trained 12 M.S. students, 10 Ph.D. candidates, and 12 postdoctoral associates.

His research areas include mRNA processing in eukaryotes, hormonal regulation of barley α-amylase genes, insect chitinolytic genes and their regulation, and plant pathogenesis-related proteins. His laboratory has isolated and characterized several chitinase and thaumatin-like protein cDNAs from fungus-infected cereal plants. In collaboration with Dr. S. K. Datta, some of these genes have been introduced into rice plants to enhance their resistance to sheath blight. He has authored over 80 publications.

Contributors

Holger Bohlmann, Ph.D.
Swiss Federal Institute of Technology
Institüt für Pflanzenwissenschaften
Zurich, Switzerland

Willem Broekaert, Ph.D.
F.A. Janssens Laboratory of Genetics
Katholieke Universiteit Leuven
Heverlee-Leuven, Belgium

Annemarie S. Buchel, Ph.D.
Institute of Molecular Plant Sciences
Gorlaeus Laboratories
Leiden University
The Netherlands

C. Campargue, Ph.D.
Centre de Biologie et Physiologie Végétale
Université Paul Sabatier
Toulouse, France

Bushra Chaudhry, Ph.D.
Center for Excellence in Molecular Biology
Lahore, Pakistan

Jaishree M. Chittoor, Ph.D.
Department of Plant Pathology
Throckmorton Plant Sciences Center
Kansas State University
Manhattan, Kansas, U.S.A.

Alexandre da Silva Conceicão, Ph.D.
F.A. Janssens Laboratory of Genetics
Katholieke Universiteit Leuven
Heverlee-Leuven, Belgium

Karabi Datta, Ph.D.
Plant Breeding, Genetics, and Biochemistry Division
IRRI
Manila, Philippines

Swapan K. Datta, Ph.D.
Plant Breeding, Genetics, and Biochemistry Division
IRRI
Manila, Philippines

Marie-Thérèse Esquerré Tugayé, Ph.D.
CNRS–Université Paul Sabatier
Pole de Biotechnologie Végétale
Castanet
Toulouse, France

Bernard Fritig, Ph.D.
Institut de Biologie Moleculaire des Plantes
CNRS
Strasbourg, France

Pierette Geoffroy, Ph.D.
Institut de Biologie Moleculaire des Plantes
CNRS
Strasbourg, France

Thierry Heitz, Ph.D.
Institut de Biologie Moleculaire des Plantes
CNRS
Strasbourg, France

Anders Bøgh Jensen, Ph.D.
Molecular Biology Institute
Copenhagen University
Denmark

Jan E. Leach, Ph.D.
Department of Plant Pathology
Throckmorton Plant Sciences Center
Kansas State University
Manhattan, Kansas, U.S.A.

Robert Leah, Ph.D.
Carlsberg Research Laboratory
Valby, Denmark

Michael Legrand, Ph.D.
Institut de Biologie Moleculaire des Plantes
CNRS
Strasbourg, France

Gerhard Leubner-Metzger, Ph.D.
Friedrich Miescher-Institut
Basel, Switzerland

Huub J.M. Linthorst, Ph.D.
Institute of Molecular Plant Sciences
Gorlaeus Laboratories
Leiden University
The Netherlands

D. Mazau, Ph.D.
Centre de Biologie et Physiologie Végétale
Université Paul Sabatier
Toulouse, France

Frederick Meins, Jr., Ph.D.
Friedrich Miescher-Institut
Basel, Switzerland

John Mundy, Ph.D.
Molecular Biology Institute
Copenhagen University
Denmark

Subbaratnam Muthukrishnan, Ph.D.
Department of Biochemistry
Kansas State University
Manhattan, Kansas, U.S.A.

Jean-Marc Neuhaus, Ph.D.
Laboratoire de Biochimie
Université de Neuchatel
Switzerland

Leendert Cornelis van Loon, Ph.D.
Department of Plant Ecology & Evolutionary Biology
Utrecht University
The Netherlands

Rethinasamy Velazhahan, Ph.D.
Department of Plant Pathology
Tamil Nadu Agricultural University
Coimbatore, India

Frank F. White, Ph.D.
Department of Plant Pathology
Throckmorton Plant Sciences Center
Kansas State University
Manhattan, Kansas, U.S.A.

Jian-Min Zhou, Ph.D.
Department of Plant Pathology
Throckmorton Plant Sciences Center
Kansas State University
Manhattan, Kansas, U.S.A.

Preface

Since the discovery of pathogenesis-related (PR) proteins nearly 30 years ago in tobacco plants reacting hypersensitively to infection by tobacco mosaic virus, a great deal of research has focused on the isolation, characterization, and regulation of expression of this unique class of defense proteins in a variety of plants including rice. The studies on PR and PR-like proteins have literally ballooned in the past 10 years. The major impetus for the acute interest in this area has come from the realization that several of the PR proteins had antimicrobial or insecticidal activity which could be demonstrated in the laboratory. While the early studies on transgenic plants in which PR proteins were over-expressed represented a mixed bag of positive and negative results, more recent studies have led to the understanding that PR proteins, especially in combinations, can delay the progression of diseases caused by several pathogens belonging to diverse genera. Although there can be some selectivity in the interaction between a PR protein and its intended target pathogen, the notion that PR proteins represent generalized plant defense responses for broad-range, albeit incomplete, protection against a range of pathogens is gaining ground. Indeed, many groups are developing transgenic plants with multigenic resistance based on combinations of PR proteins and other antimicrobial or insecticidal proteins. In fact, we are at the threshold of a very exciting period where constitutive (or inducible) expression of PR proteins at effective levels could be used as a tool to enhance or stabilize yield in areas where pathogens and pests are endemic. For this reason we felt that it was timely to write a book devoted entirely to PR proteins in plants which would serve as a textbook for students and researchers and also as a convenient reference source for investigators in this field.

Along with the practical aspects of employing PR proteins for plant protection, basic studies have led to some very useful insights into the mode of action of at least some PR proteins. In addition to a possible role as the first or last line of defense against pathogens and pests, PR-like proteins which are expressed in apparently healthy tissues during normal plant growth, such as seed development and flowering, may have additional unsuspected roles in morphogenesis or in symbiosis. Some of these aspects are also dealt with in various chapters in this book.

A few words about the content and organization of the chapters are in order. The first chapter by Dr. van Loon describes the distribution and classification of PR proteins, which has eliminated a great deal of confusion and brought some order to the nomenclature of PR proteins. A general summary of the known and potential role of PR proteins is also presented in this chapter. Chapters 2 through 7 deal with individual PR proteins belonging to PR-1 to PR-11 with the exception of class 10. We apologize for our inability to recruit someone to write about this important class of PR proteins. However, in our opinion, this deficiency is more than compensated by the inclusion of several chapters (10 through 12) dealing with other plant defense proteins including thionins, defensins, ribosome-inactivating proteins, and plant cell wall hydroxyproline-rich glycoproteins. Chapter 9, by Dr. Jian-Min Zhou, deals with signal transduction mechanisms in PR protein synthesis and provides several useful insights into resistance mechanisms and plant defense responses. Finally, Chapter 13, by Dr. Karabi Datta et al., summarizes work in which PR-proteins have been utilized to enhance plant resistance to diseases and insects.

We have had our ups and downs in bringing out this volume. Throughout this endeavor, we have received enthusiastic support from our contributors. Without their patience and willingness to put up with some unexpected delays, this volume would not have been feasible. We are very fortunate to have received contributions from so many pioneers who have continued to be at the cutting edge of research in this field. They have done an outstanding job in presenting the data in a readable and thoughtful manner. We are confident you will find this book a pleasure to read. Our grateful thanks go to all the authors.

Our thanks go also to John Sulzycki of CRC Press, who decided to publish this book on short notice after our original publisher could not fulfil their agreement. We believe that this might turn out to be a fortunate outcome because we can expect a more widespread circulation of this book by CRC Press. Our special thanks to Lina Torrizo, Michelle Viray, and Dr. Karabi Datta at the International Rice Research Institute who helped assemble, revise, and reformat the manuscripts.

Swapan K. Datta
Subbaratnam Muthukrishnan

Contents

Chapter 1
Occurrence and Properties of Plant Pathogenesis-Related Proteins........................ 1
Leendert Cornelis van Loon

Chapter 2
PR-1: A Group of Plant Proteins Induced Upon Pathogen Infection 21
Annemarie S. Buchel and Huub J.M. Linthorst

Chapter 3
Functions and Regulation of Plant β-1,3-Glucanases (PR-2) 49
Gerhard Leubner-Metzger and Frederick Meins, Jr.

Chapter 4
Plant Chitinases (PR-3, PR-4, PR-8, PR-11)... 77
Jean-Marc Neuhaus

Chapter 5
The PR-5 Family: Thaumatin-Like Proteins.. 107
Rethinasamy Velazhahan, Swapan K. Datta, and Subbaratnam Muthukrishnan

Chapter 6
The PR-6 Family: Proteinase Inhibitors in Plant–Microbe and Plant–Insect Interactions... 131
Thierry Heitz, Pierette Geoffroy, Bernard Fritig, and Michael Legrand

Chapter 7
The Response of Plant Cell Wall Hydroxyproline-Rich Glycoproteins to Microbial Pathogens and Their Elicitors 157
Marie-Thérèse Esquerré-Tugayé, C. Campargue, and D. Mazau

Chapter 8
Induction of Peroxidase During Defense Against Pathogens.............................. 171
Jaishree M. Chittoor, Jan E. Leach, and Frank F. White

Chapter 9
Signal Transduction and Pathogen-Induced PR Gene Expression 195
Jian-Min Zhou

Chapter 10
The Role of Thionins in the Resistance of Plants 207
Holger Bohlmann

Chapter 11
Ribosome-Inactivating Proteins: Structure, Function, and Engineering 235
Anders Bøgh Jensen, Robert Leah, Bushra Chaudhry, and John Mundy

Chapter 12
Plant Defensins 247
Alexandre da Silva Conceicão and Willem F. Broekaert

Chapter 13
Expression and Function of PR-Protein Genes in Transgenic Plants 261
Karabi Datta, Subbaratnam Muthukrishnan, and Swapan K. Datta

Index 279

1 Occurrence and Properties of Plant Pathogenesis-Related Proteins

Leendert Cornelis van Loon

CONTENTS

1.1 Pathogenesis-Related Proteins Are Plant Proteins Induced in Pathological or Related Situations 1
1.2 Occurrence and Properties of PRs and PR-Like Proteins 5
1.3 Possible Functions of PRs 9
References 12

1.1 PATHOGENESIS-RELATED PROTEINS ARE PLANT PROTEINS INDUCED IN PATHOLOGICAL OR RELATED SITUATIONS

Pathogenesis-related proteins were first observed as new protein components induced by tobacco mosaic virus (TMV) in hypersensitively reacting tobacco.[1,2] This discovery was a direct consequence of the introduction of polyacrylamide gel electrophoresis, which allowed complex mixtures of proteins to be separated with unprecedented resolution on the basis of their combination of size and charge.[3,4] Soluble protein extracts from tobacco (*Nicotiana tabacum*) leaves were resolved into over 30 bands, with additional separation being obtained by varying the concentration of acrylamide in the gel.[5] In extracts from TMV-diseased leaves, one new band consisting of free viral coat protein oligomers was identified, but no additional qualitative changes were apparent.[1,6] In contrast, development of necrotic local lesions in *N*-gene-containing tobacco cultivars in response to TMV infection was associated with the appearance of at least four new bands, designated as new protein components I, II, III, and IV in order of increasing electrophoretic mobility,[1] or b_1, b_2, b_3, and b_4 in order of decreasing mobility.[2] Figure 1.1 illustrates the original observation, depicting densitometer tracings of the protein patterns of leaves from healthy and infected plants, as first described by Van Loon and Van Kammen.[1] In the subsequent paper

0-8493-0697-3/99/$0.00+$.50

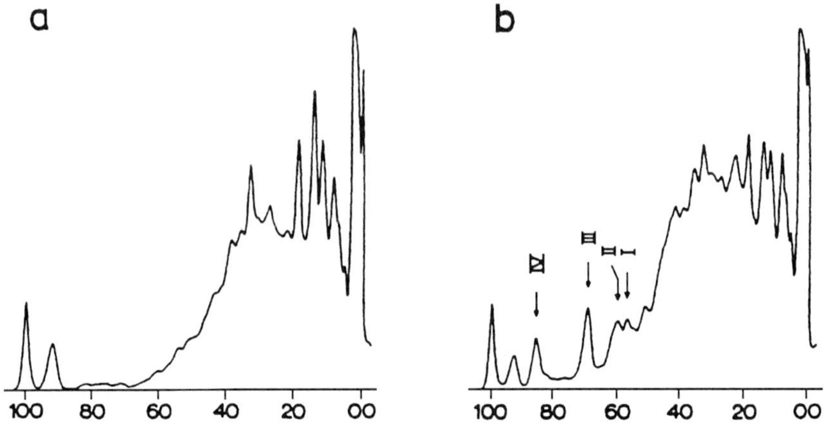

FIGURE 1.1 The first representation of pathogenesis-related proteins: densitometer tracings of electrophoretic patterns of leaf extracts from (a) water-inoculated and (b) 7-day TMV-infected Samsun NN tobacco, as described by Van Loon and Van Kammen.[1] The peaks denoted IV, III, II, and I correspond to PR-1a, -1b, -1c, and -2a, respectively.

by Gianinazzi et al.[2] the presence of additional b-proteins (b_5–b_8) was inferred, but those remained essentially unstudied for over a decade.

The same proteins were induced in tobacco by different viruses, whereas a different protein appeared upon TMV infection in *N. glutinosa*,[1,7,8] indicating that these new proteins were host specific. Substantial amounts accumulated in host–virus combinations giving rise to a hypersensitive reaction resulting in local necrotic lesions or to systemic necrosis, whereas none or only small amounts were induced under conditions where plants reacted with chlorotic or mosaic-type symptoms.[8,9] Thus, it was demonstrated that changes were connected with the type of symptoms produced, rather than with the genetic constitution of the host plant.[9]

Because these new protein components were found in small amounts in senescing leaves of flowering plants[10] and in progressively larger quantities when necrosis was more severe, it was assumed by many that these polypeptides were stable proteolytic breakdown products of larger leaf proteins. According to this hypothesis, their accumulations of up to 1% each of the soluble protein present would have required degradation of one or more abundant leaf proteins. The main leaf protein, ribulosebisphosphate carboxylase (Rubisco), which makes up about 50% of the protein in photosynthetically active tissues, decreased substantially in mosaic-diseased leaves, but none of the new protein components appeared under those conditions.[1] In necrotizing tissues, Rubisco hardly decreased, if at all. Moreover, when purification of the proteins allowed their amino acid composition to be determined, components IV, III, and II (b_1, b_2, b_3) were shown to have similar molecular weights and amino acid compositions, the latter differing from that of Rubisco in being particularly rich in acidic amino acids and tyrosine.[11] Component I (b_4) clearly differed from the other three proteins. This suggested that IV, III, and II constituted a protein family with structural, and possibly functional, relationships, whereas component I represented a different type of protein.

TABLE 1.1
Plant Species in Which Pathogenesis-Related Proteins Have Been Identified

Family	Species
Amaranthaceae	*Gomphrena globosa*
Chenopodiaceae	*Chenopodium amaranticolor, C. quinoa,* sugar beet
Compositae	*Gynura aurantiaca,* sunflower
Cruciferae	*Arabidopsis thaliana, Brassica nigra,* oilseed rape, radish
Cucurbitaceae	Cucumber
Gramineae	Barley, maize, oats, rice, wheat
Malvaceae	Cotton
Papilionaceae	Alfalfa, asparagus bean, bean, chickpea, cowpea, groundnut, lima bean, *Lablab purpureus,* pea, soybean, white lupin
Pinaceae	*Picea abies*
Rutaceae	Citron, *Citrus sinensis*
Solanaceae	Pepper, petunia, potato, *Solanum demissum, S. dulcamara, tobacco, Nicotiana debneyi, N. glutinosa, N. langsdorfii, N. plumbaginifolia, N. rustica, N. sylvestris, N. tomentosiformis,* tomato
Umbelliferae	Celery, parsley
Vitaceae	Grapevine

Recognition of the accumulation of the newly induced proteins in hypersensitively reacting tobacco prompted research into the more general occurrence of such proteins, their biochemical properties, induction mechanism, genetic diversity, and functional significance in infected plants. Aided by selective extraction methods,[12-15] it was established soon that not only viruses, but also fungi and bacteria were able to induce similar new protein components in various plant species, particularly in incompatible combinations resulting in hypersensitive necrosis (Table 1.1).[16] Although referred to by different names and symbols, these proteins resembled each other in that they apparently consisted of one or more families of host-coded proteins, were induced by different types of pathogens and abiotic stresses, and were most often of relatively low molecular weight, preferentially extracted at low pH, highly resistant to proteolytic degradation, and localized predominantly in the intercellular space of the leaf.[16,17] In order to unify the naming of these proteins, to describe the full set of novel proteins in the reaction of a plant to a pathogen, and to facilitate their study and explore their role in pathology and host defense, the term *pathogenesis-related proteins* (abbreviated: PRs) was introduced in 1980 to designate "proteins coded for by the host plant but induced only in pathological or related situations."[11]

According to this definition, any host protein induced by any type of infectious agent or comparable condition would qualify as a PR. Hence, the term *pathological situations* is taken to refer to all types of infected states, not just to resistant, hypersensitive responses in which PRs are most common; they also include parasitic attack by nematodes, phytophagous insects, and herbivores. Historically, abiotic pathological conditions have not been considered, but noninfectious ("physiological") types of necrosis or toxin-induced chlorosis and necrosis have been found to induce PRs and, thus, can be categorized either as a pathological or as a related situation. Although the term *pathogenesis-related proteins* by itself may give the impression that (all) proteins common to pathological processes in a given plant species can be considered PRs, proteins that are constitutively present in low, yet readily detectable amounts, but which are increased substantially by pathogen infection, such as oxidative and hydrolytic enzymes or general enzymes of aromatic biosynthesis, have been excluded. However, if specific isoforms of such enzymes do not normally occur but are induced as a result of infection, such isoenzymes do belong to the PRs. The major criterion for inclusion among the PRs is that the protein concerned is newly expressed upon infection, although not necessarily in all pathological conditions. These considerations imply that the characteristics of the induction of PRs take priority over other identifying characteristics, such as chemical properties and cellular localization.[18]

Proteins that are induced exclusively during disease development in compatible host–pathogen combinations have hardly been considered as PRs. Rather, the "classical" PRs are a collective set of novel proteins associated with host defense mainly in incompatible interactions and, thus, are related primarily to a special type of pathogenesis, i.e., one culminating in the impediment of further pathogen progress. In compatible host–pathogen interactions, pathogenesis, symptom expression, and reduction of growth are generally more severe, and enzyme activities may be increased to high levels, but novel host proteins have been observed only occasionally and remain poorly characterized. Proteins expressed under such conditions seem to be rather specific, being induced under the influence of particular pathogens but not others. This situation contrasts with the massive, apparently coordinated, induction of a common set of PRs in virtually any incompatible interaction, which in some cases can amount to 10% of the total soluble protein of the leaf. For this reason, PRs have been implicated in the resistance response and, more specifically, in the acquired resistance that is induced accompanying necrotic lesion formation in various plant species. When plants are locally infected with a pathogen inducing hypersensitive necrosis, they acquire systemically enhanced resistance to subsequent infection by various types of pathogens, and this systemic acquired resistance (SAR) is associated with the induction of PRs in tissues distant from the original inoculation site.[1,7,8,19] Such observations suggested that PRs are defense proteins functioning in limiting pathogen multiplication and/or spread. Among the many metabolic alterations characteristic of a hypersensitive response, induction of PRs is a relatively late event,[14] and their contribution to resistance against the initial infection is likely to be limited. In contrast, PRs induced in noninfected, distant leaves as a result of primary infection might afford an enhanced level of protection, resulting in reduced size or number of lesions, or even in reduced systemic symptom severity, upon further infection.

The "related situations" in which PRs were found to be induced, seem to prove the point: application of chemicals that mimic the effect of pathogen infection or induce some aspects of the host response, as well as wound responses that give rise to proteins that are also induced during infections, can induce both PRs and acquired resistance. This is the case for salicylic acid,[20,21] for example, which has since been shown to be an essential intermediate in the signal–transduction pathway leading to the induction of PRs in tobacco.[22,23] Similarly, wound-induced proteinase inhibitors in tomato and potato[24] inevitably accompany infections resulting in tissue damage. Hybrids of *N. glutinosa* and *N. debneyi* express a PR constitutively and are exceptionally resistant to virus infection.[25]

1.2 OCCURRENCE AND PROPERTIES OF PRs AND PR-LIKE PROTEINS

On the basis of similarities in molecular weights, amino acid composition, and serological properties, and confirmed by nucleotide sequencing of corresponding cDNAs, the 10 major acidic PRs of tobacco were grouped into five families, designated PR-1 to -5.[26,27] This classification has set a convenient standard for other plant species, in which PRs with properties homologous to the tobacco PRs are now similarly designated by these family numbers. To accommodate further classes of PRs with different properties, additional families were adopted. In 1994 a unifying nomenclature was proposed based on the grouping of PRs into plant-wide families sharing amino acid sequences, serological relationship, and/or enzymatic or biological activity. Eleven families have been recognized (Table 1.2) and classified for tobacco and tomato.[28] The current nomenclature of the PRs serves as an example that is being extended to other plant species, e.g., *Arabidopsis,* barley, bean, cucumber, and maize. The families are numbered and the different members within each family are assigned letters according to the order in which they are described. Thus, the same designation for a PR in different plant species does not necessarily imply that they are the same protein. They must belong to the same PR-family (number), but the lettering only reflects how many proteins of that family had been identified within those plant species previous to their discovery.

PR-genes are designated as *ypr* (Table 1.2), followed by the same suffix, in accordance with the recommendations of the Commission for Plant Gene Nomenclature. Because PRs are generally defined by their occurrence as protein bands on

TABLE 1.2
Recognized Families of Pathogenesis-Related Proteins

Family	Type member	Properties	Gene symbols
PR-1	Tobacco PR-1a	Unknown	*ypr1*
PR-2	Tobacco PR-2	β-1,3-glucanase	*ypr2, [gns2]*
PR-3	Tobacco P, Q	Chitinase type I, II, IV, V, VI, VII	*ypr3, chia*
PR-4	Tobacco R	Chitinase type I, II	*ypr4, chid*
PR-5	Tobacco S	Thaumatin-like	*ypr5*
PR-6	Tomato inhibitor I	Proteinase-inhibitor	*ypr6, pis ('pin')*
PR-7	Tomato P_{69}	Endoproteinase	*ypr7*
PR-8	Cucumber chitinase	Chitinase type III	*ypr8, chib*
PR-9	Tobacco "lignin-forming peroxidase"	Peroxidase	*ypr9*
PR-10	Parsley "PR1"	"Ribonuclease-like"	*ypr10*
PR-11	Tobacco class V chitinase	Chitinase, type I	*ypr11, chic*

gels and classified within each family in the order in which new members are described, cDNA or genomic sequences without information on the corresponding protein cannot be fitted into the adopted nomenclature. Conversely, homologies at the cDNA or genomic level may be encountered without information on the expression or characteristics of the encoded protein. Thus, for naming, it is necessary to gather information at both the nucleic acid and the protein level when dealing with a stress-related protein falling within the definition of PRs. A type member has been defined within each PR family (Table 1.2) the nucleotide sequence of the mRNA of which can be used in the search for homologs in the same or in different plant species. Newly defined mRNAs (cDNAs) may be added to the existing families when shown to be induced by pathogens or specific elicitors. Thionins[29] and defensins,[30] both families of small, basic, cysteine-rich polypeptides, qualify for inclusion as new families of PRs.

Coordinate expression of PR-mRNAs in response to infection may also involve other, similarly inducible, mRNAs, the corresponding proteins of which have not (yet) been observed. For example, an mRNA of about 650 nucleotides was induced in hypersensitively reacting tobacco and deduced to code for a glycine-rich protein.[31] The protein could not be detected, however, and was assumed to become incorporated into cell walls. If identified, it would have been classified along with the other PR-families, but because information on the protein itself is lacking, it cannot be: induction of the mRNA only is not sufficient for inclusion.

Demonstration that the PR-2 family are β-1,3-endoglucanases[32] and the PR-3, -4, -8, and -11 families consist of chitinases with (PR-8) or without lysozyme activity,[33-36] immediately suggested that these PRs are directed against cell walls of fungi and bacteria. Homology of the thaumatin-like proteins of the PR-5 family with a bifunctional α-amylase/trypsin inhibitor from maize seeds[37] seemed consistent with a role in protection against phytophagous insects. However, no proteinase-inhibitor activity has been demonstrated for PR-5 proteins. Resistance to insect attack is taken to be conferred primarily by wound-inducible proteinase inhibitors, which have now been grouped into the family PR-6.[28] Because acquired resistance is active against all types of pathogens, it seemed logical to assume that one or more of the further PR-families might be involved in resistance to viruses and viroids.

Localization of the major, acidic PRs in the intercellular space of the leaf seems to guarantee contact with invading fungi or bacteria before these are able to penetrate. However, in spite of numerous investigations involving injection of plants with purified proteins, *in vitro* tests for inhibition of pathogen growth, or *in vivo* infection of transgenic plants, few of the inducible acidic PRs associated with SAR have been shown to possess significant anti-pathogenic activity.[38,39] Based on these observations, their functioning as general defense proteins now appears less likely. In contrast, hypersensitive resistance to viruses is a cellular phenomenon,[14] acting not at the level of virus entry but limiting viral multiplication in the cytoplasm, and viral spread through symplastic connections. It is difficult to see how extracellular proteins might selectively influence intracellular viral replication and transport. It could be that PRs make cells less conducive, but any such evidence is entirely lacking. Van Loon and Antoniw[21] observed an association between the presence of PRs and decreased TMV multiplication and spread in inoculated leaves of Samsun tobacco, which reacts to the virus with systemic mosaic symptoms. However, under other conditions, the presence of PRs did not reduce viral multiplication or impede systemic spreading.[40-42] Moreover, in tobacco, synthesis of PRs is initiated only after resistance reactions have become operative. There is no quantitative relationship between the amount of PR-1 proteins, for example, and resistance in primarily inoculated leaves, and systemic acquired resistance appears to develop well before significant amounts of PRs are detectable in distant leaves.[43] Also, transformed tobacco plants constitutively expressing PR-1 or PR-5 were not more resistant to TMV than untransformed controls,[44] and neither were they resistant to the leaf-eating insects *Spodoptera exigua* and *Heliotis virescens.*[45] Reports that members of the PR-2 and PR-5 family have antiviral activity[46] have not been substantiated. On the contrary, callose deposited in reaction to virus infection and acting as a physical barrier to virus spread is a substrate for β-1,3-glucanase, and induction of PR-2 might reduce callose accumulation and promote virus replication and spread.[47]

In tomatoes infected with the fungus *Cladosporium fulvum,* the major PR-1 protein accumulates early in incompatible combinations, but induction also occurs in compatible interactions. Although in the latter case the protein appears later and is synthesized more slowly, it may eventually reach levels even higher than in the incompatible situation.[48] Similar observations were made repeatedly by other workers and strongly suggest that PRs do not determine the resistance response, and at most have only an accessory role. Recently, a rhizobacterially mediated induced systemic resistance phenotypically similar to SAR was demonstrated in radish and *Arabidopsis* that was not accompanied by an accumulation of PRs.[49,50] Thus, inducible PRs are not required for the enhanced defensive capacity characteristic of the state of induced resistance.[39]

PRs of the families -7 to -11 have been less extensively studied. Whereas chitinase families PR-3 and -4 correspond to the *chia* and *chid* gene families, PR-8 and -11 have now been equated with *chib* and *chic,* respectively (Table 1.2). The PR-7 endoproteinase is a major pathogen-inducible protein in tomatoes.[51] Specific peroxidase isoenzymes recognized as PRs in tobacco,[52] that are identical to or homologous with a lignin-forming peroxidase, have been classified as PR-9. The parsley "PR-1" protein[53] was taken as the type member of the PR-10 family. This protein turned out to belong to an extensive family of intracellular defense-related proteins with structural homology to ribonucleases.[54] The latter finding suggested that PR-10 proteins might be the long-sought inhibitor associated with SAR against viruses. Although some PR-10 type proteins are enzymatically active, the parsley protein did not display RNase activity and no PR-10-like homologs were detectable in tobacco with SAR induced by TMV.

Elucidation of the biochemical properties of the major, pathogen-inducible PRs of tobacco and subsequent cloning of their cDNAs and/or genes revealed that proteins with substantial similarity to the classical PRs can be present in healthy plants. The term PR-like proteins was proposed to accommodate proteins homologous to PRs as deduced from their amino acid sequence or on the basis of the nucleotide sequence of their corresponding cDNA or gene, but induced principally in a developmentally controlled, tissue-specific manner.[18] In contrast to the classical PRs, which are mostly acidic and extracellular proteins, the homologous counterparts

TABLE 1.3
Developmentally Regulated Induction of Pathogenesis-Related Proteins in Tobacco

PR	Organs	References
Acidic PR-1	Sepals	Lotan et al., 1989[57]
	Senescing leaves	Fraser, 1981[10]
Basic PR-1	Roots, flowers	Memelink et al., 1990[58]
Acidic PR-2	Pistil	Lotan et al., 1989[57]
	Sepals	Coté et al., 1991[114]
Basic PR-2	Roots, maturing leaves	Shinshi et al., 1987[56]
	Stem, flowers	Memelink et al., 1990[58]
	Seeds	Vögeli-Lange et al., 1994[112]
Acidic PR-3a	Pedicel, sepals, anthers, ovary	Lotan et al., 1989[57]
PR-3b	Sepals, anthers	Lotan et al., 1989[57]
Basic PR-3	Roots, maturing leaves	Shinshi et al.,1987[56]
	Stem, flowers	Memelink et al.,1987[58]
Basic PR-5	Roots, flowers	Neale et al., 1990[86]

are mostly basic and localized intracellularly in the vacuole.[27] As far as it has been possible to deduce, they possess the same type of enzymatic activities, but their substrate specificity and specific activity may be rather different.[32,33]

In the last few years, the distinction between PRs and PR-like proteins has become difficult to maintain. In several situations it is difficult to distinguish PRs from related proteins/mRNAs that are present in some organs or appear during specific developmental stages. As has been shown clearly in tobacco, the same proteins can be expressed in a developmentally controlled manner and also be induced in response to infection in the same organs, for example, basic (type I) glucanase and chitinase in mature leaves.[55,56] Or PRs and PR-like proteins may be expressed constitutively in some organs and be inducible by pathogens in other organs, i.e., acidic and basic glucanase and chitinase in floral organs and leaves, respectively.[57,58] Table 1.3 lists both the acidic and basic proteins of tobacco that, besides being coordinately induced by infection, are present in healthy plants in developmentally controlled, organ-specific manners. Because infection is necessary to trigger the appearance of the proteins in other organs or in the same organs at a different stage of development, those PR-like proteins are now also regarded as PRs.

Meanwhile, the term *PRs* or *PR-proteins* has been widely adopted to designate sequences from various plant species that share sequence homology with established PRs, or proteins with some of the characteristics of PRs, even though induction by pathogens or related stresses has not been investigated. Such proteins cannot be considered to be "induced only in pathological or related situations," and the term *PR-like proteins* can be appropriately retained for the latter. Thus, homologous proteins/mRNAs in healthy tissues in which no induction by pathogen infection has (yet) been demonstrated are to be termed PR-like proteins (PRLs) and their genes *yprl*. Many seed-specific proteins fall within this category. In other cases, occurrence of PR sequences has only been shown at the mRNA (cDNA) level, and the presence of the protein itself remains to be demonstrated.

In tobacco, the developmentally controlled, basic PRs appear to share all characteristics with the acidic ones, except for their cellular localization. However, in contrast to the acidic PRs, their induction by pathogens is not systemic,[59] and they are more prone to being induced by various stress conditions, particularly the stress hormone ethylene. Since ethylene is produced as an early response during incompatible reactions,[60,61] the vacuolar PRs were proposed to function as a last line in primary defense, to be released when the attacked host cells lyse.[62]

Various basic PRs have indeed been shown to reduce infection by selected pathogens.[39] Both basic tobacco PR-1g and basic PR-5c have antifungal activity against selected oomycetes.[63-65] Notably, simultaneous expression of the basic tobacco glucanase PR-2e and chitinase PR-3d in tomatoes affords protection against *Fusarium oxysporum* f.sp. *lycopersici*.[66] Similarly, co-expression of glucanase and chitinase genes in tobacco enhanced resistance against *Cercospora nicotianae*.[67,68] Moreover, tobacco chitinase has been reported to synergistically enhance the antifungal action of a nettle lectin *in vitro*.[69] However, it is difficult to understand how intravacuolar PRs could exert antifungal activity at an early stage of infection, because fungal pathogens would only be exposed to these PRs after defense reactions that lead to cellular disruption have been initiated. Therefore, it is questionable whether the vacuolar proteins have a general defensive role. Increased levels of resistance to TMV, comparable to resistance obtained by previous infection with the virus, were found in UV-irradiated leaves but not upon wounding or treatment with ethylene, both of which induced high levels of basic PR-mRNAs.[59] Similarly, transformation of *N. sylvestris* with a basic chitinase gene did not substantially increase resistance to the chitin-containing fungus *C. nicotianae*.[70]

On the other hand, several PRs inhibit growth of selected fungi *in vitro*. Tests for plant proteins with antimicrobial activity have revealed various PR-like proteins, as well as thionins and defensins. PR-like chitinases and thaumatin-like proteins isolated from seeds of wheat, barley, oats, sorghum, and maize were found to be antifungal[71-73] and to display synergism.[72] The latter observation suggests that, while each PR may have an intrinsically small defensive capacity, in the reaction of a plant to fungi, and perhaps bacteria, coordinated induction of the different families of PRs may indeed confer a level of resistance considerably broader than any protein by itself. Nevertheless, most of the phytopathogenic fungi that have been tested *in vitro* are resistant to glucanases and chitinases. Apparently, fungal pathogens can secrete metabolites that inhibit selective plant enzymes.[74] Combination of several PRs might be necessary to reduce pathogen activity sufficiently to reduce or prevent disease development.

1.3 POSSIBLE FUNCTIONS OF PRs

The occurrence of homologous PRs as small multigene families in various plant species belonging to different families (Table 1.1), their tissue-specific expression during development and consistent localization in the apoplast as well as in the vacuolar compartment, and their differential induction by endogenous and exogenous signaling compounds suggest that PRs may have important functions extending beyond their apparently limited role in plant defense. It is interesting to note that PR-1 proteins show substantial sequence similarity to a mammalian androgen-dependent sperm-coating glycoprotein,[75] the major insect venom allergen of the hornet wasp,[76] proteins secreted by a fungus during formation of fruiting bodies,[77] and a protein expressed in mouse B cells under the control of the Oct2 transcription factor.[78] The functions of these proteins are unknown, but their occurrence in such diverse organisms is suggestive of some wider significance.

The cellular damage and death occurring during a hypersensitive reaction is a major stress to the plant, as exemplified by high increases in abscisic acid (ABA) and ethylene.[14] It is possible, therefore, that PRs are stress proteins directed to alleviate harmful effects of cellular degradation products on hitherto untouched neighboring cells. Both acidic and basic PRs may be induced by high concentrations of ethylene[59,79] or physiological necrosis,[80] plasmolysis (osmotic stress),[81,82] or wounding.[83,84] Typically, tobacco basic PR-5c (osmotin) is induced in leaves, stems, and roots by drought, high salt, or ABA,[85] as well as in leaves by wounding,[86] UV light, ethylene,[59] or jasmonic acid.[87] Such induction in the absence of pathogenic attack might be taken to indicate protection of cellular structures, either physically to stabilize sensitive membranes or macromolecules, or chemically to keep potentially harmful saprophytic microorganisms on tissue surfaces or in intercellular spaces in check. In virtually any natural stress condition, e.g., heat, cold, drought, osmotic stress, water logging, anaerobiosis, metal toxicity, etc., plants are known to react by the synthesis of novel, and sometimes partly overlapping, sets of proteins.[88,89] The various conditions under which PRs occur are reminiscent of those under which heat-shock proteins (HSP) are induced. These proteins are ubiquitous in living organisms and associated with the acquisition of thermotolerance to otherwise lethal temperatures, but a causal connection is not evident.[90] HSP may display coordinate induction by heat shock, i.e., sudden transfer to 5 to 10°C above the normal growing temperature, variable induction by such stress conditions as osmotic stress, high salt, dinitrophenol, arsenite, cadmium, anaerobiosis, or high auxin,[91] and developmentally controlled induction such as during oogenesis and metamorphosis in insects and amphibia, and glucocorticoid-dependent processes in animals.[92] At least some of the HSP have been shown to act as molecular chaperones that bind other polypeptides to facilitate folding into their native structure, formation of oligomers, or localization to their correct intracellular compartment.[93] Interestingly, the promoters of all three tobacco PR-1 genes that are expressed, as well as of a different type of PR in parsley, contain a heat-shock regulatory element,[94,95] but the proteins are not induced to detectable levels by heat shock.[96] Nevertheless, PRs might have an analogous, though quite different, chaperonin-like function: unlike PRs, HSP are intercellular proteins that do not accumulate during heat shock. However, the specific occurrence of individual PRs in some floral organs, but not in others, points to other, more specific roles.

The relative ineffectiveness of PRs in determining resistance to pathogens does not preclude an involvement in defense. As first proposed by Mauch and Staehelin,[97] acidic, extracellular PRs might be involved in recognition processes, releasing defense-activating signal molecules from the walls of invading pathogens. This would hold particularly for chitinases and glucanases that could liberate elicitor-type carbohydrate molecules from fungal and bacterial cell walls. Thus, a β-1,3-glucanase induced in soybean seedlings by infection or chemical stress releases elicitor-active fragments from cell wall preparations of the fungus *Phytophthora megasperma* f.sp. *glycinea*.[98,99] Such elicitors could help stimulate defense responses in adjacent cells and thus accelerate and enhance these reactions, as well as induce acquired resistance to further infection.

It is well known that during the hypersensitive response many other biochemical alterations are coordinately induced that can contribute to, or may even determine the resistance attained, i.e., stimulation of aromatic biosynthesis, leading to the synthesis of phytoalexins and phenolic compounds that, through the action of increased oxidative enzyme activities, result in accelerated lignification, as well as further cell wall changes involving structural hydroxyproline-rich glycoproteins and glycine-rich proteins, that contribute to barrier formation.[100] Moreover, antiviral proteins inhibiting viral multiplication have been reported to be induced, which may or may not resemble the ribosome-inactivating proteins described to be present in the sap of some plant species that have antiviral activity.[101]

Exogenous elicitors cannot play a role in the induction of resistance by viruses. Nevertheless, the same physiological and biochemical changes occur during a hypersensitive reaction to fungi, bacteria, and viruses. Occurrence of endogenous substrates of PRs in plants might allow these proteins to function similarly through the release of endogenous elicitor-type signaling compounds. Specific oligosaccharides have been implicated as signal molecules in plant development and reactions to biotic and abiotic factors.[102] A role of PRs as specific internal signal-generating enzymes would be consistent both with their occurrence in specific organs and with their induction during development and in response to stressful pathogenic infections.

In the last few years, chitinase has been implicated as a factor regulating developmental processes. In the yeast *Saccharomyces cerevisiae,* a specific chitinase is secreted into the growth medium that is required for cell separation after division has taken place.[103] This activity hydrolyzes chitin in the septum that forms the temporary junction between mother and daughter cell. Dot matrix comparison has indicated that the catalytic domain of the yeast chitinase is highly homologous to the PR-8 type III chitinase of cucumber. Two regions are also conserved in several bacterial chitin-hydrolyzing enzymes and chitinase-like glycosidases, as well as the killer toxin of the yeast *Kluyveromyces lactis.* Since the latter acts on other micro-organisms, it was suggested that perhaps the secreted yeast chitinase plays a dual role: enabling cell separation to occur and suppressing the growth of other micro-organisms through hydrolysis of cell wall chitin or related polysaccharides.[103] Interestingly, the major chitinase of bean leaves, first described by Boller et al.[104] to be induced by ethylene and located in the vacuole, appears to also be induced in abscission zones at the stem–petiole junction,[105,106] together with a PR-1-like protein, two isoforms of β-1,3-glucanase, other chitinases, and a thaumatin-like protein.[106]

The specificity of *Rhizobium*–legume interactions may be in part due to differential inactivation of bacterial nodulation (nod) factors by root chitinases.[107] Thus,

PR-like chitinases can be implicated in the turnover of signal molecules acting as morphogenetic factors. In embryogenic cell suspensions of carrot, a specific chitinase with homology to PR-4 is necessary for somatic embryogenesis to proceed beyond the globular stage. A different chitinase homologous to PR-3 has similar activity.[108,109] These chitinases are only minor components compared with the total chitinase activity secreted into the growth medium and apparently have a special function, i.e., to release signal molecules from the cell walls. Rhizobial nod-factors can substitute for these signal molecules, supporting the notion of chitin oligomer signaling as a more general mechanism in plants.[110,111] However, the natural substrates for chitinases in higher plant cell walls remain to be determined. The tobacco PR-2 glucanases vary 250-fold in specific activity on laminarin,[32] and their relative activities on different substrates vary greatly, suggesting that their normal actions may be diverse. Expression studies of PR-2d in transgenic tobacco suggest that this protein functions developmentally in seed germination by weakening the endosperm, thus allowing the radicle to protrude.[112]

The occurrence of almost all types of PRs in various floral tissues (Table 1.3) also suggests specific physiological functions during flower development rather than a role in general defense against pathogen invasion. This notion is supported by the presence in floral organs of additional PR-like proteins, glucanases,[58,113,114] and thaumatin-like proteins.[115] In petunia flowers, chitinase activity is localized in the petals (about 15%) and stigma (about 85%). In the stigma, it increases about five-fold following anther dehiscence, strongly suggesting that the chitinase has a specific function in reproduction.[116]

If these cell-wall polymer-hydrolyzing enzymes release specific oligosaccharides, those might function as signal molecules: as morphogenetic factors during plant development and/or exogenous or endogenous elicitors after pathogen infection. Other PRs might be involved in protection of cellular structures and might contribute to the reduction of multiplication or spread of invading pathogens. This raises the question of whether PR-genes evolved primarily to limit damage inflicted by invading pathogens or were adapted from other functions to serve accessory protective roles. The occurrence of small multigene families appears to preclude the use of mutants to solve this problem. However, genetic engineering techniques using antisense RNAs to block PR-gene expression may provide some answers. It is not inconceivable, however, that other proteins that are either constitutively present or induced, provide a certain level of redundancy, and that a full understanding will have to await a better insight into the regulation of developmental processes and induced responses at the cellular level.

REFERENCES

1. Van Loon, L. C. and Van Kammen, A., Polyacrylamide disc electrophoresis of the soluble leaf proteins from *Nicotiana tabacum* var. "Samsun" and "Samsun NN" II. Changes in protein constitution after infection with tobacco mosaic virus, *Virology,* 40, 199, 1970.

2. Gianinazzi, S., Martin, C., and Vallée, J.-C., Hypersensibilité aux virus, température et protéines solubles chez le *Nicotiana Xanthi* n.c. Apparition de nouvelles macromolécules lors de la répression de la synthèse virale, *C.R. Acad. Sci. Paris,* 270D, 2383, 1970.
3. Ornstein, L., Disc electrophoresis I. Background and theory, *Ann. N. Y. Acad. Sci.,* 121, 321, 1964.
4. Davis, B. J., Disc electrophoresis II. Method and application to human serum proteins, *Ann. N. Y. Acad. Sci.,* 121, 404, 1964.
5. Van Loon, L. C. and Van Kammen, A., Polyacrylamide disc electrophoresis of the soluble leaf proteins from *Nicotiana tabacum* var. "Samsun" and "Samsun NN" I., *Phytochemistry,* 7, 1727, 1968.
6. Gianinazzi, S. and Vallée, J.-C., Température et synthèse de matériel protéique viral chez le Nicotiana Xanthi n.c. infecté par le virus de la mosaïque du tabac, *C.R. Acad. Sci. Paris,* 269D, 593, 1969.
7. Kassanis, B., Gianinazzi, S., and White, R. F., A possible explanation of the resistance of virus-infected tobacco to second infection, *J. Gen. Virol.,* 23, 11, 1974.
8. Van Loon, L. C., Polyacrylamide disc electrophoresis of the soluble leaf proteins from *Nicotiana tabacum* var. "Samsun" and "Samsun NN" IV. Similarity of qualitative changes of specific proteins after infection with different viruses and their relationship to acquired resistance, *Virology,* 67, 566, 1975.
9. Van Loon, L. C., Polyacrylamide disk electrophoresis of the soluble leaf proteins from *Nicotiana tabacum* var. "Samsun" and "Samsun NN" III. Influence of temperature and virus strain on changes induced by tobacco mosaic virus, *Physiol. Plant Pathol.,* 6, 289, 1975.
10. Fraser, R. S. S., Evidence for the occurrence of the "pathogenesis-related" proteins in leaves of healthy tobacco plants during flowering, *Physiol. Plant Pathol.,* 19, 69, 1981.
11. Antoniw, J. F., Ritter, C. E., Pierpoint, W. S., and Van Loon, L. C., Comparison of three pathogenesis-related proteins from plants of two cultivars of tobacco infected with TMV, *J. Gen. Virol.,* 47, 79, 1980.
12. Van Loon, L. C., Specific soluble leaf proteins in virus-infected tobacco plants are not normal constituents, *J. Gen. Virol.,* 30, 375, 1976.
13. Gianinazzi, S., Pratt, H. M., Shewry, P. R., and Miflin, B. J., Partial purification and preliminary characterization of soluble proteins specific to virus infected tobacco plants, *J. Gen. Virol.,* 34, 345, 1977.
14. Van Loon, L. C., Mechanisms of resistance in virus-infected plants, in *The Dynamics of Host Defense,* Bailey, J. A. and Deverall, B.J., Eds., Academic Press, North Ryde, NSW, Australia, 1983, 123.
15. Parent, J. G. and Asselin, A., Detection of pathogenesis-related proteins (PR or *b*) and of other proteins in the intercellular fluid of hypersensitive plants infected with tobacco mosaic virus, *Can. J. Bot.,* 62, 564, 1984.
16. Redolfi, P., Occurrence of pathogenesis-related (b) and similar proteins in different plant species, *Neth. J. Plant Pathol.,* 89, 245, 1983.
17. Van Loon, L. C., Stress proteins in infected plants, in *Plant-Microbe Interactions, Molecular and Genetic Perspectives,* Vol. 3, Kosuge, T. and Nester, E. W., Eds., McGraw-Hill Book Co., New York, 1989, 198.
18. Van Loon, L. C., The nomenclature of pathogenesis-related proteins, *Physiol. Mol. Plant Pathol.,* 37, 229, 1990.

19. Ward, E. R., Uknes S. J., Williams, S. C., Dincher, S. S., Wiederhold, D. L., Alexander, D. C., Ahl-Goy, P., Métraux, J.-P., and Ryals, J. A., Coordinate gene activity in response to agents that induce systemic acquired resistance, *Plant Cell,* 3, 1085, 1991.
20. White, R. F., Acetylsalicylic acid (aspirin) induces resistance to tobacco mosaic virus in tobacco, *Virology,* 99, 410, 1979.
21. Van Loon, L. C. and Antoniw, J. F., Comparison of the effects of salicylic acid and ethephon with virus-induced hypersensitivity and acquired resistance in tobacco, *Neth. J. Plant Pathol.,* 88, 237, 1982.
22. Yalpani, N., Silverman, P., Wilson, T. M. A., Kleier, D. A., and Raskin, I., Salicylic acid is a systemic signal and an inducer of pathogenesis-related proteins in virus-infected tobacco, *Plant Cell,* 3, 809, 1991.
23. Gaffney, T., Friedrich, L., Vernooij, B., Negrotto, D., Nye, G., Uknes, S., Ward, E., Kessmann, H., and Ryals, J., Requirement of salicylic acid for the induction of systemic acquired resistance, *Science,* 261, 754, 1993.
24. Green, T. R. and Ryan, C. A., Wound-induced proteinase inhibitor in plant leaves: a possible defense mechanism against insects, *Science,* 175, 776, 1972.
25. Ahl, P. and Gianinazzi, S., *b*-Protein as a constitutive component in highly (TMV) resistant interspecific hybrids of *Nicotiana glutinosa* x *Nicotiana debneyi, Plant Sci. Lett.,* 26, 173, 1982.
26. Van Loon, L. C., Gerritsen, Y. A. M., and Ritter, C. E., Identification, purification, and characterization of pathogenesis-related proteins from virus-infected Samsun NN tobacco leaves, *Plant Mol. Biol.,* 9, 593, 1987.
27. Linthorst H. J. M., Pathogenesis-related proteins of plants, *Crit. Rev. Plant Sci.,* 10, 123, 1991.
28. Van Loon, L. C., Pierpoint, W. S., Boller, T., and Conejero, V., Recommendations for naming plant pathogenesis-related proteins, *Plant Mol. Biol. Reporter,* 12, 245, 1994.
29. Bohlmann, H., The role of thionins in plant protection, *Crit. Rev. Plant Sci.,* 13, 1, 1994.
30. Broekaert, W. F., Terras, F. R. G., Cammue, B. P. A., and Osborn, R. W., Plant defensins: novel antimicrobial peptides as components of the host defense system, *Plant Physiol.,* 108, 1353, 1995.
31. Van Kan, J. A. L., Cornelissen, B. J. C., and Bol, J. F., A virus-inducible tobacco gene encoding a glycine-rich protein shares putative regulatory elements with the ribulose bisphosphate carboxylase small subunit gene, *Mol. Plant-Microbe Interact.,* 1, 107, 1988.
32. Kauffmann, S., Legrand, M., Geoffroy, P., and Fritig, B., Biological function of "pathogenesis-related" proteins: four PR proteins of tobacco have 1,3-β-glucanase activity, *EMBO J.,* 6, 3209, 1987.
33. Legrand, M., Kauffmann, S., Geoffroy, P., and Fritig, B., Biological function of pathogenesis-related proteins: four tobacco pathogenesis-related proteins are chitinases, *Proc. Natl. Acad. Sci U.S.A.,* 84, 6750, 1987.
34. Ponstein, A. S., Bres-Vloemans, S. A., Sela-Buurlage, M. B., Van den Elzen, P. J. M., Melchers, L. S., and Cornelissen, B. J. C., A novel pathogen- and wound-inducible tobacco (*Nicotiana tabacum*) protein with antifungal activity, *Plant Physiol.,* 104, 109, 1994.
35. Métraux, J.-P., Burkhart, W., Moyer, M., Dincher, S., Middlesteadt, W., Williams, S., Payne, G., Carnes, M., and Ryals, J., Isolation of a complementary DNA encoding a chitinase with structural homology to a bifunctional lysozyme chitinase, *Proc. Natl. Acad. Sci. U.S.A.,* 86, 896, 1989.

36. Melchers, L. S., Apotheker-de Groot, M., Van der Knaap, J. A., Ponstein, A. S., Sela-Buurlage, M. B., Bol, J. F., Cornelissen, B. J. C., Van den Elzen, P. J. M., and Linthorst, H. J. M., A new class of tobacco chitinases homologous to bacterial exo-chitinases displays antifungal activity, *Plant J.*, 5, 469, 1994.
37. Richardson, M., Valdez-Rodriguez, S., and Blanco-Labra, A., A possible function for thaumatin and a TMV-induced protein suggested by homology to a maize inhibitor, *Nature*, 327, 432, 1987.
38. Lawton, K., Uknes, S., Friedrich, L., Gaffney, T., Alexander, D., Goodman, R., Métraux, J.-P., Kessmann, H., Ahl Goy, P., Gut Rella, M., Ward, E., and Ryals, J., The molecular biology of systemic acquired resistance, in *Mechanisms of Plant Defense Responses,* Fritig, B. and Legrand, M., Eds., Kluwer Acad. Publ., Dordrecht, 1993, 422.
39. Van Loon, L. C., Induced resistance in plants and the role of pathogenesis-related proteins, *Eur. J. Plant Pathol.,* 103, 753, 1997.
40. Kassanis, B. and White, R. F., Effect of polyacrylic acid and b proteins on TMV multiplication in tobacco protoplasts, *Phytopath. Z.,* 91, 269, 1978.
41. Dumas, E. and Gianinazzi, S., Pathogenesis-related (b) proteins do not play a central role in TMV localization in *Nicotiana rustica, Physiol. Mol. Plant Pathol.,* 28, 243, 1986.
42. Hooft van Huijsduijnen, R. A. M., Alblas, S. W., De Rijk, R. H., and Bol, J. F., Induction by salicylic acid of pathogenesis-related proteins and resistance to alfalfa mosaic virus infection in various plant species, *J. Gen. Virol.,* 67, 2135, 1986.
43. Fraser, R. S. S., Are "pathogenesis-related" proteins involved in acquired systemic resistance of tobacco plants to tobacco mosaic virus?, *J. Gen. Virol.,* 58, 305, 1982.
44. Linthorst, H. J. M., Meuwissen, R. L. J., Kauffmann, S., and Bol, J. F., Constitutive expression of pathogenesis-related proteins PR-1, GRP, and PR-S in tobacco has no effect on virus infection, *Plant Cell,* 1, 285, 1989.
45. Linthorst, H. J. M., Cornelissen, B. J. C., Van Kan, J. A. L., Van de Rhee, M., Meuwissen, R. L. J., Gonzalez Jaén, M., and Bol, J. F., Induction of host genes by the hypersensitive response of tobacco to virus infection, in *Recognition and Response in Plant–Virus Interactions,* Fraser, R. S. S., Ed., Springer Verlag, New York, 1990, 361.
46. Edelbaum, O., Sher, N., Rubinstein, M., Novick, D., Tal N., Moyer, M., Ward, E., Ryals, J., and Sela, I., Two antiviral proteins, gp35 and gp22, correspond to β-1,3-glucanase and an isoform of PR-5, *Plant Mol. Biol.,* 17, 171, 1991.
47. Beffa, R. S. and Meins, F., Pathogenesis-related functions of plant β-1,3-glucanases investigated by antisense transformation — a review, *Gene*, 179, 97, 1996.
48. De Wit, P. J. G. M. and Bakker, J., Differential changes in soluble tomato leaf proteins after inoculation with virulent and avirulent races of *Cladosporium fulvum* (syn. *Fulvia fulva*), *Physiol. Plant Pathol.,* 17, 121, 1980.
49. Hoffland, E., Pieterse, C. M. J., Bik, L., and Van Pelt, J. A., Induced systemic resistance in radish is not associated with accumulation of pathogenesis-related proteins, *Physiol. Mol. Plant Pathol.,* 46, 309, 1995.
50. Pieterse, C. M. J., Van Wees, S. C. M., Hoffland, E., Van Pelt, J. A., and Van Loon, L. C., Systemic resistance in *Arabidopsis* induced by biocontrol bacteria is independent of salicylic acid accumulation and pathogenesis-related gene expression, *Plant Cell,* 8, 1225, 1996.
51. Vera, P. and Conejero, V., Pathogenesis-related proteins of tomato. P-69 as an alkaline endoproteinase, *Plant Physiol.,* 87, 58, 1988.

52. Stintzi, A., Heitz, T., Prasad, V., Wiedemann-Merdinoglu, S., Kauffmann, S., Geoffroy, P., Legrand, M., and Fritig, B., Plant "pathogenesis-related" proteins and their role in defense against pathogens, *Biochimie*, 75, 687, 1993.
53. Somssich, I. E., Schmelzer, E., Bollmann, J., and Hahlbrock, K., Rapid activation by fungal elicitor of genes encoding "pathogenesis-related" proteins in cultured parsley cells, *Proc. Natl. Acad. Sci. U.S.A.*, 83, 2427, 1986.
54. Moiseyev, G., Beintema, J. J., Fedoreyeva, L. I., and Yakovlev, G. I., High sequence similarity between a ribonuclease from ginseng calluses and fungus-elicited proteins from parsley indicates that intracellular pathogenesis-related (IPR) proteins are ribonucleases, *Planta*, 193, 470, 1994.
55. Felix, G. and Meins, F., Developmental and hormonal regulation of β-1,3-glucanase in tobacco, *Planta*, 167, 206, 1986.
56. Shinshi, H., Mohnen, D., and Meins, F., Regulation of a plant pathogenesis-related enzyme: inhibition of chitinase and chitinase mRNA accumulation in cultured tobacco tissues by auxin and cytokinin, *Proc. Natl. Acad. Sci. U.S.A.*, 84, 89, 1987.
57. Lotan, T., Ori, N., and Fluhr, R., Pathogenesis-related proteins are developmentally regulated in tobacco flowers, *Plant Cell*, 1, 881, 1989.
58. Memelink, J., Linthorst, H. J. M., Schilperoort, R. A., and Hoge, J. H. C., Tobacco genes encoding acidic and basic isoforms of pathogenesis-related proteins display different expression patterns, *Plant Mol. Biol.*, 14, 119, 1990.
59. Brederode, F. T., Linthorst, H. J. M., and Bol, J. F., Differential induction of acquired resistance and PR gene expression in tobacco by virus infection, ethephon treatment, UV light and wounding, *Plant Mol. Biol.*, 17, 1117, 1991.
60. De Laat, A. M. M. and Van Loon, L. C., Regulation of ethylene biosynthesis in virus infected tobacco leaves. II. Time course of levels of intermediates and *in vivo* conversion rates, *Plant Physiol.*, 69, 240, 1982.
61. Boller, T., Ethylene in pathogenesis and disease resistance, in *The Plant Hormone Ethylene*, Mattoo, A. K. and Suttle, J. C., Eds., CRC Press, Boca Raton, FL, 1991, 293.
62. Mauch, F. and Staehelin, L. A., Functional implications of the subcellular localization of ethylene-induced chitinase and β-1,3-glucanase in bean leaves, *Plant Cell*, 1, 447, 1989.
63. Vigers, A., J., Wiedema, S., Roberts, W. K., Legrand, M., Selitrennikoff, C. P., and Fritig, B., Thaumatin-like pathogenesis-related proteins are antifungal, *Plant Sci.*, 83, 155, 1992.
64. Niderman, T., Genetet, I., Bruyère, T., Gees, R., Stintzi, A., Legrand, M., Fritig, B., and Mösinger, E., Pathogenesis-related PR-1 proteins are antifungal; isolation and characterization of three 14-kilodalton proteins of tomato and of a basic PR-1 of tobacco with inhibitory activity against *Phytophthora infestans*, *Plant Physiol.*, 108, 17, 1995.
65. Zhu, B. L., Chen, T. H. H., and Li, P. H., Analysis of late-blight disease resistance and freezing tolerance in transgenic potato plants expressing sense and antisense genes for an osmotin-like protein, *Planta*, 198, 70, 1996.
66. Jongedijk, E., Tigelaar, H., Van Roekel, J. S. C., Bres-Vloemans, S. A., Dekker, I., Van den Elzen, P. J. M., Cornelissen, B. J. C., and Melchers, L. S., Synergistic activity of chitinases and β-1,3-glucanases enhances fungal resistance in transgenic tomato plants, *Euphytica*, 85, 173, 1995.
67. Zhu, Q., Maher, E. A., Masoud, S., Dixon, R. A., and Lamb, C. J., Enhanced protection against fungal attack by constitutive co-expression of chitinase and glucanase genes in transgenic tobacco, *Biotechnology*, 12, 807, 1994.

68. Jach, G., Görnhardt, B., Mundy, J., Logemann, J., Pinsdorf, E., Leah, R., Schell, J., and Maas, C., Enhanced quantitative resistance against fungal disease by combinatorial expression of different barley antifungal proteins in transgenic tobacco, *Plant J.,* 8, 97, 1995.
69. Broekaert, W. F., Van Parijs, J., Leyns, F., Joos, H., and Peumans, W. J., A chitin-binding lectin from stinging nettle rhizomes with antifungal properties, *Science*, 245, 1100, 1989.
70. Neuhaus, J. M., Ahl-Goy, P., Hinz, U., Flores, S., and Meins, F. J., High-level expression of a tobacco chitinase gene in *Nicotiana sylvestris:* susceptibility of transgenic plants to *Cercospora nicotianae* infection, *Plant Mol. Biol.,* 16, 141, 1991.
71. Roberts, W. K. and Selitrennikoff, C. P., Plant and bacterial chitinases differ in antifungal activity, *J. Gen. Microbiol.,* 134, 169, 1988.
72. Hejgaard, J., Jacobsen, S., and Svendsen, I., Two antifungal thaumatin-like proteins from barley grain, *FEBS Lett.,* 291, 127, 1991.
73. Vigers, A. J., Roberts, W. K., and Selitrennikoff, C. P., A new family of plant antifungal proteins, *Mol. Plant-Microbe Interact.,* 4, 315, 1991.
74. Ham, K. S., Wu, S. C., Darvill, A. G., and Albersheim, P., Fungal pathogens secrete an inhibitor protein that distinguishes isoforms of plant pathogenesis-related endo-β-1,3-glucanases, *Plant J.,* 11, 169, 1997.
75. Brooks, D. E., Means, A. R., Wright, E. J., Singh, S. P., and Tiver, K. K., Molecular cloning of the cDNA for androgen-dependent sperm-coating glycoproteins secreted by the rat epididymis, *Eur. J. Biochem.,* 161, 13, 1986.
76. Fang, K. S. Y., Vitale, M., Fehlner, P., and King, T. P., cDNA cloning and primary structure of a white-face hornet venom allergen, antigen 5, *Proc. Natl. Acad. Sci. U.S.A.,* 85, 895, 1988.
77. Schuren, F. H. J., Ásgeirsdóttir, S. A., Kothe, E. M., Scheer, J. M. J., and Wessels, J. G. H., The *Sc7/Sc14* gene family of *Schizophyllum commune* codes for extracellular proteins specifically expressed during fruit-body formation, *J. Gen. Microbiol.,* 139, 2083, 1993.
78. Pfisterer, P., König, H., Hess, J., Lipowsky, G., Haendler, B., Schleuning, W.-D., and Wirth, T., CRISP-3, a protein with homology to plant defense proteins, is expressed in mouse B cells under the control of Oct. 2, *Mol. Cell. Biol.,* 16, 6160, 1996.
79. Van Loon, L. C., Induction by 2-chloroethylphosphonic acid of viral-like lesions, associated proteins, and systemic resistance in tobacco, *Virology,* 80, 417, 1977.
80. Edreva, A. M., Induction of "pathogenesis-related" proteins in tobacco leaves by physiological (non-pathogenic) disorders, *J. Exp. Bot.,* 41, 701, 1990.
81. Wagih, E. E. and Coutts, R. H. A., Comparison of virus-elicited and other stresses on the soluble protein fraction of cucumber cotyledons, *Phytopath. Z.,* 104, 364, 1982.
82. White, R. F., Dumas, E., Shaw, P., and Antoniw, J. F., The chemical induction of PR (b) proteins and resistance to TMV infection in tobacco, *Antiviral Res.,* 6, 177, 1986.
83. Coutts, R. H. A., Alterations in the soluble protein patterns of tobacco and cowpea leaves following inoculation with tobacco necrosis virus, *Plant Sci. Lett.,* 12, 189, 1978.
84. Szczepanski, M. and Redolfi, P., Changes in the proteins of bean leaves infected with tobacco necrosis or alfalfa mosaic viruses, *Phytopath. Z.,* 113, 57, 1985.
85. Singh, N. K., Nelson, D. E., Kuhn, D., Hasegawa, P. M., and Bressan, R. A., Molecular cloning of osmotin and regulation of its expression by ABA and adaptation to low water potential, *Plant Physiol.,* 90, 1096, 1989.

86. Neale, A. D., Wahleithner, J. A., Lund, M., Bonnett, H. T., Kelly, A., Meeks-Wagner, D. R., Peacock, W. J., and Dennis, E. S., Chitinase, β-1,3-glucanase, osmotin, and extensin are expressed in tobacco explants during flower formation, *Plant Cell,* 2, 673, 1990.
87. Xu, Y., Chang, P. F. L., Liu, D., Narasimhan, M. L., Raghothama, K. G., Hasegawa, P. M., and Bressan, R. A., Plant defense genes are synergistically induced by ethylene and methyl jasmonate, *Plant Cell,* 6, 1077, 1994.
88. Ho, T. H. D. and Sachs, M. M., Stress-induced proteins: characterization and the regulation of their synthesis, in *The Biochemistry of Plants, A Comprehensive Treatise,* Stumpf, P. K. and Conn, E. E., Eds., Vol. 15: *Molecular Biology,* Marcus A., Ed., Academic Press, New York, 1989, 347.
89. Wasternack, C. and Parthier, B., Jasmonate-signalled plant gene expression, *Trends Plant Sci.,* 2, 302, 1997.
90. Smith, B. J. and Yaffe, M. P., Uncoupling thermotolerance from the induction of heat shock proteins, *Proc. Natl. Acad. Sci. U.S.A.,* 88, 11091, 1991.
91. Nagao, R. T., Kimpel, J. A., Vierling, E., and Key, J. L., The heat shock response: a comparative analysis, *Oxford Surveys Plant Mol. Cell. Biol.,* 3, 384, 1986.
92. Schlesinger, M. J., Aliberti, G., and Kelley, P. M., The response of cells to heat shock, *Trends Biochem. Sci.,* 1, 222, 1982.
93. Gething, M. J. and Sambrook, J., Protein folding in the cell, *Nature*, 355, 33, 1992.
94. Pfitzner, U. M., Pfitzner, A. J. P., and Goodman, H. M., DNA sequence analysis of a *PR-1a* gene from tobacco: molecular relationship of heat shock and pathogen responses in plants, *Mol. Gen. Genet.,* 211, 290, 1988.
95. Somssich, I. E., Schmelzer, E., Kawalleck, P., and Hahlbrock, K., Gene structure and in situ transcript localisation of pathogenesis-related protein 1 in parsley, *Mol. Gen. Genet.,* 213, 93, 1988.
96. Matsuoka, M. and Ohashi, Y., Induction of pathogenesis-related proteins in tobacco leaves, *Plant Physiol.,* 80, 505, 1986.
97. Mauch, F. and Staehelin, L. A., Functional implications of the subcellular localization of ethylene-induced chitinase and β-1,3-glucanase in bean leaves, *Plant Cell,* 1, 447, 1989.
98. Takeuchi, Y., Yoshikawa, M., Takeba, G., Tanaka, K., Shibata, D., and Horino, O., Molecular cloning and ethylene induction of mRNA encoding a phytoalexin elicitor-releasing factor, β-1,3-endoglucanase, in soybean, *Plant Physiol.,* 93, 673, 1990.
99. Ham, K. S., Kauffmann, S., Albersheim, P., and Darvill, A. G., Host-pathogen interactions XXXIX. A soybean pathogenesis-related protein with β-1,3-glucanase activity releases phytoalexin elicitor-active heat stable fragments from fungal cell walls, *Mol. Plant-Microbe Interact.,* 4, 545, 1991.
100. Ye, Z. H. and Varner, J. E., Tissue-specific expression of cell wall proteins in developing soybean tissues, *Plant Cell,* 3, 23, 1991.
101. Ready, M. P., Brown, D. T., and Robertus, J. D., Extracellular localization of pokeweed antiviral protein, *Proc. Natl. Acad. Sci. U.S.A.,* 83, 5053, 1986.
102. Darvill, A. G. and Albersheim, P., Phytoalexins and their elicitors — a defense against microbial infection in plants, *Annu. Rev. Plant Physiol.,* 35, 243, 1984.
103. Kuranda, M. J. and Robbins, P. W., Chitinase is required for cell separation during growth of *Saccharomyces cerevisiae, J. Biol. Chem.,* 266, 19758, 1991.
104. Boller, T., Gehri, A., Mauch, F., and Vögeli, U., Chitinase in bean leaves: induction by ethylene, purification, properties and possible function, *Planta*, 157, 22, 1983.

105. Kelly, P., Trewavas, A. J., Lewis, L. N., Durbin, M. L., and Sexton, R., Translatable mRNA changes in ethylene induced abscission zones of *Phaseolus vulgaris* (Red Kidney), *Plant Cell Environ.,* 10, 11, 1987.
106. Del Campillo, E. and Lewis, L. N., Identification and kinetics of accumulation of proteins induced by ethylene in bean abscission zones, *Plant Physiol.,* 98, 955, 1992.
107. Staehelin, C., Schultze, M., Kondorosi, E., Mellor, R. B., Boller, T., and Kondorosi, A., Structural modifications in *Rhizobium meliloti* Nod factors influence their stability against hydrolysis by root chitinases, *Plant J.,* 5, 319, 1994.
108. De Jong, A. J., Cordewener, J., Lo Schiavo, F., Terzi, M., Vandekerckhove, J., Van Kammen, A., and De Vries, S. C., A carrot somatic embryo mutant is rescued by chitinase, *Plant Cell,* 4, 425, 1992.
109. Kragh, K. M., Hendriks, T., De Jong, A. J., Lo Schiavo, F., Bucherna, N., Hojrup, P., Mikkelsen, J. D., and De Vries, S. C., Characterization of chitinases able to rescue somatic embryos of the temperature-sensitive carrot variant *ts11, Plant Mol. Biol.,* 31, 631, 1996.
110. De Jong, A. J., Heidstra, R., Spaink, H. P., Hartog, M. V., Meijer, E. A., Hendriks, T., Lo Schiavo, F., Terzi, M., Bisseling, T., Van Kammen, A., and De Vries, S. C., Rhizobium lipooligosaccharides rescue a carrot somatic embryo mutant, *Plant Cell,* 5, 615, 1993.
111. Röhrig, H., Schmidt, J., Walden, R., Czaja, I., Miklasevics, E., Wieneke, U., Schell, J., and John, M., Growth of tobacco protoplasts stimulated by synthetic lipo-chitooligosaccharides, *Science*, 269, 841, 1995.
112. Vögeli-Lange, R., Fründt, C., Hart, C. M., Beffa, R., Nagy, F., and Meins, F., Jr., Evidence for a role of β-1,3-glucanase in dicot seed germination, *Plant J.,* 5, 273, 1994.
113. Ori, N., Sessa, G., Lotan, T., Himmelhoch, S., and Fluhr, R., A major stylar matrix polypeptide (sp41) is a member of the pathogenesis-related proteins superclass, *EMBO J.,* 9, 3429, 1990.
114. Coté, F., Cutt, J. R., Asselin, A., and Klessig, D. F., Pathogenesis-related acidic β-1,3-glucanase genes of tobacco are regulated by both stress and developmental signals, *Mol. Plant-Microbe Interact.,* 4, 173, 1991.
115. Richard, L., Arró, M., Hoebeke, J., Meeks-Wagner, D. R., and Van, K. T. T., Immunological evidence of thaumatin-like proteins during tobacco floral differentiation, *Plant Physiol.,* 98, 337, 1992.
116. Leung, D. W. M., Involvement of plant chitinase in sexual reproduction of higher plants, *Phytochemistry,* 31, 1899, 1992.

2 PR-1: A Group of Plant Proteins Induced Upon Pathogen Infection

Annemarie S. Buchel and Huub J.M. Linthorst

CONTENTS

2.1 Introduction....................21
2.2 Characterization....................22
2.2.1 Acidic PR-1 Proteins....................22
2.2.2 Basic PR-1 Proteins....................23
2.2.3 PR-1 Proteins in Other Organisms....................24
2.2.4 The Function of PR-1 Proteins....................25
2.3 Expression/Induction....................25
2.3.1 Pathogens/Wounding....................26
2.3.2 Salicylic Acid....................26
2.3.3 Ethylene and Other Hormones....................29
2.3.4 UV-Light....................31
2.3.5 Developmental Stimuli....................31
2.4 PR-1 Promoter Analysis....................32
2.4.1 *CIS* Regulatory Regions in the Promoters of Acidic *PR-1* Genes....................32
2.4.2 *CIS* Regulatory Regions in the Promoters of Basic *PR-1* Genes....................34
2.4.3 Protein Factors Interacting with Acidic *PR-1* Promoter Sequences....................34
2.4.4 Protein Factors Interacting with Basic *PR-1* Promoter Sequences....................36
References....................36

2.1 INTRODUCTION

Infection of tobacco plants containing the *N*-resistance gene from *Nicotiana glutinosa* with tobacco mosaic virus (TMV) leads to a hypersensitive response, which prevents virus spread by the development of necrotic lesions at the site of virus entry. In addition, the localized virus-infection induces a long-lasting, systemic

0-8493-0697-3/99/$0.00+$.50

acquired resistance (SAR) to a variety of viral, bacterial, and fungal pathogens.[1-4] Van Loon et al.[5] showed that the accumulation of a group of so-called "pathogenesis-related" (PR) proteins correlates with the onset of SAR. PR proteins are defined as plant proteins that are induced in pathological or related situations. Pathological situations refer to all types of infected states, including the resistant hypersensitive response and parasitic attack by nematodes, phytophagous insects, and herbivores. Related situations can occur after the application of chemicals that (partly) mimic the effect of pathogen infection (e.g., salicylate), as well as wound responses that give rise to proteins that are also induced during infections.[6] In addition to its activity in tobacco, SAR has been demonstrated in many other plant species. The mechanism behind the establishment of SAR in plants is not known.

PR proteins were discovered in 1970 in TMV-infected tobacco plants.[7,8] Since then, these proteins have been found in a large number of plant species. For tobacco Xanthi nc and Samsun NN, the first identified PR proteins were grouped into the PR-1 and PR-2 families. The PR proteins were named PR-1a, -1b, and -1c in order of decreasing electrophoretic mobility in a native polyacrylamide gel. The fourth component was designated PR-2. Van Loon et al.[6] proposed that any protein whose amino acid sequence resembled a characterized PR-protein from tobacco be designated similarly. New members within a family will be assigned a letter according to the order in which they are described. Five well-established families of PR proteins have been described in tobacco. Most families contain basic and acidic isoforms.[9,10] Six additional families of PR proteins (four in tobacco) have been submitted for adoption.[6] The "type member" of the PR-10 family is parsley "PR1." Since it belongs to another family, this "PR1" protein shares no sequence homology with and is not related to the tobacco PR-1 family.[11]

2.2 CHARACTERIZATION

PR-1, the most abundant of the PR proteins, is induced 10,000-fold in infected tissue and accumulates to 1 to 2% of the total leaf protein.[12] Therefore the expression of *PR-1* genes can serve as an indicator of the activation of plant defense response. The following overview deals with the current knowledge on PR-1 and the genes encoding these proteins. Since PR-1 proteins from tobacco have been studied in most detail, the data presented here will primarily cover this biological system.

2.2.1 Acidic PR-1 Proteins

The genome of *Nicotiana tabacum* Samsun NN contains at least eight genes encoding acidic PR-1 proteins. Clones corresponding to the genes for acidic PR-1a, -1b, and -1c proteins were isolated from a genomic library of Samsun NN tobacco.[13-18] Pseudogenes have been found for different tobacco cultivars.[16,19,20] *Nicotiana tabacum* var. Samsun NN is amphidiploid, containing 2n chromosomes from *N. sylvestris* and 2n chromosomes from *N. tomentosiformis*. The genes for the acidic PR-1a and -1c proteins are derived from *N. sylvestris,* while the PR-1b gene originates from *N. tomentosiformis.*[19] The acidic *PR-1* genes do not contain introns.[19,21,22]

At the 3′ end of the *PR-1a* gene, alternate polyadenylation sites have been identified, resulting in 194 bp and 270 bp long noncoding regions.[15] The length of the open reading frame is 504 nucleotides; the primary structure is strictly conserved among the acidic PR-1 proteins.[21] The 138 amino acid long mature proteins are synthesized on membrane-bound ribosomes as higher molecular weight precursors containing an N-terminal 30 amino acid hydrophobic signal peptide, which is cleaved to yield 15 kDa mature proteins.[8,13,15,23,24] Lucas et al.[25] reported that no amino sugars can be detected in the PR-1a protein. The PR-1a,-1b, and -1c proteins were shown to belong to a serological group in tobacco and to be serologically related to PR-proteins from other plant species.[26-30] The three acidic isoforms of the tobacco PR-1 family share more than 90% amino acid sequence identity, while minor sequence variations exist between the homologous PR-1 proteins from *N. tabacum* cvs. Samsun NN, Wisconsin 38, and Xanthi nc.[31]

The acidic PR-1 proteins remain soluble at pH 3, whereas most other plant proteins are denatured under these conditions.[32,33] The relative resistance to the action of endogenous and exogenous proteases, suggests that PR-proteins have a low turnover rate.[34,35] The estimated half-life of the acidic PR-1 proteins is 40 to 70 hours.[31] The high resistance of PR-proteins to acidic pH and to proteolytic attack appears to be a good adaptation to the conditions of the extracellular environment. Acidic PR-1 proteins were detected in the extracellular spaces and xylem elements of TMV-infected tobacco leaves, using biochemical and immuno-localization experiments.[36-39] In addition, Dixon et al.[40] described the accumulation of PR-1a, -1b, and -1c after TMV infection of *N. tabacum* Xanthi in the central vacuoles of specialized leaf cells, known as crystal idioblasts. There are indications for the existence of special proteolytic enzymes that are responsible for the turnover of PR-proteins. Rodrigo et al.[41] described an extracellular aspartyl protease in tomato which is able to degrade PR proteins.

The acidic *PR-1* genes do not encode a known targeting peptide sequence for vacuolar targeting. Therefore, the synthesis and accumulation within crystal idioblasts of acidic PR-1 proteins, indicates that these cells sort the proteins in a unique manner.[40] In other plants, extracellular proteins corresponding to the acidic PR-1 proteins of tobacco do not necessarily have a low pI. For example, the tomato P4 and P6 proteins purified from the apoplastic fluid are basic.[42]

2.2.2 Basic PR-1 Proteins

cDNAs and genomic clones corresponding to basic proteins, which are approximately 65% similar to the acidic PR-1a, -1b, and -1c proteins, have been characterized from TMV-infected leaves.[19,43,44] So far there are indications for two basic PR-1 proteins in tobacco although hybridization studies revealed that the genome contains at least eight (pseudo) genes.[19] Also, the genes encoding the basic PR-1 proteins do not appear to contain introns. Antibodies against the tobacco acidic PR-1 proteins did not react with the basic proteins. Nidermann and co-workers[45] isolated a 17 kDa basic PR-1 protein (PR-1g) from tobacco. Partial amino acid sequencing (59 amino acids out of 179) indicated that this protein had 94% sequence identity to the protein encoded by the genomic clone of Payne et al.,[43] 97% to the genomic clone of Eyal et al.,[44] and 97% sequence identity to the cDNA clone described by Cornelissen et al.[19] Sessa et al.[46] showed that the product of the basic dark inducible *PR-1* gene *PRB-1b,* characterized by Eyal et al.,[44] was a 16 kDa protein. Although the *PRB-1b* coding and noncoding region appear to be identical to the previously published basic-type *PR-1* DNA,[47] an additional "C" present at position 476 in this gene results in a frame shift relative to the described basic-type *PR-1* DNA sequence. The *PRB-1b* open reading frame codes for a 179 amino acid protein which shares 87% homology with the protein encoded by the genomic clone of Payne et al.[43]

The basic PR-1 proteins, like the acidic PR-1 proteins, contain a hydrophobic N-terminal region of 30 amino acids which probably functions as a signal peptide for translocation to the ER.[43] In contrast to the acidic PR-1 proteins, there is an extra domain present in the basic PR-1. Based on the deduced amino acid sequence, this domain was found as an extension of 36 amino acids encoded in the cDNA clone,[19] while the two genomic clones encode extensions of 18 amino acids.[43,46]

These C-terminal extensions in the basic-type PR proteins are thought to represent vacuolar targeting signals.[48,49] Sessa et al.[46] showed the subcellular localization of the basic PRB-1b proteins in the vacuoles of ethylene-treated tobacco leaves. The preliminary data of Nidermann et al.[45] indicate that the basic PR-1g protein is at least partially present in the extracellular space. This suggests that, like their acidic counterparts, basic PR-1 proteins may have a dual localization, irrespective of specific targeting signals present in the protein sequence.

2.2.3 PR-1 Proteins in Other Organisms

The genes encoding the acidic and basic PR-1 proteins in tobacco and other plant species probably arose relatively early during evolution. The existence of *PR-1*-like genes/sequences not only in many plant species but also in fungi, insects, mammals, and yeast supports this idea.[50-53] The function of the proteins encoded by the *PR-1*-like genes present in these organisms is not known. Immunoblot analysis using antibodies against the acidic tobacco PR-1b revealed the existence of cross-reacting proteins in a number of infected dicots and monocots, including tomato, potato, cowpea, maize, sunflower, barley, *Gomphrena globosa, Chenopodium amaranticolor,* and *Solanum demissum.*[30,54] cDNA clones with sequences similar to tobacco *PR-1* sequences have been isolated from *Brassica napus, Solanum tuberosum, Medicago truncatula,* and maize.[55-58] PR-1 proteins have been purified from tomato, barley, and *Arabidopsis,* as well as from tobacco.[42,59,60] Phylogenetic analysis of protein sequence data described by Devereux et al.[61] showed, for example that within the Solanaceae, the acidic PR-1 proteins from tobacco are more closely related to the acidic PR-1a1 protein from tomato than to the tobacco basic PR-1g and PRB-1b.[62] The existence of distinct basic and acidic *PR-1* genes in plants suggests an early duplication of *PR-1* genes during evolution. PR-1 proteins (PR-1a-g) have been detected in different tobacco cultivars and species, e.g., in *Nicotiana tabacum* cvs. Samsun NN, Xanthi nc., Wisconsin-38, White Burley, and *Nicotiana* sp. *tomentosiformis, sylvestris, debneyi, glutinosa.*[7,15,23,59,63-65]

2.2.4 THE FUNCTION OF PR-1 PROTEINS

Since PR-1 proteins were first discovered in response to virus infection, it was believed that they were involved in virus resistance or in virus localization. Treatments, such as spraying plants with salicylic acid or UV-light irradiation, that induce PR-1 production also efficiently induce resistance. Ahl and Gianinazzi[66] reported that the interspecific hybrid of *Nicotiana glutinosa* and *N. debneyi,* which constitutively produces PR-1 proteins, was highly resistant to TMV. However, transgenic tobacco plants, constitutively expressing chimeric genes composed of the cauliflower mosaic virus (CaMV) 35S promoter fused to the *PR-1a* or *-1b* coding region, did not show higher levels of resistance to local or systemic infection with TMV or alfalfa mosaic virus.[12,67,68] This lack of resistance was not due to a possible failure in targeting, since both the PR-1a and -1b proteins were correctly secreted into the extracellular space. Apparently, despite the high production after virus infection, the PR-1a and -1b proteins do not function as antiviral factors.

PR-1 proteins do not play a role in resistance to leaf-eating insects, because it was demonstrated that larvae of *Spodoptera exigua* and *Heliotis virescens* fed with leaf material derived from PR-1a transgenic tobacco did not show any apparent difference in development, morphology, or reproductive capabilities compared to the control larvae fed with leaves of nontransgenic tobacco.[69,70] In addition, Klessig et al.[71] showed that transgenic tobacco plants producing PR-1b proteins are as susceptible to *Manduca sexta* as nontransformed control plants.

There is increasing evidence that PR-1 proteins may play a role in the resistance to fungal infection. In 1993, Alexander et al.[12] reported that transgenic *N. tabacum* cv Xanthi nc, which constitutively produce PR-1a protein, showed an increased tolerance to two oomycete pathogens. Significant delays of infection and substantial reduction of disease symptoms were found on transgenic plants inoculated with *Phytophthora parasitica* var. *nicotianae* (causal agent of black shank disease) and *Peronospora tabacina* (causal agent of blue mold disease). The transgenic tobacco plants remained susceptible to infection with TMV, potato virus Y (PVY), *Cercopspora nicotianae,* and *Pseudomonas syringae* pv *tabaci.* More recently, Niderman et al.[45] tested the antifungal activity of purified tobacco and tomato PR-1 proteins against *Phythopthora infestans.* They measured the degree of inhibition of zoospore germination *in vitro* and scored the reduction of infected leaf surface *in vivo.* The basic PR-1 proteins PR-1g (tobacco) and P14c (tomato), which show considerable sequence similarity, turned out to be most effective in the assays. The acidic tobacco PR-1 proteins were only slightly inhibitory.

2.3 EXPRESSION/INDUCTION

PR-1 genes are known to be expressed in response to various external stimuli, including pathogens, wounding, chemical elicitors, hormones, and UV-light.[72,73] In addition, the genes also appear to be under developmental control. The induction of PR proteins is not dependent on the presence of the *N*-resistance gene. After infection with tobacco necrosis virus (TNV), Samsun nn plants, which lack the *N*-gene, induce the same set of PR proteins as Samsun NN tobacco plants.[74] Genes encoding basic and acidic isoforms of PR-1 are differentially regulated.[75]

2.3.1 Pathogens/Wounding

Although *PR-1* genes are induced by a plethora of different pathogens, their induction by TMV has been studied in most detail. In healthy Samsun NN tobacco, the mRNA for the acidic PR-1a, -1b, and -1c proteins accumulates at a very low level. Upon TMV infection, this level is highly increased.[75,76] The mRNA for the basic PR-1 proteins is also induced by TMV infection.[75] The expression of acidic *PR-1* genes in the local leaves reaches its maximum at five days after TMV inoculation, while the basic genes reach a maximal expression after three days. Only the acidic genes are systemically induced. mRNA levels for acidic PR-1 proteins are detectable in systemic leaves at three days after TMV inoculation of the bottom leaves; the maximum expression is reached at nine to fifteen days.[73]

In contrast to Ohshima et al.,[77] Uknes et al.,[78] Malamy et al.,[79] and Van de Rhee et al.[80] reported that acidic *PR-1* genes do not respond to wounding. Wounding, however, results in enhanced accumulation of basic PR-1 mRNA, but this does not result in increased resistance to TMV.[73]

2.3.2 Salicylic Acid

An ever-increasing body of evidence suggests that salicylic acid (SA) plays an important role in the activation of defense responses against pathogen attack in plants. In 1979, White[81] reported that spraying tobacco plants with salicylic acid (SA) induced local resistance to TMV. Salicylate induces the expression of both acidic and basic *PR-1* genes.[31] After pathogen infection, the endogenous plant hormone SA accumulates to high levels in tobacco, cucumber, and *Arabidopsis*.[82-86]

Experiments with the protein synthesis inhibitor cycloheximide indicated that protein synthesis is required for salicylate-dependent, acidic PR-1 mRNA accumulation.[78,87] Levels of SA and salicylic acid-β-glucoside (SAG) increase after infection of resistant (Xanthi nc) but not susceptible (Xanthi) tobacco cultivars.[82,88] SAG possibly functions as a storage form of metabolically active SA. SAG is as active as SA in induction of *PR-1* gene expression. The induction of *PR-1* by SAG is preceded by a transient release of SA, which occurs in the extracellular space.[89] When Xanthi nc (genotype NN) plants are inoculated with TMV and incubated at temperatures higher than 28°C, the replication and spread of the virus is not restricted, necrotic lesions are not formed, the increase in SA and SAG levels is blocked, and the *PR*-genes are not induced. However, when the infected plants are moved to lower temperatures (22°C), SA and SAG levels rise dramatically, PR proteins accumulate, and the HR is rapidly activated.[8,82,84,88,90,91]

A variety of *Arabidopsis* mutants (e.g., *cpr, lsd, acd*), which constitutively express *PR-1* genes, exhibit enhanced resistance to pathogens, while also showing elevated levels of SA.[92-94] Other *Arabidopsis* mutants that are defective in SA signal transduction (e.g., *npr, nim,* and *sai*) fail to activate *PR* gene expression after treatment with SA and exhibit enhanced susceptibility to pathogens.[95-97] Recently,

Ryals et al.[98] reported that the *Arabidopsis nim1* gene product (NIM1) shares significant homology to the mammalian IκB subclass of transcription inhibitors. The IκB protein functions in signal transduction by binding to the transcription factor NF-κB and preventing it from entering the nucleus. In animals, the transcription factor NF-κB is post-transcriptionally activated by reactive oxygen species (ROS), such as H_2O_2, and mediates expression of many genes involved in acute phase, immune, and inflammatory responses.[99] When the signal transduction pathway is activated, phosphorylation and ubiquitination of IκB results in transport of the NF-κB/IκB complex through the proteosome. IκB is degraded, and NF-κB is released to the nucleus where it stimulates transcription.[100,101] The homology between NIM1 and IκB suggests that NIM1 may interact with an NF-κB-related transcription factor in plants to induce SAR gene expression and trigger disease resistance. Since mutations in the *nim1* gene result in inhibition of the SAR signal, the transcription factor targeted by the NIM1 protein possibly serves as a repressor of SAR gene expression and disease resistance.[98] Transgenic tobacco and *Arabidopsis* plants expressing a salicylate hydroxylase gene (*NahG*) from *Pseudomonas putida* fail to accumulate SA upon pathogen attack. The substrate-specific salicylate hydroxylase enzyme uses NADH as a cofactor to decarboxylate and hydroxylate SA to biologically inactive catechol. The *NahG* plants do not display *PR-1* gene induction and show no local and systemic resistance.[102,103]

Histochemical analysis performed by Ohshima et al.[77] after SA induction of the *PR-1a* promoter-GUS transgene in tobacco showed induced GUS-activity in all tissues of the leaf blade and vascular bundle, especially the phloem, cortical parenchyma, pith parenchyma, and the leaf base of the petiole. This induced gene expression reached the highest levels in young rigid leaves. However, it must be noted that this pattern of expression was obscured by the heterologous CaMV 35S enhancer present in the constructs used. Transgenic plants with high GUS activity did not express endogenous *PR-1* genes. Tissue specificity of GUS expression from the *PR-1a* promoter was identical to that obtained with the CaMV 35 S promoter.[104]

During the past several years, different studies have been initiated to elucidate the mechanism of SA action in plant disease resistance. However, the precise role of SA in this context has yet to be determined. The function of SA as a primary long-distance signal in SAR is questionable. In contrast to the data published by Rasmussen et al.[86] and Vernooij et al.,[105] Shulaev et al.[106] reported that SA may be the major translocated SAR-inducing signal in tobacco.

In 1991, Chen and Klessig identified a soluble (SA)-binding protein in tobacco which appeared to be a catalase.[107] Many compounds that induce *PR-1* gene expression and disease resistance in plants (e.g., SA, 2,6-dichloroisonicotinic acid and their biologically active analogues) inactivate catalase directly or indirectly.[108,109] Among different plant species, there is a large variety in the levels of SA-binding activity and SA-sensitivity of catalase activity.[110] The other key H_2O_2-scavenging enzyme, ascorbate peroxidase (APX), is also shown to be inhibited by SA.[111] Lipid peroxides are able to induce *PR* genes.[112] Conrath et al.[113] suggested that the interaction of SA with catalase and APX leads to elevated H_2O_2 levels and lipid peroxidation. Since SA treatment of tobacco results in a moderate increase in H_2O_2, Chen et al.[109,114]

proposed that elevated levels of H_2O_2 or other reactive oxygen species (ROS) derived from H_2O_2 are involved in the activation of plant defenses. Analyses of transgenic tobacco plants that expressed tobacco catalase 1 and 2 in an antisense orientation indicated that severe reduction of catalase levels was accompanied by the induction of *PR-1* genes, the development of necrosis/chlorosis, and enhanced resistance to TMV. Despite severely depressed catalase levels, the systemic leaves of the antisense catalase plants did not exhibit *PR-1* gene expression.[115] Similarly, Chamnong-pol et al.[116] observed *PR-1* gene activation in catalase-deficient transgenic tobacco plants only under high light conditions which cause necrosis in these plants. Apparently, elevated H_2O_2 due to the inhibition of catalase *per se* is not sufficient for induction of *PR-1* genes. Bi et al.[117] showed that H_2O_2 and the catalase inhibitor 3-amino-1,2,4-triazole (3AT) were weak inducers of *PR-1a* gene expression in wild-type tobacco. Both H_2O_2 and 3AT were unable to induce *PR-1a* in *Nah G* plants, suggesting that their action depends on the accumulation of SA.[117,118] After infection of tobacco with *P. syringae* pv. *syringae* or pretreatment of leaf discs with SA, no significant changes in catalase activity could be detected.[117] Neither was there an increase in H_2O_2 levels during the onset of SAR.[118] Increasing concentrations of H_2O_2 result in a dose-dependent accumulation of total SA and activation of benzoic acid-2-hydroxylase, the enzyme that converts benzoic acid to SA.[118-120] Taken together, these data suggest that H_2O_2 affects *PR-1* gene expression at a step in signal transduction upstream of SA.

Activation of defense responses may occur through the interaction of SA with other cellular factors, rather than, or in addition to, catalase and APX. The contention that plant SA-sensitive catalases play a role only in certain aspects of SA-mediated responses is supported by experiments described by Chen et al.[121] The authors found an inverse correlation between SA levels and the sensitivity of catalase to SA both among plant species and in different tissues of *Oryza sativa.*

H_2O_2 could be just one of several stimuli that induces *PR-1* through SA. Green and Fluhr[87] showed that increases in PR-1 protein levels by SA are unaffected by pretreatment of the tobacco leaves with antioxidants. Vernooij et al.[105] reported that even very slight increases in SA levels in systemic tissue can induce PR-gene expression via a pathway that appears not to involve H_2O_2.

Recently, Du and Klessig[112] identified a new soluble SA-binding protein (SABP2) in tobacco. It reversibly binds SA (and its active analogues) with an affinity that is 150-fold higher than that between SA and catalase. SABP2's relatively high affinity for SA and its low abundance in leaves suggests that it may function as a receptor for SA.

Chivasa et al.[122] showed that treatment of tobacco with SA interferes with the normal function of the TMV RdRp complex. The pathway leading to this inhibition of virus replication is required for SAR. It appears to be distinct from that leading to PR-1 protein gene activation in that it requires the activity of alternative oxidase (AOX) or some other salicylic hydroxamic (SHAM)-sensitive activity. SHAM is an inhibitor of the SA-induced terminal oxidase of the alternative respiratory pathway in plants.[122]

Conrath et al.[113] reported that dephosphorylation of serine/threonine residues of two or more phosphoproteins is involved in induction of acidic *PR-1* genes. At least one of the target phosphoproteins acts downstream of SA, since the protein phosphatase inhibited okadaic acid, and calyculin A blocked SA-mediated induction of *PR-1*.

In 1996, Yang and Klessig[123] reported the isolation of a tobacco Myb1 protein which appears to function as a signaling component in the pathway leading to *PR-1* gene expression downstream of SA. In animals, *myb* genes encode transcription factors involved in cell cycling, proliferation, and differentiation. So far in plants (e.g., barley, maize, potato, petunia, *Arabidopsis, Anthirrhinum*), homologs of *myb* oncogenes have been found to play a role in flavonoid biosynthesis, epidermal cell differentiation, salt stress, dehydration, and gibberellin response.[124-133] The tobacco *myb1* gene is activated systemically by incompatible pathogens, SA, 2,6-dichloroisonicotinic acid (INA), and their biologically active analogs. The *myb1* gene activation precedes *PR-1* gene expression.

Certain chemicals induce the accumulation of PR-1 proteins and enhance virus resistance in tobacco. These compounds include polyacrylic acid (PAA), thiamine, L-α-aminobutyric acid, barium chloride, the antiviral agent 2-thiouracil, β-1,4-endoxylanase, benzothiadiazole (BTH), INA, and cholera toxin.[134-140] Since all these chemical elicitors except BTH and INA induce the accumulation of SAG, it is suggested that they are involved in the SA signaling pathway leading to *PR-1* gene expression and SAR.[79,89,141] Studies performed by Conrath et al.[108] and Lawton et al.[141] indicated that INA and BTH activate the SAR pathway downstream of SA. In contrast to INA, BTH is able to bind to SABP2.[112]

2.3.3 Ethylene and Other Hormones

Hormones play a central role in regulating many aspects of plant growth, development, and stress response. The concentration of hormones present in plant tissue influences the expression of "stress"-related genes like *PR-1*. Basic PR-1 mRNA accumulates in hormone-starved tobacco callus.[75] In tissues that develop abnormally due to endogenously overproduced or exogenously applied hormones, both basic and acidic *PR-1* genes are activated.[75]

Cytokinins and auxins represent two important groups of plant hormones. Cytokinins are plant hormones which delay senescence, inhibit HR, and induce susceptibility to pathogens.[142] Auxins are growth-promoting hormones regulating different processes in plant development (e.g., cell division, elongation, and differentiation). Tobacco *PR-1* genes are not induced by the cytokinins kinetin and 6-benzylaminopurine (BAP, see references 75 and 78). Different plant species show a different response to the exogenous application of plant hormones. The tobacco *PR-1a* promoter does not significantly respond to treatment with the auxin 1-naphthaleneacetic acid (NAA). In contrast, Jung et al.[143] reported a strong induction of sunflower PR-1 proteins by NAA, whereas the auxin 2,4D (2,4-dichlorophenoxyacetic acid) does not induce PR-1 protein expression in this species.

A phytohormone that is able to induce both basic and acidic *PR-1* genes is ethylene.[44,75] Ethylene is produced in plants in response to stress conditions.[144] During the normal life cycle, ethylene production is induced at certain stages of plant growth (e.g., during germination, fruit ripening, flower senescence). Ethylene is not directly involved in the signal transduction pathway that leads to SA accumulation.[145] Treatment of tobacco with the ethylene-releasing compound ethephon results in a strong induction of basic *PR-1* genes, while acidic *PR-1* genes are hardly expressed. Basic PR-1 transcripts accumulate 1 to 2 days after ethephon treatment.[73,75] Inhibitors of protein kinase C have been shown to block the ethylene-induced transcription of *PR*-genes.[146] Ethylene-dependent *PR-1* gene induction requires light.[44] Histochemical analysis by Eyal et al.[147] showed ethylene inducible expression of the tobacco basic *PRB-1b* promoter/GUS transgene in the leaf abscission zone, the vascular tissues of the stem, and the petiole. In all organs except the leaf, inducible GUS expression was limited to vascular tissues, mainly the inner phloem. In sunflower, floating leaf discs on ethephon results in a strong induction of the acidic *PR-1* genes.[143] Ethephon treatment does not induce resistance to TMV in tobacco.[73] Lawton et al.[148] demonstrated that in *Arabidopsis*, SAR requires SA but not ethylene, while ethylene-insensitive mutants show the same biological induction of SAR as wild-type *Arabidopsis*. However, recent experiments by Knoester et al. (personal communication) indicated that ethylene may play a role in resistance. Tobacco plants carrying the mutant *etr1-1* ethylene receptor gene from *Arabidopsis thaliana* are insensitive to ethylene. These plants no longer display TMV-inducible basic *PR-1* gene expression in the infected leaf, while their acidic *PR-1* gene expression is not altered. These ethylene insensitive plants are severely compromised in basal resistance to fungal pathogens, and initial studies revealed that induced resistance (SAR) may also be diminished.

The precursor of ethylene is methionine. The two key enzymes in the ethylene biosynthesis pathway are 1-aminocyclopropane-1-carboxylic acid (ACC) synthase and ACC-oxidase. ACC-synthase catalyzes the conversion of *S*-adenosyl-L-methionine (SAM) to ACC. ACC-oxidase catalyzes the conversion of ACC to ethylene.[144] ACC-synthase and ACC-oxidase are the rate-limiting enzymes in ethylene biosynthesis.[149] The elicitor β-1,4-endoxylanase (EIX) isolated from *Trichoderma viride* stimulates accumulation of ACC-oxidase and ACC-synthase transcripts, and elevated enzyme activity in Xanthi tobacco leaves.[150] EIX induces ethylene biosynthesis, accumulation of PR proteins and phytoalexins, electrolyte leakage, and hypersensitive responses in *N. tabacum* cv Xanthi.[136,151,152] The accumulation of basic PRB-1b transcript shows complex regulation with regard to ethylene and EIX. EIX treatment stimulates accumulation of PRB-1b transcript within 4 hours. Ethylene treatment of Xanthi tobacco leaves for 8 to 16 hours induces a much higher accumulation of PRB-1b transcript. However, EIX treatment caused a decline in PRB-1b transcript levels in leaves pretreated (16 hours) with ethylene.[153] Recently, a high-affinity binding site for EIX was found on tobacco membranes. This protein may function as the receptor that mediates the hypersensitive response induced by EIX binding.[154]

Xu et al.[155] suggested that signal combinations may synergistically hyperinduce plant defense genes and that such synergistic signals may be more specifically related to gene function than any single inductive signal. The authors showed that the acidic *PR-1b* gene was synergistically induced by ethylene/methyl jasmonate combinations. The plant-growth regulator jasmonic acid (JA) and its methyl ester methyl jasmonate (MeJa) are widely distributed in the plant kingdom. Jasmonates, the key molecules of the octadecanoid signaling pathway, have been proposed to be stress-related compounds. MeJa in particular is a signal molecule that is released in plants in response to various stresses, such as wounding and pathogen attack. In dicotyledonous species, defense-related genes are usually induced either by wounding and jasmonates or by pathogen attack and SA.[156-159] Jasmonates may be embedded in a signaling network together with other pathogen induced pathways. All the different plant responses to jasmonates appear to be correlated with alterations in gene expression. Different plant species show distinct reaction patterns to exogenously applied jasmonates. In contrast to treatment with SA, application of JA to rice plants induces the accumulation of PR-1-like proteins.[160] However, in tobacco, the acidic *PR-1b* gene is not induced by MeJa, while SA treatment of these plants results in high levels of *PR-1b* expression. A combination of MeJa and SA synergistically hyper-induces *PR-1b* in tobacco.[155] The difference in response to SA and JA might be explained by the fact that the basal levels of these hormones vary among plant species. For example, in contrast to tobacco, high basal levels of nonconjugated SA are found in rice leaves.[161]

2.3.4 UV-Light

Exposure of tobacco plants to UV-C light (200 to 280 nm) induces basic and acidic *PR-1* genes. Like TMV-infection and SA-treatment, UV-C irradiation induces virus resistance as measured by the reduced size of TMV local lesions.[73] However, the *PR-1* gene induction by UV-C is correlated with extensive leaf damage, which itself might induce a pathogenesis response.[73] Green and Fluhr[87] showed that UV-B (280 to 320 nm) irradiation results in accumulation of PR-1a, -1b, and -1c proteins in Samsun NN tobacco, in a fluence-dependent manner. The experiments were performed using UV-B fluences which were well below the range of fluences that caused detectable damage. Accumulation of PR-1 transcripts by continuous UV-B irradiation could be detected after 10 hours. PR-1 protein accumulation occurred in all irradiated parts of the leaf. Irradiation with UV-A (320 to 360 nm) did not cause induction of *PR-1* genes.[87] UV-light responsiveness of *PR-1* genes appears to be species specific. Exposure of sunflower leaf discs to UV-light strongly induces PR-3 and PR-5 protein synthesis but fails to induce PR-1 protein production.[143] The signal transduction pathway between UV-B and *PR-1* gene expression in tobacco requires active oxygen species and protein synthesis since antioxidants and cycloheximide are able to block the induction of *PR-1* by UV-B. Treatment of leaves with a generator of reactive oxygen species (ROS) causes the accumulation of PR-1 protein. The ROS involved in this pathway are not generated during photosynthesis.[87]

2.3.5 Developmental Stimuli

Acidic PR-1 proteins are not detectable in young, healthy plants.[7,65,75,162] However, Fraser[163] detected PR-1a and -1b proteins in healthy untreated tobacco Xanthi nc leaves as the plants began to flower and senesce. Tobacco PR-1a, -1b, and -1c proteins are present in the sepal tissue of the flower and the mesophyll and epidermal tissues of the leaf blade. No accumulation of PR-1 proteins is found in petals, filaments, and seed capsules.[78,164,165] Fraser[163] suggested that the induction of the PR-1 proteins in mature plants depends on the presence of both the senescing lower leaves and the developing inflorescence. In contrast, Grüner and Pfitzner[165] showed that expression of the endogenous PR-1 proteins and the GUS reporter gene under the control of the *PR-1a* promoter is readily detected in vegetative tobacco plants which do not yet carry any flower buds. The expression starts, without any prior inductive treatment, 12 weeks after germination in the lowest leaves of the plants. The lowest leaves do not display any signs of senescence at this time. The expression develops into distinct gradients from the bottom to the top in the green leaves of the plants. Recently, Hanfrey et al.[55] identified a *PR-1a*-like gene from *Brassica napus* that is expressed early in leaf senescence, before any yellowing of the leaves is visible. The expression of this gene appears to be triggered by a developmental process, such as flowering or silique development.

High levels of basic PR-1 mRNA have been reported to occur constitutively in tobacco roots.[75] Irrespective of microbial attack or other stress conditions, tobacco basic PR-1 proteins are present in abscission zones of leaves and inflorescences. Vascular tissue of stems and petioles also shows a low level of constitutive basic *PR-1* gene expression.[147]

2.4 PR-1 PROMOTER ANALYSIS

Several lines of evidence suggest that regulation of PR protein biosynthesis occurs primarily at the level of gene transcription. Possibly, the regulation of acidic and basic *PR-1* gene expression depends on the binding of certain transcription factors to particular regions of the *PR-1* promoter. Indeed, important *cis*-regulatory regions have been characterized in a number of plant *PR*-genes.[166] In addition, several protein factors have been identified which interact with *PR-1* promoter sequences

2.4.1 *Cis* Regulatory Regions in the Promoters of Acidic *PR-1* Genes

A number of studies have been performed to characterize upstream regions *PR-1* genes for a possible role in the induction of gene expression. These studies all made use of transgenic plants transformed with gene constructs in which various *PR-1* promoter regions, alone or in combination with the "core" 35S promoter of CaMV (nt –90 to + 1), were fused to reporter genes like the bacterial β-glucuronidase gene (GUS). Different results were obtained when a series of promoter deletions was tested for TMV- and SA-induced expression in transgenic tobacco plants.

A *PR-1a* promoter fragment containing nucleotides –902 to +29 is able to confer high levels of induced expression to the GUS reporter gene in transgenic tobacco.[80] When this *PR-1a* promoter fragment is gradually truncated from the 5′ end to position –689, the expression level is lowered several times.[167]

According to experiments described by Ohshima et al.,[77] the first 300 bp of the *PR-1a* promoter are sufficient to drive TMV- and SA-dependent reporter gene expression. In contrast, Uknes et al.,[78] Van de Rhee et al.,[80] and Beilmann et al.[168] failed to observe TMV or SA inducibility in studies using 300 bp *PR-1a* promoter fragments fused to the GUS gene. Possibly, these conflicting results can be ascribed to the presence of a 35S enhancer element immediately upstream of the *PR-1* promoter/GUS transgene. Beilmann et al.[104] showed that the *PR-1a* promoter can be influenced by a heterologous enhancer located upstream or downstream of the transcription start site. However, more recent results of Ohashi and Ohshima[169] with an enhancer that is not SA-responsive confirmed the sufficiency of the 300 bp *PR-1a* promoter for the induction of the GUS reporter gene by SA and TMV. While Ohshima et al.[77] marked a 50% increase in the level of GUS activity as "induced," Van de Rhee et al.[80] and Uknes et al.[78] used a level of 500% as the lower-level limit for induction. Similar conclusions would probably have been drawn when all groups had used the same criteria. Van de Rhee et al.[80] and Uknes et al.[78] reported that, respectively, 689 and 661 bp of the *PR-1a 5′*-flanking sequence are required for SA/TMV-induced gene expression. Since constructs with 643 bp or smaller of the *PR-1a* upstream sequence were inactive, Van de Rhee et al.[80] suggested that the region between position –689 and –643 is important for gene induction by SA treatment or TMV infection.

Van de Rhee and Bol[167] showed that the *PR-1a* promoter contains at least four regulatory elements which function in a context-dependent way. These interacting elements are located between nucleotides –902 and –691 (element 1), –689 and –643 (element 2), –643 and –287 (element 3), and –287 and +29 (element 4). All four elements are required for maximum induction of the reporter gene by TMV or SA. The authors did not detect promoter activity with each fragment alone. Elements 1 to 3 positively regulate the *PR-1a* promoter; element 4 is suggested to be important for a correct spacing between the other three elements and the transcription start site.

Beilmann et al.[168] suggest that the *PR-1a* promoter consists of at least two functional domains. One is located upstream of position –335 and contains a strong positive regulatory element. The other domain resides within the region between –71 and +28 of the *PR-1a* gene. This region is sufficient to support considerable transcription from the *PR-1a* promoter in transient assays. In stable transformants, this region is nearly inactive. However, activation to high levels occurs in transgenic plants by heterologous enhancers like the CaMV 35S enhancer.

Uknes et al.[78] described data indicating the presence of *PR-1a* promoter sequences responsible for high-level constitutive expression that lie approximately at position –300 and a possible negative regulatory element that contributes to constitutive expression near position –150. DNA sequence analysis of the *PR-1a* promoter of Wisconsin 38 tobacco revealed a consensus TATA box at position 34 and an 11 bp long imperfect direct repeat at positions –116 and –135 with respect to the transcription start site. At position –57, the sequence CTAATTTCTG is found. This sequence differs from the consensus heat shock regulatory element (HSE) only by the insertion of one extra nucleotide. Identical or very similar structural elements were found at the same positions in two acidic *PR-1* pseudogenes.[21] The HSE is a common motif in the flanking region of all heat-inducible genes. It has been demonstrated to have a positive effect on transcription.[170] Although the induction of *PR-1* genes by TMV and SA is known to be temperature-dependent, heat shock treatment does not induce the synthesis of PR-1 proteins.[19,90] Taken together, these data suggest that the *PR-1a* promoter contains a mixture of *cis* regulatory regions with either a positive or negative effect on gene expression.

2.4.2 *Cis* Regulatory Regions in the Promoters of Basic *PR-1* Genes

Promoter analysis of the basic *PRB-1b* gene of tobacco showed that the promoter sequence from position –213 to the transcription start site was sufficient to direct ethylene responsiveness. Constructs using 863 bp of the basic *PRB-b1* promoter fused to the GUS gene display 25 to 50% higher GUS activity compared to the 213 bp basic *PRB-1b* promoter/GUS construct.[147] Studies by Meller et al.[171] indicated that regulatory sequences necessary for activation of the basic *PRB-1b* gene by ethylene are located in the promoter between positions –213 and –142. Within this region two protein binding sites were mapped: region G (–200 to –178) and region Y (–179 to –154). The G region contains a G-box consensus core sequence at position –187. G-box motifs, which are commonly found in plant promoters, bind a subclass of bZIP DNA-binding proteins.[44,172-175] Besides the G-box motif, the G region also contains the 11 bp sequence, TAAGAGCCGCC (GCC-box), which is highly conserved in promoters of ethylene-induced genes of *Nicotiana* species. Ohme-Takagi and Shinshi[176] showed that the GCC-box is essential for ethylene responsiveness when incorporated into a heterologous promoter (e.g., CaMV 35S promoter).

2.4.3 Protein Factors Interacting with Acidic *PR-1* Promoter Sequences

In 1993, Hagiwara et al.[177] demonstrated the binding of nuclear proteins from healthy tobacco to a region in the *PR-1a* promoter between positions –68 and –51 (which contains the HSE-like sequence) and a region between positions –184 and –172. Although this binding was sequence specific, the proteins were not able to distinguish one binding site from the other. No protein binding to these two regions was found using nuclear proteins derived from the interspecific hybrid of *Nicotiana glutinosa* x *Nicotiana debneyi,* which constitutively produces the PR-1a protein. Gel retardation assays performed by Buchel et al.,[178] using *PR-1a* promoter fragments comparable to those of Hagiwara et al.,[177] also showed the formation of a slow migrating protein-DNA complex with nuclear proteins from healthy Samsun NN tobacco. In contrast to the results obtained with proteins from the interspecific hybrid, Buchel et al.[178] detected the same shift when proteins from tobacco plants were used which display SA- or TMV-induced *PR-1* expression. Extending the gel shift analysis to more upstream *PR-1a* promoter regions showed that the same nuclear factor(s) bind(s) various promoter fragments with different affinity. The factors involved in these interactions are probably GT-1-like factors.[178] GT-1 is a nuclear factor which interacts with DNA sequences in light-responsive and other inducible plant gene promoters like the pea *rbc S-3A* promoter and the rice *phyA* promoter. GT-1 probably binds to the DNA in a homotetrameric form.[179] The binding sites are usually positioned in tandem. The spacing between the neighboring sites appears to be critical for promoter activity but not for GT-1 binding *in vitro*.[180] The *PR-1a* promoter contains a number of putative GT-1 binding sites distributed over the entire 900 bp upstream region. In comparison to extracts from untreated plants, nuclear protein preparations from SA-treated and TMV-infected tobacco showed a severely reduced GT-1 binding activity to the upstream *PR-1a* promoter region, which was earlier shown to be important for inducible gene expression (–906 to –656). Mutation of four of the putative GT-1 binding sites in this part of the *PR-1a* promoter resulted in a reduced GT-1 binding *in vitro* (Buchel et al., unpublished results). *In vivo* experiments with a mutated *PR-1a* promoter fused to the GUS reporter gene will help to elucidate the role of GT-1 binding in *PR-1a* gene regulation.

TGA1a is a transcription factor from tobacco that specifically binds to the SA-responsive motifs *as-1* and *ocs*.[173,181,182] The elements *as-1* and *ocs* are present in the CaMV 35S promoter and the promoter of the octopine synthase gene of *Agrobacterium tumefaciens,* respectively. TGA1a, which is highly expressed in roots, belongs to the basic-leucine-zipper (bZIP) class of transcriptional factors. Southern blot and genomic cloning analyses have shown that the genome of tobacco contains at least four closely related genes corresponding to TGA1a.[183] The N-terminus of TGA1a is important for transactivation, while the DNA-binding domain resides in the bZIP region, present in the central part of the protein.[184,185] Homo- and/or heterodimer formation is essential for DNA binding. The C-terminal region of TGA1a is involved in the stabilization of dimers.[184] Strompen and Pfitzner[186] expressed the transcription factor TGA1a in bacteria. Although TGA1a specifically binds to an *as-1*-like sequence in the *PR-1a* promoter (present at position –593), the role of this factor in transcriptional activation of the *PR-1a* gene is unclear.

Several phytohormones are known to enhance the activity of promoters containing *as-1* type elements (e.g., CaMV 35S promoter and nopaline synthase promoter). These include SA, MeJa, and auxins, but not cytokinin and abscisic acid. Other stress-related treatments, such as heat shock and high concentrations of cadmium, have little or no effect.[187] As mentioned before, a number of (hormone) treatments have the same effect on induction of the *PR-1* genes. It is possible that the effects of these different hormones on the expression of the *PR-1a* gene may be at least partly mediated through binding of TGA1a to the *as-1*-like element present in the *PR-1a* promoter.

Yang and Klessig[123] reported the specific binding to the *PR-1a* promoter of the product of a *myb1* oncogene homolog from tobacco. The *PR-1a* promoter contains several Myb1 binding sites in the region between positions –643 and –169. Strong binding of recombinant Myb1 protein was found to a binding site at position –520 to –514. Exogenous SA treatment rapidly (within 15 min) activated the expression of tobacco *myb1* in both resistant and susceptible cultivars. The subsequent induction of *PR-1* genes occurred several hours later (between 6 and 12 hours). This delay suggests that, while *myb1* gene activation by SA may be necessary, it is probably not sufficient for the induction of the *PR-1* genes.[123] Myb1 proteins are known to be involved in activation of gene expression in combination with other transcription factors.[188] Therefore, the tobacco Myb1 protein may act in concert with additional factors during the activation of *PR-1* genes.[123]

2.4.4 Protein Factors Interacting with Basic *PR-1* Promoter Sequences

Eyal et al.[147] studied the binding of nuclear proteins to the promoter of the basic *PRB-1b* gene in tobacco. Gel retardation experiments with subfragments of the 863 bp promoter region detected four specific protein-DNA complexes within the regions –642 to –369 and –240 to –141. No differences in the number or intensity of the interactions were observed using extracts from untreated or ethylene-induced plants. Meller et al.[171] fractionated crude tobacco nuclear extract by heparin–agarose chromatography and analyzed the binding of proteins to the ethylene responsive *PRB-1b* promoter region (–213 and –142). Two binding activities were separated which show differential specificity toward the protein binding sites G (–200 to –178) and Y (–179 to –154). The first factor interacts with both sequences G and Y. The other factor specifically binds to the region G. Mutations in the G region, which alter the GCC-box or the G-box motif, disrupt factor binding and abolish the ethylene inducibility of the minimum –213 bp basic *PRB-1b* promoter/GUS transgene. Ohme-Takagi and Shinshi[176] cloned and characterized cDNAs corresponding to four ethylene-responsive element binding proteins (EREBPs) from tobacco that specifically interact with the GCC box. These EREBPs are novel DNA binding proteins that are unrelated to bZIP and zinc finger protein families. They exhibit no sequence homology with known transcription factors or DNA binding proteins.

So far, a number of binding activities/transcription factors have been identified which show interaction with particular sequences present in the promoter regions of different *PR-1* genes. Future experiments will help to elucidate the role of these protein–DNA interactions with respect to the induction and expression of *PR-1* genes in plants.

REFERENCES

1. Ross, A. F., Localized acquired resistance to plant virus infection in hypersensitive hosts, *Virology*, 14, 329, 1961.
2. Ross, A. F., Systemic acquired resistance induced by localized virus infections in plants, *Virology*, 14, 340, 1961.
3. Collinge, D. B. and Slusarenko, A. J., Plant gene expression in response to pathogens, *Plant Mol. Biol.*, 9, 389, 1987.
4. Kuc, J., Induced immunity to plant disease, *Bioscience*, 32, 854, 1982.
5. Van Loon, L. C. and Antoniw, J. F., Comparison of the effects of salicylic acid and ethephon with virus-induced hypersensitivity and acquired resistance in tobacco, *Neth. J. Plant Path.*, 88, 237, 1982.
6. Van Loon, L. C., Pierpoint, W. S., Boller, T. H., and Conejero, V., Recommendations for naming plant pathogenesis-related proteins, *Plant Mol. Biol. Rep.*, 12, 245, 1994.
7. Van Loon, L. C. and Van Kammen, A., Polyacrylamide disc electrophoresis of the soluble leaf proteins from *Nicotiana tabacum* var. "Samsun" and "Samsun NN." II. Changes in protein constitution after infection with TMV, *Virology*, 40, 199, 1970.

8. Gianinazzi, S., Martin, C., and Vallee, J.C., Hypersensibilité aux virus, température et protéines soluble chez le *Nicotiana* Xanthi n.c. Apparition de novelles macromolécules lors de la répression de la synthèse virale, *C.R. Acad. Sci. (Ser. D),* 270, 2383, 1970.
9. Linthorst, H. J. M., Pathogenesis-related proteins of plants, *Crit. Rev. Plant Sci.,* 10, 123, 1991.
10. Stintzi, A., Heitz, T., Prasad, V., Weidmann-Merdinoglu, S., Kauffman, S., Geoffrey, P., Legrand, M., and Fritig, B., Plant pathogenesis-related proteins and their role in defense against pathogens, *Biochimie,* 75, 687, 1993.
11. Sommsich, I. E., Schmelzer, E., Bollman, J., and Hahlbrock, K., Rapid activation by fungal elicitor of genes encoding "pathogenesis-related" proteins in cultured parsley cells, *Proc. Natl. Acad. Sci. U.S.A.,* 83, 2427, 1986.
12. Alexander, D., Goodman, R. M., Gut-Rella, M., Glascock, C., Weymann, K., Friedrich, L., Maddox, D., Ahl-Goy, P., Luntz, T., Ward, E., and Ryals, J., Increased tolerance to 2 oomycete pathogens in transgenic tobacco expressing pathogenesis-related protein-1a, *Proc. Natl. Acad. Sci. U.S.A.,* 90, 7327, 1993.
13. Cornelissen, B. J. C., Hooft Van Huijsduijnen, R. A. M., Van Loon, L. C., and Bol, J. F., Molecular characterization of messenger RNAs for "pathogenesis-related" proteins 1a, 1b and 1c, induced by TMV infection of tobacco, *EMBO J.,* 5, 37, 1986.
14. Payne, G., Parks, T. D., Burkhart, W., Dincher, S., Carnes, M., and Ryals, J., Isolation of the genomic clone for pathogenesis-related protein 1a from *Nicotiana tabacum* cv. Xanthi-nc, *Plant Mol. Biol.,* 11, 89, 1988.
15. Pfitzner, U. M. and Goodman, H. M., Isolation and characterization of cDNA clones encoding pathogenesis-related proteins from tobacco mosaic virus infected tobacco plants, *Nucleic Acids Res.,* 15, 4449, 1987.
16. Ohshima, M., Matsuoka, M., Yamamoto, N., Tanaka, Y., Kano-Murakami, Y., Ozeki, Y., Kato, A., Harada, N., and Ohashi, Y., Nucleotide sequence of the PR-1 gene of *Nicotiana tabacum, FEBS Lett.,* 225, 243, 1987.
17. Ohshima, M., Harada, N., Matsuoka, M., and Oyashi, Y., The nucleotide sequence of pathogenesis-related *(PR) 1b* protein gene of tobacco, *Nucleic Acids Res.,* 18, 181, 1990.
18. Ohshima, M., Harada, N., Matsuoka, M., and Oyashi, Y., The nucleotide sequence of pathogenesis-related *(PR) 1c* protein gene of tobacco, *Nucleic Acids Res.,* 18, 182, 1990.
19. Cornelissen, B. J. C., Horowitz, J., Van Kan, J. A. L., Goldberg, R. B., and Bol, J. F., Structure of tobacco genes encoding pathogenesis-related proteins from the PR-1 group, *Nucleic Acids Res.,* 15, 6799, 1987.
20. Pfitzner, A. J. P., Pfitzner, U. M., and Goodman, H. M., Nucleotide sequences of two *PR-1* pseudogenes from *Nicotiana tabacum* cv. Wisconsin 38, *Nucleic Acids Res.,* 18, 3404, 1990.
21. Pfitzner, U. M., Pfitzner, A. J. P., and Goodman, H. M., DNA sequence analysis of a *PR-1a* gene from tobacco: molecular relationship of heat shock and pathogen responses in plants, *Mol. Gen. Genet.,* 211, 290, 1988.
22. Matsuoka, M., Yamamoto, N., Kano-Murakami, Y., Tanaka, Y., Ozeki, Y., Hirano, H., Kagawa, H., Oshima, M., and Ohashi, Y., Classification and structural comparison of full-length cDNAs for pathogenesis-related proteins, *Plant Physiol.,* 85, 942, 1987.
23. Ahl, P., Cornu, A., and Gianinazzi, S., Soluble proteins as genetic markers in studies of resistance and phylogeny in *Nicotiana, Phytopathology,* 72, 80, 1982.

24. Carr, J. P., Dixon, D. C., and Klessig, D. F., Synthesis of pathogenesis-related proteins in tobacco is regulated at the level of mRNA accumulation and occurs on membrane bound polysomes, *Proc. Natl. Acad. Sci. U.S.A.*, 82, 7999, 1985.
25. Lucas, J., Camacho Henriquez, A., Lottspeich, F., Henschen, A., and Sänger, H. L., Amino acid sequence of the "pathogenesis-related" leaf protein p14 from viroid-infected tomato reveals a new type of structurally unfamiliar proteins, *EMBO J.*, 4, 2745, 1985.
26. Antoniw, J. F., White, R. F., Barbara, D. J., Jones, P., and Longey, L., The detection of PR(b) protein and TMV by Elisa in systemic and localized virus infections of tobacco, *Plant Mol. Biol.*, 4, 55, 1985.
27. Matsuoka, M. and Ohashi, Y., Biochemical and serological studies of pathogenesis-related proteins of *Nicotiana* species, *J. Gen. Virol.*, 65, 2209, 1984.
28. Van Loon, L. C., Gerritsen, Y. A. M., and Ritter, C. E., Identification, purification and characterization of pathogenesis-related proteins from virus-infected Samsun NN tobacco leaves, *Plant Mol. Biol.*, 9, 593, 1987.
29. Nassuth, A. and Sänger, H. L., Immunological relationships between "pathogenesis-related" leaf proteins from tomato, tobacco and cowpea, *Virus Res.*, 4, 229, 1986.
30. White, R. F., Rybicki, E. P., Von Wechmar, M. B., Dekker, J. L., and Antoniw, J. F., Detection of PR-1 type proteins in Amaranthaceae, Chenopodiaceae, Gramineae and Solanaceae by immunoelectroblotting, *J. Gen. Virol.*, 68, 2043, 1987.
31. Bol, J. F., Linthorst, H. J. M., and Cornelissen, B. J. C., Plant pathogenesis-related proteins induced by virus infection, *Annu. Rev. Phytopathol.*, 28, 113, 1990.
32. Van Loon, L. C., Specific soluble leaf proteins in virus-infected tobacco plants are not normal constituents, *J. Gen. Virol.*, 30, 375, 1976.
33. Gianinazzi, S., Pratt, H. M., Shewry, P. R., and Miflin, B. J., Partial purification and preliminary characterization of soluble proteins specific to virus infected tobacco plants, *J. Gen. Virol.*, 34, 345, 1977.
34. Van Loon, L. C., Regulation of changes in proteins and enzymes associated with defence against virus infection, in *Active Defense Mechanisms in Plants,* Wood, R. K. S., Ed., Plenum Press, New York, 1982, 247.
35. Van Loon, L. C. and Gerritsen, Y. A. M., Protease activity and pathogenesis-related proteins in virus-infected Samsun nn tobacco leaves, *Plant Sci.*, 63, 141, 1989.
36. Parent, J. G. and Asselin, A., Detection of pathogenesis-related proteins (PR or b) and of other proteins in the intercellular fluid of hypersensitive plants infected with tobacco mosaic virus, *Can. J. Bot.*, 62, 564, 1984.
37. Dumas, E., Lherminier, J., Gianninazzi, S., White, R. F., and Antoniw, J. F., Immunocytochemical location of pathogenesis-related b1 protein induced in tobacco mosaic virus-infected or polyacrylic acid-treated tobacco plants, *J. Gen. Virol.*, 69, 2687, 1988.
38. Ohashi, Y. and Matsuoka, M., Localization of pathogenesis-related proteins in the epidermis and intercellular spaces of tobacco leaves after their induction by potassium salicylate or tobacco mosaic virus infection, *Plant Cell Physiol.*, 28, 1227, 1987.
39. Carr, J. P., Dixon, D. C., Nikolau, B. J., Voelkerding, K. V., and Klessig, D. F., Synthesis and localization of pathogenesis-related proteins in tobacco, *Mol. Cell. Biol.*, 7, 1580, 1987.
40. Dixon, D. C., Cutt, J. R., and Klessig, D. F., Differential targeting of the tobacco PR-1 pathogenesis-related proteins to the extracellular space and vacuoles of crystal idioblasts, *EMBO J.*, 10, 1317, 1991.

41. Rodrigo, I., Vera, P., Van Loon, L. C., and Conejero, V., Degradation of tobacco pathogenesis-related proteins. Evidence for conserved mechanisms of degradation of pathogenesis-related proteins in plants, *Plant Physiol.*, 95, 616, 1991.
42. Joosten, M. H. A. J., Bergmans, C. J. B., Meulenhoff, E. J. S., Cornelissen, B. J. C., and De Wit, P. J. G. M., Purification and serological characterization of three basic 15-kilodalton pathogenesis-related proteins from tomato, *Plant Physiol.*, 94, 585, 1990.
43. Payne, G., Middlesteadt, W., Desai, N., Williams, S., Dincher, S., Carnes, J., and Ryals, J., Isolation and sequence of a genomic clone encoding the basic form of pathogenesis-related protein 1 from *Nicotiana tabacum, Plant Mol. Biol.*, 12, 595, 1989.
44. Eyal, Y., Sagee, O., and Fluhr, R., Dark-induced accumulation of a basic pathogenesis-related (PR-1) transcript and a light requirement for its induction by ethylene, *Plant Mol. Biol.*, 19, 589, 1992.
45. Niderman, T., Genetet, I., Bruyère, T., Gees, R., Stinzi, A., Legrand, M., Fritig, B., and Mösinger, E., Pathogenesis-related PR-1 proteins are antifungal: Isolation and characterization of three 14 kilodalton proteins of tomato and of a basic PR-1 of tobacco with inhibitory activity against *Phytophthora infestans, Plant Physiol.*, 108, 17, 1995.
46. Sessa, G., Yang, X. Q., Raz, V., Eyal, Y., and Fluhr, R., Dark-induction and subcellular localization of the pathogenesis-related PRB-1b protein, *Plant Mol. Biol.*, 28, 537, 1995.
47. Van Kan, J. A. L., Structure and expression of virus-inducible plant genes. Ph.D. thesis, State University of Leiden, The Netherlands, 1988.
48. Melchers, L. S., Sela-Buurlage, M. B., Vloemans, S. A., Woloshuk, C. P., Van Roekel, J. S. C., Van den Elzen, P. J. M., and Cornelissen, B. J. C., Extracellular targeting of the vacuolar tobacco proteins AP24, chitinase and β-1,3-glucanase in transgenic plants, *Plant Mol. Biol.*, 21, 583, 1993.
49. Neuhaus, J. M., Sticher, L., Meins, F., Jr., and Boller, T. A short C-terminal sequence is necessary and sufficient for the targeting of chitinases to the plant vacuole, *Proc. Natl. Acad. Sci. U.S.A.*, 88, 10362, 1991.
50. Schuren, F. H. J., Asgeirsdottic, S. A., Kothe, E. M., Scheer, J. M. J., and Wessels, J. G. H., The *Sc7/Sc14* gene family of *Schizophyllum commune* codes for extracellular proteins specifically expressed during fruit-body formation, *J. Gen. Microbiol.*, 139, 2083, 1993.
51. Miosga, T., Schaaff-Gerstenschlager, I., Chalwatzis, N., Baur, A., Boles, E., Fournier, C., Schmitt, S., Velten, C., Wilhelm, N., and Zimmermann, F. K., Sequence analysis of a 33.1 kb fragment from the left arm of *Sacharomyces cervisiae* chromosome X, including putative proteins with leucine zippers, a fungal Zn(II)2-Cys-6 binuclear cluster domain and a putative alpha 2-SCB-alpha 2 binding sit, *Yeast*, 11(7), 681, 1995.
52. Fang, K. S. F., Vitale, M., Fehlner, P., and King, T. P., cDNA cloning and primary structure of a white-face hornet venom allergen, antigen 5, *Proc. Natl. Acad. Sci. U.S.A.*, 85, 895, 1988.
53. Murphy, E. V., Zhang, Y., Zhu, W. J., and Biggs, J., The human glioma pathogenesis-related protein is structurally related to plant pathogenesis-related proteins and its gene is expressed specifically in brain tumors, *Gene*, 159, 131, 1995.
54. Jung, J. L., Fritig, B., and Hahne, G., Sunflower (*Helianthus annuus* L.) pathogenesis-related proteins: induction by aspirin (acetylsalicylic acid) and characterization, *Plant Physiol.*, 101, 873, 1993.

55. Hanfrey, C., Fife, M., and Buchanan-Wollaston, V., Leaf senescence in *Brassica napus:* expression of genes encoding pathogenesis-related proteins, *Plant Mol. Biol.,* 30, 597, 1996.
56. Casacuberta, J. M., Puigdomènech, P., and San Segundo, B., A gene coding for a basic pathogenesis-related (PR-like) protein from *Zea mays*. Molecular cloning and induction by a fungus (*Fusarium moniliforme*) in germinating maize seeds, *Plant Mol. Biol.,* 16, 527, 1991.
57. Van Eldik, G. J., Wingens, M., Ruiter, R. K., Van Herpen, M. M. A., Schrauwen, J. A. M., and Wullems, G. J., Molecular analysis of a pistil-specific gene expressed in the stigma and stylar cortex of *Solanum tuberosum, Plant Mol. Biol.,* 30, 171, 1996.
58. Szybiak-Strozycka, U., Lescure, N., Cullimore, J. V., and Gamas, P., A cDNA encoding a PR-1-like protein in the model legume *Medicago truncatula, Plant Physiol.,* 107, 273, 1995.
59. Uknes, S., Mauch-Mani, B., Moyer, M., Potter, S., Williams, S., Dincher, S., Chandler, D., Slusarenko, A., Ward, E., and Ryals, J., Acquired resistance in *Arabidopsis, Plant Cell,* 4, 645, 1992.
60. Bryngelsson, T., Sommer-Knudsen, J., Gregersen, P. L., Collinge, D. B., Ek, B., and Thordal-Christensen, H., Purification, characterization, and molecular cloning of basic PR-1-type pathogenesis-related proteins from barley, *Mol. Plant-Microbe Interact.,* 7, 267, 1994.
61. Devereux, J., Haeberli, P., and Smithies, O., A comprehensive set of sequence analysis programs for the VAX, *Nucleic Acid Res.,* 12, 387, 1984.
62. Tornero, P., Conejero, V., and Vera, P., A gene encoding a novel isoform of the PR-1 protein family from tomato is induced upon viroid infection, *Mol. Gen. Genet.,* 243, 47, 1994.
63. Ahl, P., Antoniw, J. F., White, R. F., and Gianinazzi, S., Biochemical and serological characterization of b-proteins from *Nicotiana* species, *Plant Mol. Biol.,* 4, 31, 1985.
64. Gianinazzi, S., Ahl, P., and Cornu, A., b-Protein variation in virus-infected intraspecific tobacco hybrids, *Acta Phytopathol. Acad. Sci. Hung.,* 16, 73, 1981.
65. Antoniw, J. F., Ritter, C. E., Pierpoint, W. S., and Van Loon, L. C., Comparison of three pathogenesis-related proteins from plants of two cultivars of tobacco infected with TMV, *J. Gen. Virol.,* 47, 79, 1980.
66. Ahl, P. and Gianinazzi, S., b-Protein as a constitutive component in highly (TMV) resistant interspecific hybrids of *Nicotiana glutonosa* x *Nicotiana debneyi, Plant Sci. Lett.,* 26, 173, 1982.
67. Linthorst, H. J. M., Meuwissen, R. L. J., Kauffmann, S., and Bol, J. F., Constitutive expression of pathogenesis-related proteins PR-1, GRP, and PR-S in tobacco has no effect on virus infection, *Plant Cell,* 1, 285, 1989.
68. Cutt, J. R., Harpster, M. H., Dixon, D. C., Dunsmuir, P., and Klessig, D. F., Disease response to tobacco mosaic virus in transgenic tobacco plants that constitutively express the pathogenesis related *PR-1b* gene, *Virology,* 173, 89, 1989.
69. Linthorst, H. J. M., Cornelissen, B. J. C., Van Kan, J. A. L., Van de Rhee, J., Meuwissen, R. L., Gonzalez-Jaén, M. T., and Bol, J. F., Induction of plant genes by compatible and incompatible virus-plant interactions in *Recognition and Response in Plant-Virus Interactions,* Fraser, R. S. S., Ed., Springer Verlag, Wien/New York, 1990, 361.
70. Bol, J. F., Van Rossum, C. M. A., Cornelissen, B. J. C., and Linthorst, H. J. M., Induction of host genes by the hypersensitive response of tobacco to virus infection, in *Viral Genes and Plant Pathogenesis,* Pirone, T. and Shaw, J. G., Eds., Springer Verlag, New York, 1990, 1.

71. Klessig, D. F., Cutt, J. R., Dixon, D. C., Carr, J. P., Malamy, J., Metzler, M., Harpster, M. H., Dunsmuir, P., Beachy, R. N., and Raskin, I., The PR-1 proteins — their synthesis, location and possible function (Abstract), in *Workshop on Pathogenesis Related Proteins in Plants,* Conejero, V. and Van Loon, L. C., Eds., Fundación Juan March, Madrid, Spain, 1989, 61.
72. Van Loon, L. C., Pathogenesis-related proteins, *Plant Mol. Biol.,* 4, 111, 1985.
73. Brederode, F. T., Linthorst, H. J. M., and Bol, J. F., Differential induction of acquired resistance and PR gene expression in tobacco by virus infection, ethephon treatment, UV light and wounding, *Plant Mol. Biol.,* 17, 1117, 1991.
74. Antoniw, J. F. and White, R. F., The effects of aspirin and polyacrylic acid on soluble leaf proteins and resistance to virus infection in five cultivars of tobacco, *Phytopath. Z.,* 98, 331, 1980.
75. Memelink, J., Linthorst, H. J. M., Schilperoort, R. A., and Hoge, J. H. C., Tobacco genes encoding acidic and basic isoforms of pathogenesis-related proteins display different expression patterns, *Plant Mol. Biol.,* 14, 119, 1990.
76. Hooft van Huijsduijnen, R. A. M., Cornelissen, B. J. C., Van Loon, L. C., Van Boom, J. H., Tromp, M., and Bol, J. F., Virus-induced synthesis of messenger RNAs for precursors of pathogenesis-related proteins in tobacco, *EMBO J.,* 4, 2167, 1985.
77. Ohshima, M., Itoh, H., Matsuoka, M., Murakami, T., and Ohashi, Y., Analysis of stress-induced or salicylic acid-induced expression of the pathogenesis-related *1a* protein gene in transgenic tobacco, *Plant Cell,* 2, 95, 1990.
78. Uknes, S., Dincher, S., Friedrich, L., Negrotto, D., Williams, S., Thompson-Taylor, H., Potter, S., Ward, E., and Ryals, H., Regulation of pathogenesis-related protein-*1a* gene expression in tobacco, *Plant Cell,* 5, 159, 1993.
79. Malamy, J., Sánchez-Casas, P., Hennig, J., Guo, A., and Klessig, D. F., Dissection of the salicylic acid signaling pathway in tobacco, *Mol. Plant Microbe Interact.,* 9, 474, 1996.
80. Van de Rhee, M. D., Van Kan, J. A. L., Gonzalez-Jaén, M. T., and Bol, J. F., Analysis of regulatory elements involved in the induction of two tobacco genes by salicylate treatment and virus infection, *Plant Cell,* 2, 357, 1990.
81. White, R. F., Acetylsalicylic acid (aspirin) induces resistance to tobacco mosaic virus in tobacco, *Virology,* 99, 410, 1979.
82. Malamy, J., Carr, J. P., Klessig, D. F., and Raskin, I., Salicylic acid: a likely endogenous signal in the resistance response of tobacco to viral infection, *Science,* 250, 1002, 1990.
83. Métraux, J.-P., Signer, H., Ryals, J., Ward, E., Wyss-Benz, M., Gaudin, J., Raschdorf, K., Schmid, E., Blum, W., and Inverardi, B., Increase in salicylic acid at the onset of systemic acquired resistance in cucumber, *Science,* 250, 1004, 1990.
84. Yalpani, N., Silverman, P., Wilson, T. M. A., Kleier, D. A., and Raskin, I., Salicylic acid is a systemic signal and an inducer of pathogenesis-related proteins in virus-infected tobacco, *Plant Cell,* 3, 809, 1991.
85. Uknes, S., Winter, A. M., Delaney, T., Vernooij, B., Morse, A., Friedrich, L., Nye, G., Potter, S., Ward, E., and Ryals, J., Biological induction of systemic acquired resistance in *Arabidopsis, Plant Cell,* 4, 645, 1993.
86. Rasmussen, J. B., Hammerschmidt, R., and Zook, M. N., Systemic induction of salicylic acid accumulation in cucumber after inoculation with *Pseudomonas syringae* pv syringae, *Plant Physiol.,* 97, 1342, 1991.
87. Green, R. and Fluhr, R., UV-B-induced PR-1 accumulation is mediated by active oxygen species, *Plant Cell,* 7, 203, 1995.

88. Enyedi, A. J., Yalpani, N., Silverman, P., and Raskin, I., Localization, conjugation, and function of salicylic acid in tobacco during the hypersensitive reaction to tobacco mosaic virus, *Proc. Natl. Acad. Sci. U.S.A.*, 89, 2480, 1992.
89. Hennig, J., Malamy, J., Grynkiewicz, G., Indulski, J., and Klessig, D. F., Interconversion of the salicylic acid signal and its glucoside in tobacco, *Plant J.*, 4, 593, 1993.
90. Malamy, J., Hennig, J., and Klessig, D. F., Temperature-dependent induction of salicylic acid and its conjugates during the resistance response to tobacco mosaic virus infection, *Plant Cell*, 4, 359, 1992.
91. Kassanis, B., Some effects of high temperatures on the susceptibility of plants to infection with viruses, *Ann. Appl. Biol.*, 39, 358, 1952.
92. Bowling, S. A., Guo, A., Cao, H., Cao, H., Gordon, S., Klessig, D. F., and Dong, X., A mutation in *Arabidopsis* that leads to constitutive expression of systemic acquired resistance, *Plant Cell*, 6, 1845, 1994.
93. Dietrich, R. A., Delaney, T. P., Uknes, S. J., Ward, E. R., Ryals, J. A., and Dangl, J. L., *Arabidopsis* mutants simulating disease resistance response, *Cell*, 77, 565, 1994.
94. Greenberg, J. T., Guo, A. L., Klessig, D. F., and Ausubel, F. M., Programmed cell death in plants: a pathogen-triggered response activated coordinately with multiple defense functions, *Cell*, 77, 551, 1994.
95. Cao, H., Bowling, S. A., Gordon, A. S., and Dong, X., Characterization of an *Arabidopsis* mutant that is nonresponsive to inducers of systemic acquired resistance, *Plant Cell*, 6, 1583, 1994.
96. Delaney, T. P., Friedrich, L., and Ryals, J. A., *Arabidopsis* signal transduction mutant defective in chemically and biologically induced disease resistance, *Proc. Natl. Acad. Sci. U.S.A.*, 92, 6602, 1995.
97. Shah, J., Tsui, F., and Klessig, D. F., Characterization of a salicylic acid-insensitive mutant (*sai1*) of *Arabidopsis thaliana*, identified in a selective screen utilizing the SA-inducible expression of the *tms2* gene, *Mol. Plant Microbe Interact.*, 10, 69, 1997.
98. Ryals, J., Weymann, K., Lawton, K., Friedrich, L., Ellis, D., Steiner, J. Y., Johnson, J., Delaney, T. P., Jesse, T., Vos, P., and Uknes, S., The *Arabidopsis* NIM1 protein shows homology to the mammalian transcription factor inhibitor IκB, *Plant Cell*, 9, 425, 1997.
99. Schreck, R., Rieber, P., and Baeuerle, P. A., Reactive oxygen intermediates as apparently widely used messengers in the activation of the NF-κB transcription factor and HIV-1, *EMBO J.*, 10, 2247, 1991.
100. Baeuerle, P. and Baltimore, D., NF-κB: Ten years after, *Cell*, 87, 13, 1996.
101. Baldwin, A., The NF-κB and IκB proteins: New discoveries and insights, *Annu. Rev. Immunol.*, 14, 649, 1996.
102. Gaffney, T., Friedrich, L., Vernooij, B., Negrotto, D., Nye, G., Uknes, S., Ward, E., Kessmann, H., and Ryals, J., Requirement of salicylic acid for the induction of systemic acquired resistance, *Science*, 261, 754, 1993.
103. Friedrich, L., Vernooij, B., Gaffney, T., Morse, A., and Ryals, J., Characterization of tobacco plants expressing a bacterial salicylate hydroxylase gene, *Plant Mol. Biol.*, 29, 959, 1995.
104. Beilmann, A., Albrecht, K., Schultze, S., Wanner, G., and Pfitzner, U. M., Activation of a truncated *PR-1* promoter by endogenous enhancers in transgenic plants, *Plant Mol. Biol.*, 18, 65, 1992.
105. Vernooij, B., Friedrich, L., Morse, A., Reist, R., Kolditz-Jawhar, R., Ward, E., Uknes, S., Kessman, H., and Ryals, J., Salicylic acid is not the translocated signal responsible for inducing systemic acquired resistance but is required in signal transduction, *Plant Cell*, 6, 959, 1994.

106. Shulaev, V., Leon, J., and Raskin, I., Is salicylic acid a translocated signal of systemic acquired resistance in tobacco?, *Plant Cell,* 7, 1691, 1995.
107. Chen, Z. and Klessig, D. F., Identification of a soluble salicylic acid binding protein that may function in signal transduction in plant disease resistance response, *Proc. Natl. Acad. Sci. U.S.A.,* 88, 8179, 1991.
108. Conrath, U., Chen, Z., Ricigliano, J. R., and Klessig, D. F., Two inducers of plant defense responses, 2,6-dichloroisonicotinic acid and salicylic acid, inhibit catalase activity in tobacco, *Proc. Natl. Acad. Sci. U.S.A.,* 92, 7143, 1995.
109. Chen, Z. X., Malamy, J., Hennig, J., Conrath, U., Sanchez-Casas, P., Silva H., Ricigliano, J., and Klessig, D. F., Induction, modification, and transduction of the salicylic acid signal in plant defense responses, *Proc. Natl. Acad. Sci. U.S.A.,* 92, 4134, 1995.
110. Sanchez-Casas, P. and Klessig, D. F., A salicylic acid-binding activity and a salicylic acid-inhibitable catalase activity are present in a variety of plant species, *Plant Physiol.,* 106, 1675, 1994.
111. Durner, J. and Klessig, D. F., Inhibition of ascorbate peroxidase by salicylic acid and 2,6-dichloroisonicotinic acid, two inducers of plant defense responses, *Proc. Natl. Acad. Sci. U.S.A.,* 92, 11312, 1995.
112. Du, H. and Klessig, D. F., Identification of a soluble, high affinity salicylic acid-binding protein in tobacco, *Plant Physiol.,* 113, 1319, 1997.
113. Conrath, U., Silva, H., and Klessig, D. F., Protein dephosphorylation mediates salicylic acid-induced expression of *PR-1* genes in tobacco, *Plant J.,* 11, 747, 1997.
114. Chen, Z. X., Silva, H., and Klessig, D. F., Active oxygen species in the induction of plant systemic acquired resistance by salicylic acid, *Science*, 262, 1883, 1993.
115. Takahashi, H., Chen, Z., Du, H., Liu, Y., and Klessig, D. F., Development of necrosis and activation of disease resistance in transgenic tobacco plants with severely reduced catalase levels, *Plant J.,* 11, 993, 1997.
116. Chamnongpol, S., Willekens, H., Langebartels, C., Van Montagu, M., Inzé, D., and Van Camp, W., Transgenic tobacco with a reduced catalase activity develops necrotic lesions and induces pathogenesis-related expression under high light, *Plant J.,* 10, 491, 1996.
117. Bi, Y. M., Kenton, P., Mur, L., Darby, R., and Draper, J., Hydrogen peroxide functions upstream of salicylic acid in the induction of PR protein expression, *Plant J.,* 8, 235, 1995.
118. Neuenschwander, U., Vernooij, B., Friedrich, L., Uknes, S., Kessmann, H., and Ryals, J., Is hydrogen peroxide a second messenger of salicylic acid in systemic acquired resistance?, *Plant J.,* 8, 227, 1995.
119. Léon, J., Lawton, M. A., and Raskin, I., Hydrogen peroxide stimulates salicylic acid biosynthesis in tobacco, *Plant Physiol.,* 108, 1673, 1995.
120. Summermatter, K., Sticher, L., and Métraux, J.-P., Systemic responses in *Arabidopsis thaliana* infected and challenged with *Pseudomonas syringae* pv syringae, *Plant Physiol.,* 108, 1379, 1995.
121. Chen, Z., Iyer, S., Caplan, A., Klessig, D. F., and Fan, B., Differential accumulation of salicylic acid and salicylic acid-sensitive catalase in different rice tissues, *Plant Physiol.,* 114, 193, 1997.
122. Chivasa, S., Murphy, A. M., Naylor, M., and Carr, J. P., Salicylic acid interferes with tobacco virus replication via a novel salicylhydroxamic acid-sensitive mechanism, *Plant Cell,* 9, 547, 1997.
123. Yang, Y. and Klessig, D. F., Isolation and characterization of a tobacco virus-inducible *myb* oncogene homolog from tobacco, *Proc. Natl. Acad. Sci. U.S.A.,* 93, 14972, 1996.

124. Grotewold, E., Drummond, B. J., Bower, B., and Peterson, T., The *myb*-homologous *P* gene controls phlobaphene pigmentation in maize floral organs by directly activating a flavonoid biosynthetic gene subset, *Cell*, 76, 543, 1994.
125. Avila, J., Nieto, C., Canas, L., Benito, M. J., and Paz-Ares, J., *Petunia hybrida* genes related to the maize regulatory *C1* gene and to animal *Myb* proto-oncogenes, *Plant J.*, 3, 553, 1993.
126. Solano, R., Nieto, C., Avila, J., Cañas, L., and Diaz, I., Dual DNA binding specificity of a petal epidermis specific MYB transcription factor (MYB.Ph3) from *Petunia hybrida, EMBO J.*, 14, 1773, 1995.
127. Baranowskij, N., Frohberg, C., Prat, S., and Willmitzer, L., A novel DNA binding protein with homology to MYB oncoproteins containing only one repeat can function as a transcriptional activator, *EMBO J.*, 13, 5383, 1994.
128. Jackson D., Culianez-Macia, F. A., Prescott, A. G., Roberts, K., and Martin, C., Expression pattern of *myb* genes from *Anthirrhinum* flowers, *Plant Cell,* 3, 115, 1991.
129. Sablowski, R. W. M., Moyano, E., Culianez-Macia, F. A., Schuch, W., Martin, C., and Bevan, M. A., A flower-specific Myb protein activates transcription of phenylpropanoid biosynthetic genes, *EMBO J.*, 13, 128, 1994.
130. Oppenheimer, D. G., Herman, P. L., Sivakumaran, S., Esch, J., and Marks, M. D., A *myb* gene required for leaf trichome differentiation is expressed in stipales, *Cell*, 67, 483, 1991.
131. Urao, T., Yamaguchi-Shinozaki, K., Urao, S., and Shinozaki, K., An *Arabidopsis myb* homolog is induced by dehydration stress and its gene product binds to the conserved *myb* recognition sequence, *Plant Cell,* 5, 1529, 1993.
132. Gubler, F., Kalla, R., Roberts, J. K., and Jacobsen, J. V., Gibberillin-regulated of a *myb* gene in barley aleurone cells: evidence for *myb* transactivation of a high-pI α-amylase promoter, *Plant Cell,* 7, 1879, 1995.
133. Hattori, T., Vasil, V., Rosenkrans, L., Hannah, L. C., McCarty, D. R., and Vasil, I. K., The *viviparous-1* gene and abscisic acid activate the *C1* regulatory gene for anthocyanin biosynthesis during seed maturation in maize, *Gene Dev.*, 6, 609, 1992.
134. Asselin, A., Grenier, J., and Cote, F., Light-influenced extracellular accumulation of b-(pathogenesis-related) proteins in *Nicotiana* green tissue induced by various chemicals or prolonged floating on water, *Can. J. Bot.*, 63, 1276, 1985.
135. Gianinazzi, S. and Kassanis, B., Virus resistance induced in plants by polyacrylic acid., *J. Gen. Virol.*, 23, 1, 1974.
136. Lotan, T. and Fluhr, R., Xylanase, a novel elicitor of pathogenesis-related proteins in tobacco, uses a non-ethylene pathway for induction, *Plant Physiol.*, 93, 811, 1990.
137. Ward, E. R., Uknes, S. J., Williams, S. C., Dincher, S. S., Wiederhold, D. L., Alexander, D. C., Ahl-Goy, P., Métraux, J.-P., and Ryals, J. A., Coordinate gene activity in response to agents that induce systemic acquired resistance, *Plant Cell*, 3, 1085, 1991.
138. White, R. F., Dumas, E., Shaw, P., and Antoniw, J. F., The chemical induction of PR(b) proteins and resistance to TMV infection on tobacco, *Antivir. Res.*, 6, 177, 1986.
139. Friedrich, L., Lawton, K., Ruess, W., Masner, P., Specker, N., Gut-Rella, M., Meier, B., Dincher, S., Staub, T., Uknes, S., Métraux, J.-P., Kessmann, H., and Ryals, J., A benzothiadiazole derivative induces systemic acquired resistance in tobacco, *Plant J.*, 10, 61, 1996.
140. Beffa, R. S., Szell, M., Meuwly. P., Pay, A., Vögeli-Lange, R., Métraux, J.-P., Neuhaus, G., Meins, F., Jr., and Nagy, F., Cholera toxin elevates pathogen resistance and induces pathogenesis-related gene expression in tobacco, *EMBO J.*, 14, 5753, 1995.

141. Lawton, K. A., Friedrich, L., Hunt, M., Weymann, K., Delaney, T., Kessmann, H., Staub, T., and Ryals, J., Benzothiadiazole induces disease resistance in *Arabidopsis* by activation of the systemic acquired resistance signal transduction pathway, *Plant J.,* 10, 71, 1996.
142. Beckman, K. B. and Ingram, D. S., The inhibition of the hypersensitive response of potato tuber tissues by cytokinins: Similarities between senescence and plant defense responses, *Physiol. Mol. Plant Pathol.,* 45, 229, 1994.
143. Jung, J. L., Maurel, S., Fritig, B., and Hahne, G., Different pathogenesis-related proteins are expressed in sunflower (*Helianthus annuus* L.) in response to physical, chemical and stress factors, *J. Plant Physiol.,* 145, 153, 1995.
144. Yang, S. F. and Hoffmann, N. E., Ethylene biosynthesis and its regulation in higher plants, *Annu. Rev. Plant Physiol.,* 35, 155, 1984.
145. Silverman, P., Nuckles, E., Ye, X. S., Kuc, J., and Raskin, I., Salicylic acid, ethylene and pathogen resistance in tobacco, *Mol. Plant Microbe Interact.,* 6, 775, 1993.
146. Raz, V. and Fluhr, R., Ethylene signal is transduced via protein phosphorylation events in plants, *Plant Cell,* 5, 523, 1993.
147. Eyal, Y., Meller, Y., Lev Yadun, S., and Fluhr, R., A basic-type *PR-1* promoter directs ethylene responsiveness, vascular and abscission zone-specific expression, *Plant J.,* 4, 225, 1993.
148. Lawton, K., Weymann, K., Friedrich, L., Vernooij, B., Uknes, S., and Ryals, J., Systemic acquired resistance in *Arabidopsis* requires salicylic acid but not ethylene, *Mol. Plant-Microbe Interact.,* 6, 863, 1995.
149. Mattoo, A. K. and Suttle, J. C., *The Plant Hormone Ethylene,* CRC Press, Boca Raton, Florida, 1991.
150. Avni, A., Bailey, B. A., Mattoo, A. K., and Anderson, J. D., Induction of ethylene biosynthesis in *Nicotiana tabacum* by a *Trichoderma viride* xylanase is correlated to the accumulation of ACC-synthase and ACC-oxidase transcripts, *Plant Physiol.,* 106, 1049, 1994.
151. Bailey, B. A., Dean, J. F. D., and Anderson, J. D., An ethylene biosynthesis-inducing endoxylanase elicits electrolyte leakage and necrosis in *Nicotiana tabacum* cv Xanthi leaves, *Plant Physiol.,* 94, 1849, 1990.
152. Bailey, B. A., Korcak, R. F., and Anderson, J. D., Alteration in *Nicotiana tabacum* cv Xanthi cell membrane integrity following treatment with ethylene biosynthesis-inducing endoxylanase, *Plant Physiol.,* 100, 749, 1992.
153. Bailey, B., Avni, A., and Anderson, J. D., The influence of ethylene and tissue age on the sensitivity of Xanthi tobacco leaves to a *Trichoderma viride* xylanase, *Plant Cell Physiol.,* 36, 1669, 1995.
154. Hanania, U. and Avni, A., High affinity binding site for ethylene-inducing xylanase elicitor on *Nicotiana tabacum* membranes, *Plant J.,* 12, 113, 1997.
155. Xu, Y., Chang, P. F. L., Liu, D., Narasimhan, M. L., Raghothama, K. G., Hasegawa, P. M., and Bressan, R. A., Plant defense genes are synergistically induced by ethylene and methyl jasmonate, *Plant Cell,* 6, 1077, 1994.
156. Choi, D., Bostock, R. M., Avdiushko, S., and Hildebrand, D. F., Lipid-derived signals that discriminate wound- and pathogen-responsive isoprenoid pathways in plants: methyl jasmonate and the fungal elicitor arachidonic acid induce different 3-hydroxyl-3-methylglutaryl-coenzyme A reductase genes and antimicrobial isoprenoids in *Solanum tuberosum* L., *Proc. Natl. Acad. Sci. U.S.A.,* 91, 2329, 1994.
157. Doares, S. H., Syrovets, F., Weiler, E. W., and Ryan, C. A., Oligogalacturonides and chitosan activate plant defensive genes through the octadecanoid pathway, *Proc. Natl. Acad. Sci. U.S.A.,* 92, 4095, 1995.

158. Pena-Cortes, H., Albrecht, T., Prat, S., Weiler, E. W., and Willmitzer, L., Aspirin prevents wound-induced gene expression in tomato leaves by blocking jasmonic acid biosynthesis, *Planta*, 191, 123, 1993.
159. Seo, S., Okamoto, M., Seto, H., Ishizuka, K., Sano, H., and Ohashi, Y., Tobacco MAP kinase: possible mediator in wound signal transduction pathways, *Science*, 270, 1988, 1995.
160. Schweizer, P., Buchala, A., Silverman, P., Seskar, M., Raskin, I., and Métraux, J.-P., Jasmonate-inducible genes are activated in rice by pathogen attack without a concomitant increase in endogenous jasmonic acid levels, *Plant Physiol.,* 114, 79, 1997.
161. Silverman, P., Seskar, M., Kanter, D., Schweizer, P. Métraux, J.-P., and Raskin, I., Salicylic acid in rice: biosynthesis, conjugation, and possible role, *Plant Physiol.,* 103, 633, 1995.
162. Kassanis, B., Gianinazzi, S., and White, R.F., A possible explanation of the resistance of virus-infected plants to a second infection, *J. Gen. Virol.,* 23, 11, 1974.
163. Fraser, R. S. S., Evidence for the occurrence of "pathogenesis-related" proteins in leaves of healthy tobacco plants during flowering, *Physiol. Plant Pathol.,* 19, 69, 1981.
164. Lotan, T., Ori, N., and Fluhr, R., Pathogenesis-related proteins are developmentally regulated in tobacco flowers, *Plant Cell,* 1, 881, 1989.
165. Grüner, R. and Pfitzner, U. M., The upstream region of the gene for the pathogenesis-related protein 1a from tobacco responds to environmental as well as developmental signals in transgenic plants, *Eur. J. Biochem.,* 220, 247, 1994.
166. Sommsich, I. E., Regulatory elements governing pathogenesis-related (*PR*) gene expression, in *Results and Problems in Cell Differentiation,* Vol. 20, Plant Promoters and Transcription Factors, Nover, L., Ed., Springer-Verlag, Berlin/Heidelberg, 1994, 163.
167. Van de Rhee, M. D. and Bol, J. F., Induction of the tobacco *PR-1a* gene by virus infection and salicylate treatment involves an interaction between multiple regulatory elements, *Plant J.,* 3, 71, 1993.
168. Beilmann, A., Pfitzner, A. J. P., Goodman, H. M., and Pfitzner, U. M., Functional analysis of the pathogenesis-related *1a* protein gene minimal promoter region: comparison of reporter gene expression in transient and in stable transfections, *Eur. J. Biochem.,* 196, 415, 1991.
169. Ohashi, Y. and Ohshima, M., Stress-induced expression of genes for pathogenesis-related proteins in plants, *Plant Cell Physiol.,* 33, 819, 1992.
170. Pelham, H. R. B., Activation of heat-shock genes in eukaryotes, *Trends Genet.,* 1, 31, 1985.
171. Meller, Y., Sessa, G., Eyal, Y., and Fluhr, R., DNA-protein interactions on a cis-DNA element essential for ethylene regulation, *Plant Mol. Biol.,* 23, 453, 1993.
172. Giulano, G., Pichersky, E., Malik, V. S., Timko, M. P., Scolnic, P. A., and Cashmore, A. R., An evolutionary conserved protein binding sequence upstream of a plant light-regulated gene, *Proc. Natl. Acad. Sci. U.S.A.,* 85, 7089, 1988.
173. Katagiri, F., Lam, E., and Chua, N. H., Two tobacco DNA binding proteins with homology to the nuclear factor CREB, *Nature*, 340, 727, 1989.
174. Schindler, U., Beckmann, H., and Cashmore, A. R., TGA1 and G-box binding factors: two distinct classes of *Arabidopsis* leucine zipper proteins compete for the G-box-like element TGACGTGG, *Plant Cell,* 4, 1309, 1992.
175. Schulze-Lefert, P., Dangl, J. L., Becker-Andre, M., Hahlbrock, K., and Schulz, W., Inducible *in vivo* DNA footprints define sequences necessary for UV-light activation of the parsley chalcone synthase gene, *EMBO J.,* 8, 651, 1989.

176. Ohme-Takagi, M. and Shinshi, H., Ethylene-inducible DNA binding proteins that interact with an ethylene-responsive element, *Plant Cell,* 7, 173, 1995.
177. Hagiwara, H., Matsuoka, M., Ohshima, M., Watanabe, M., Hosokawa, D., and Ohashi, Y., Sequence-specific binding of protein factors to 2 independent promoter regions of the acidic tobacco pathogenesis-related-1 protein gene (*PR-1*), *Mol. Gen. Genet.,* 240, 197, 1993.
178. Buchel, A. S., Molenkamp, R., Bol, J. F., and Linthorst, H. J. M., The *PR-1a* promoter contains a number of elements that bind GT-1-like nuclear factors with different affinity, *Plant Mol. Biol.,* 30, 493, 1996.
179. Lam, E., Domain analysis of the plant DNA-binding protein GT-1a: requirement of four putative α-helices for DNA binding and identification of a novel oligomerization region, *Mol. Cell Biol.,* 15, 1014, 1995.
180. Gilmartin, P. M. and Chua, N. H., Spacing between GT-1 binding sites within a light-responsive element is critical for transcriptional activity, *Plant Cell,* 2, 447, 1990.
181. Qin, X. F., Holuigue, L., Horvath, D. M., and Chua, N. H., Immediate early transcription activation by salicylic acid via the cauliflower mosaic virus *as-1* element, *Plant Cell,* 6, 863, 1994.
182. Ulmasov, T., Hagen, G., and Guilfoyle, T., The *ocs* element in the soybean GH2/4 promoter is activated by both active and inactive auxin and salicylic acid analogues, *Plant Mol. Biol.,* 26, 1055, 1994.
183. Miao, Z. H., Liu, X., and Lam, E., TGA3 is a distinct member of the TGA family of bZIP transcription factors in *Arabidopsis thaliana, Plant Mol. Biol.,* 25, 1, 1994.
184. Katagiri, F., Seipel, K., and Chua, N. H., Identification of a novel dimer stabilization region in a plant bZIP transcription activator, *Mol. Cell Biol.,* 12, 4809, 1992.
185. Neuhaus, G., Neuhaus-Url, G., Katagiri, F., Seipel, K., and Chua, N. H., Tissue-specific expression of as-1 in transgenic tobacco, *Plant Cell,* 6, 827, 1994.
186. Strompen, G. and Pfitzner, U. M., Transcription factor TGA1a binds to the PR-1a promoter (Abstract), in *4th Int. Workshop Pathogenesis-Related Proteins in Plants: Biology and Biotechnological Potential,* Kloster Irsee, Germany, 1995.
187. Miao, Z. H. and Lam, E., Construction of a *trans*-dominant inhibitor for members of the TGA family of transcription factors in higher plants, *Plant J.,* 7, 887, 1995.
188. Martin, C. and Paz-Ares, J., Myb transcription factors in plants, *Trends in Genet.,* 13, 67, 1997.

3 Functions and Regulation of Plant β-1,3-Glucanases (PR-2)

Gerhard Leubner-Metzger and Frederick Meins, Jr.

CONTENTS

3.1 Introduction 49
3.2 Structural Classes of β-1,3-Glucanases and PR-2 Nomenclature 50
3.3 Biological Functions of β-1,3-Glucanases 52
3.3.1 Plant Reproductive Biology 52
3.3.2 Pathogenesis-Related Functions 57
3.4 Regulation of β-1,3-Glucanase Expression 60
3.4.1 Developmental and Hormonal Regulation 60
3.4.2. Pathogenesis-Related Regulation 62
3.4.3 *Cis*-Acting Elements 63
3.4.4 Signal Transduction and *Trans*-Acting Factors 64
3.5 Concluding Remarks 65
References 66

3.1 INTRODUCTION

Several classes of proteins, called pathogenesis-related (PR) proteins, are induced in response to the infection of plants with microbial pathogens.[1-3] This chapter deals with the family of PR-2 proteins, which are β-1,3-glucanases (glucan endo-1,3-β-glucosidases, E.C. 3.2.1.39) able to catalyze endo-type hydrolytic cleavage of the 1,3-β-D-glucosidic linkages in β-1,3-glucans. The β-1,3-glucanases are abundant, highly regulated enzymes widely distributed in seed-plant species.[4-7] Although the major interest in β-1,3-glucanases stems from their possible role in the response of plants to microbial pathogens, there is strong evidence that these enzymes are also implicated in diverse physiological and developmental processes in the uninfected plant, including cell division,[8,9] microsporogenesis,[10,11] pollen germination and tube growth,[12,13] fertilization,[14,15] embryogenesis,[16,16a] fruit ripening,[17] seed germination,[18,19] mobilization of storage reserves in the endosperm of cereal grains,[20] bud dormancy,[21] and responses to wounding, cold, ozone, and UV-B.[22-26]

0-8493-0697-3/99/$0.00+$.50

In this chapter we focus on progress being made in understanding the function and regulation of β-1,3-glucanases in reproductive development and pathogenesis. For more general information on plant β-1,3-glucanases, the reader is referred to the extensive reviews by Stone and Clarke,[7] Meins et al.,[4] Simmons,[6] and Høj and Fincher.[5]

3.2 STRUCTURAL CLASSES OF β-1,3-GLUCANASES AND PR-2 NOMENCLATURE

β-1,3-glucanases (βGlu) exist as multiple structural isoforms that differ in size, isoelectric point, primary structure, cellular localization, and pattern of regulation.[4] The most detailed sequence information for these isoforms is available from cDNA and genomic clones of tobacco βGlu, which form a multigene family. Based on amino acid sequence identity, the various βGlu of the genus *Nicotiana* have been classified into three structural classes.[4,27,28] The classification, nomenclature, and salient features of these βGlu are summarized in Table 3.1. Similar structural isoforms have been reported for tomato, potato, and other plant species.[4,7,29-32]

The approximately 33 kDa class I enzymes (βGlu I) of *Nicotiana tabacum,* which constitute the PR-2e subgroup of tobacco PR proteins, are basic proteins localized in the cell vacuole.[33-35] βGlu I is produced as a preproprotein with an N-terminal hydrophobic signal peptide, which is cotranslationally removed, and a C-terminal extension N-glycosylated at a single site. The proprotein is transported from the endoplasmic reticulum via the Golgi compartment to the vacuole where the C-terminal extension is removed to give the mature, approximately 33 kDa enzyme, which is not glycosylated.[35,36] There is considerable indirect evidence that, in analogy to tobacco class I chitinases[37] and barley lectin,[38] the C-terminal extension contains a signal for targeting to the vacuole.[10,39] Recent results obtained with cultured tobacco cells provide strong evidence that vacuolar class I βGlu and chitinases can be secreted into the medium via a novel pathway.[39a]

The known mature βGlu I of tobacco and Gn2 of *Nicotiana plumbaginifolia* share about 98% amino acid identity.[40] In contrast, with only about 76% similarity, the Gn1 of *N. plumbaginifolia* is structurally more distinct.[41] The tobacco βGlu I multigene family consists of very similar homeologs derived from the *N. sylvestris* and *N. tomentosiformis* progenitors of tobacco as well as recombinants of the two progenitor types.[42]

In contrast to βGlu I, the class II and III members of the PR-2 family are secreted into the extracellular space.[2,4,6] The tobacco class II βGlu PR-2a, PR-2b, PR-2c, and the class III βGlu PR-2d, also known as PR-2, PR-N, PR-O, PR-Q′, respectively,[1] are acidic proteins without the C-terminal extension present in the class I enzymes ranging in apparent size from approximately 34 to 36 kDa in denaturing gels.[43] The class II tobacco isoforms are at least 82% identical in amino acid sequence and differ from the class I enzymes at a minimum of 48.8% of the positions.[4,6,27,44] Class II also includes the two acidic 41 kDa stylar βGlu isoforms, Sp41a, and Sp41b, which are exclusively expressed in the style of tobacco flowers.[15] They do not appear to be induced by pathogen infection and, hence, are referred to as "PR-like proteins."[1]

TABLE 3.1
PR-2 Family Members (β-1,3-Glucanases) of Tobacco and Other *Nicotiana* Species

Class[a]	Member name	Trivial name	Origin[b]	MW (kDa)[c]	pI	Localization	References
I	PR-2e	Glb	Nt (T)	33	Basic	Vacuole	42, 36, 18, 19, 71, 124
I	PR-2e	Gla	Nt (S)	33	Basic	Vacuole	36, 42
I	PR-2e	Gglb50	Nt (S)		Basic	Vacuole	44, 125
I	PR-2e	Gln2	Nt (S)		Basic	Vacuole	132, 126
I		Gn2[d]	Np		Basic	Vacuole	40
I		Gn1[d]	Np	34	Basic	Vacuole	41, 127
II	PR-2a	PR-2 (I, b_4)	Nt	35	Acidic	Secreted	27
II	PR-2b	PR-N (b_5), G19	Nt	35	Acidic	Secreted	27, 44, 125
II	PR-2c	PR-O ($b_{6(b)}$, O′)	Nt	35	Acidic	Secreted	27, 142
II		PR-2d	Nt		Acidic	Secreted	128, 129
II	Stylar βGLU[e]	Sp41a	Nt	41	Acidic	Secreted	14, 15, 60
II	Stylar βGLU[e]	Sp41b	Nt	41	Acidic	Secreted	14, 15
III	PR-2d	PR-Q′	Nt	35	Acidic	Secreted	28
IV[f]	Anther βGLU[e]	Tag1	Nt	35	Acidic	Secreted	11

[a] Classification according to amino acid sequence identity of the mature proteins (reference 4).
[b] *Nicotiana tabacum* (Nt); *N. plumbaginifolia* (Np); T and S refer to the *N. tomentosiformis* and *N. sylvestris* progenitors of tobacco, respectively.
[c] Approximate molecular weight of mature protein; selected values from the literature, which might differ between publications.
[d] Amino acid sequence identity to tobacco βGlu I enzymes of the *N. plumbaginifolia* enzymes is ca. 98% for Gn2, but only ca. 76% for Gn1.
[e] Not induced by pathogens, i.e., a "PR-like protein" (reference 1).
[f] Tag1 is assigned to a new class, since it shares only 37 to 38% amino acid sequence identity to Gla, PR-2, and PR-Q′.

The acidic, approximately 35 kDa, PR-2d (PR-Q′) is the sole representative of tobacco class III βGlu and differs in sequence by at least 43% from the class I and class II enzymes.[28] Two highly homologous cDNA clones for class III βGlu have been isolated from tomato plants infected with a viroid.[31] Based on deduced amino-acid sequences, TomPR-Q′a is an acidic isoform 86.7% identical to tobacco PR-Q′ and TomPR-Q′b, is a basic isoform 78.7% identical to tobacco PR-Q′.

Tag1 appears to represent a novel class of tobacco βGlu. It is a "PR-like" protein which is expressed specifically in tobacco anthers.[11] Like the tobacco class I βGlu, Tag1 is encoded by a small gene family with at least two members derived from the *N. sylvestris* and *N. tomentosiformis* progenitors of tobacco. Based on deduced amino acid sequence, *Tag1* encodes a polypeptide with an N-terminal signal peptide, but no C-terminal extension, suggesting that the protein may be secreted. The mature form of Tag1 is an acidic, 35 kDa protein, which shares absolutely conserved sequences found in all classes of tobacco βGlu. It is 37 to 38% identical in sequence to the mature forms of tobacco class I Gla, class II PR-2, and class III PR-Q′. Based on the criteria used earlier for tobacco βGlu,[4] we have assigned Tag1 to a new class, IV (Table 3.1).

The specific enzymatic activities and substrate specificities of different βGlu vary considerably. The βGlu I and class II PR-2c appear to be 50 to 250 times more active in degrading the β-1,3-glucan substrate laminarin than the class II PR-2a and PR-2b and the class III PR-2d enzymes.[24,45] The mechanism of catalysis has been recently reviewed by Davies and Henrissat[46] and by Høj and Fincher.[5] The complete three-dimensional structure of a barley βGlu has been determined.[47]

3.3 BIOLOGICAL FUNCTIONS OF β-1,3-GLUCANASES

3.3.1 Plant Reproductive Biology

3.3.1.1 Anther β-1,3-Glucanases

Pollen development begins with the division of diploid sporophytic cells within the anther, which gives rise to tapetal cells and pollen mother cells.[48,49] The tapetum forms a single layer of cells around the anther locule in which the pollen develops. The pollen mother cells undergo meiosis to form tetrads of haploid microspores. In almost all higher plants, each individual microspore of the tetrad is surrounded by a thick callose wall composed of a β-1,3-glucan, which is laid down between the cellulose cell wall and the plasma membrane. At a critical stage of microspore development, the callose wall of the tetrads is degraded by callase activity, which is secreted by the tapetal cells. The microspores are then released into the anther locule where they develop into mature pollen grains. Although the callose wall is essential for production of fertile pollen, its function is unknown. Proposed functions include physical and chemical isolation of the developing gametes from sporophytic tissues, mechanical isolation of the meiocytes and tetrads, protection from environmental and osmotic stress, and formation of the pollen cell wall.[11]

In the anthers of petunia and lily, expression and secretion of callase activity is under strict developmental control.[50,51] The callase complex of lily consists of a 32 kDa endo-βGlu and a 62 kDa exo-type βGlu.[52] The endo-type enzyme seems to be most important for the degradation of the callosic walls, while the exo-type βGlu is involved in the further hydrolysis of released oligosaccharides. Alterations in the timing of βGlu expression, or failure to express βGlu, leads to abnormal dissolution of the tetrad callose walls, which has been shown to be a primary cause of male sterility in cytoplasmic male-sterile lines of petunia,[53] sorghum,[54] and soybean.[55]

Two βGlu genes have been identified that are expressed during microsporogenesis. The *A6* gene was originally identified as an anther-specific *Brassica napus* cDNA, which was then used to isolate genomic clones of the *Arabidopsis thaliana* homologs.[56] Based on deduced amino acid sequences and immunoblotting experiments, *A6* encodes a polypeptide with a domain similar in sequence to βGlu and a 114 amino-acid long C-terminal domain, which is not present in other known βGlu. Reporter gene studies established that *A6*-gene expression is tapetum specific and temporally correlated with the expression of callase activity. Transcripts of the class IV tobacco anther βGlu Tag1 are also expressed exclusively in the tapetum and show a callase-like pattern of expression.[11] *Tag1* mRNA is not detectable in flower buds, pistil, sepals, petals, roots, healthy leaves, and in leaves not infected with tobacco mosaic virus (TMV).

Further evidence for a role of βGlu in callose-wall dissolution and microsporogenesis has come from sense-transformation experiments. Worrall et al.[10] transformed tobacco with a gene encoding a tobacco βGlu I with the C-terminal extension deleted, which is secreted into the extracellular space. Tobacco plants that express the recombinant extracellular βGlu I under the control of the tapetum-specific promoters of the *Arabidopsis A3* and *A9* genes exhibited premature degradation of callose in microspore cell walls, production of abnormal microspores, and partial-to-complete male sterility. No male sterility was observed in transformants obtained with the extracellular βGlu I regulated by the cauliflower mosaic virus (CaMV) 35S RNA promoter. Similar results were obtained in lettuce[56a] and by expressing in transgenic tobacco the cDNA of a pathogenesis-related extracellular endo-βGlu from soybean using the rice, tapetum-specific *Osg6B* promoter.[57] Taken together, the sense transformation experiments indicate that premature callose degradation is sufficient to cause male sterility and suggest that formation of the callose cell wall and its proper developmental degradation by endo-βGlu are critical for microsporogenesis.

3.3.1.2 Stylar β-1,3-Glucanases

As part of the fertilization process, pollen tubes grow through the transmitting tissue of the style toward the ovary.[58,59] The transmitting tissue consists of elongated, secretory cells, connected end-to-end through plasmodesmata. It is believed that interactions between the transmitting tissue and the growing pollen tube are important for guiding pollen tube growth toward the ovules and successful fertilization.

The class II βGlu Sp41 of tobacco is a "PR-like" protein encoded by the two closely related *Sp41a* and *Sp41b* genes.[14,15] It is a major component — up to 20%

— of the soluble protein in the stylar transmitting tissue of the tobacco flower. The mature form of Sp41 is an approximately 41 kDa, acidic glycoprotein, which is secreted into the extracellular matrix. Measurements of Sp41 protein, Sp41 steady-state mRNA, and activity of the *Sp41a*-gene promoter indicate that Sp41 is expressed exclusively in the style in a developmentally regulated fashion.[14,15,60] Accumulation of the protein begins 4 days before anthesis and reaches a broad peak from 2 days before anthesis until style senescence. No expression is found in leaf, root, sepal, petal, anther, pollen sac, and ovary. Moreover, the protein is not induced by TMV in leaves or by treatment with abiotic elicitors of PR proteins in leaves and styles. Reporter-gene experiments have shown that 2.5 kb 5′-noncoding region of the *Sp41a* gene is sufficient to confer transmitting-tissue specific expression.

Sessa and Fluhr[60] used antisense transformation to find out if expression of Sp41 has an essential role in reproductive development. Tobacco plants were transformed with a construct containing a partial *Sp41a* cDNA in reverse orientation fused to a CaMV 35S promoter, which is active in the style, and a TMV enhancer of translation. The results are difficult to interpret. Although neither Sp41 protein nor βGlu activity was detected and fertility was reduced in 3 of 18 primary transformants, progeny obtained by outcrossing the Sp41-deficient transformants were fully fertile even though many of them had undetectable or greatly reduced levels of Sp41. Moreover, no direct effects on stylar development or pollen tube growth related to Sp41 deficiency were observed. Sessa and Fluhr[60] suggest as possible explanations that either the Sp41 deficiency is significant only under specific environmental conditions not met in their experiments; or that Sp41 has a role in defense against pathogen infection rather than in the fertilization process. Another possibility is that the plants can compensate for the deficiency by producing other proteins with Sp41-like functions as has been reported for βGlu I induced in tobacco by TMV infection.[43]

3.3.1.3 Endosperm β-1,3-Glucanases

Germination of seeds is a complex physiological process triggered by imbibition of water. Under favorable conditions, rapid growth of the embryo culminates in rupture of the covering layers and emergence of the radicle. In many species the enclosing tissues act as a physical barrier which must be overcome by the growth potential of the embryo if the seed is to complete its germination.[61,62] Little is known about the molecular basis for the rupture and physical penetration of these covering layers.

In the case of tobacco, the embryo is surrounded by three to five layers of rather thick-walled endosperm cells in the mature seed. The periphery of the endosperm is pressed against the thin seed coat (testa), consisting of cuteinized and lignified dead cells.[63] Rupture of the testa and rupture of the endosperm are separate events in the germination of tobacco seeds (Figure 3.1),[19,64] and there is strong evidence that endosperm rupture is the limiting factor in the germination of these seeds. Electron microscopic studies support the view that the endospermic hole of the germinated seed, which is always at the micropylar end, is formed by "dissolution" rather than by "pushing" action.[64] In photodormant varieties of tobacco, both the seed coat and endosperm remain intact in the 80 to 90% of seeds that do not germinate in darkness.[63,65-67] However, when the seed coat and endosperm are mechanically removed, there is radicle growth even in the absence of light.[68-70] Finally, treatment of tobacco seeds with 10 μ*M* abscisic acid (ABA) greatly delays endosperm rupture and results in the formation of a novel structure, consisting of the enlarging radicle with a sheath of greatly elongated endosperm tissue[19] (Figure 3.1).

The first hint that βGlu may play a role in tobacco seed germination was our observation that βGlu I is induced during germination.[18,19] Measurements of βGlu activity, βGlu I protein and mRNA in combination with reporter-gene experiments with the *E. coli uidA* gene (*Gus* reporter-gene) fused to the promoter of the tobacco class I β-1,3-glucanase B gene (*Glb*) established that most, if not all, of the βGlu activity is due to transcriptional induction of the βGlu I isoforms. βGlu I is induced exclusively in the micropylar region of the endosperm where the radicle will penetrate, but prior to endosperm rupture. Finally, βGlu I induction during germination is not a classical defense-type response, since chitinases (Chn) and the known acidic class II and III βGlu are not induced. Based on these findings, we proposed as a working hypothesis that βGlu I weakens the endosperm by digestion of cell-wall material and that this promotes radical protrusion to facilitate germination (Figure 3.1).

In support of this hypothesis, treatment of tobacco seeds incubated in the light with ABA does not affect seed-coat rupture but greatly delays subsequent endosperm rupture and inhibits the rate of endosperm-specific βGlu I accumulation in a concentration-dependent manner.[19] Gibberellins (GA), which can substitute for light in releasing dormancy, induced βGlu I in the dark in association with germination.[65] Recent results[71] provide evidence that ethylene is required both for high levels of βGlu I expression in the micropylar endosperm and for endosperm rupture. Although the close correlation between βGlu I induction and the onset of endosperm rupture under a variety of physiological conditions (Figure 3.1) is consistent with our working hypothesis, direct evidence is still lacking. One approach we are currently exploring is the use of sense- and antisense transformation to alter the level of expression of βGlu I during the germination process.

It is well established that 1,3;1,4-β-glucanases, which are structurally related to βGlu but differ in substrate specificity,[5,47,72] hydrolyze the 1,3;1,4-β-glucan cell walls of the starchy endosperm during the germination of cereals.[5,6,73] Less is known about the function of cereal βGlu, which are present in ungerminated grain and rise markedly in concentration during germination. However, the putative substrate, β-1,3 glucan, is not abundant in cereal grains and is restricted to small callosic deposits scattered through the starchy endosperm. Fincher[73] has proposed that the high levels of βGlu are part of a preemptive strategy to protect the grain against microbial attack. Cordero et al.[74] found that one βGlu isoform and three Chn isoforms are induced in germinating maize kernels infected by *Fusarium moniliforme*. In contrast, a second βGlu isoform is expressed in embryo and radicle tissues but is not induced by fungal infection. These findings support the view that cereal βGlu, in addition to their possible role in pathogen defense, might also be involved in embryogenesis and seed germination.

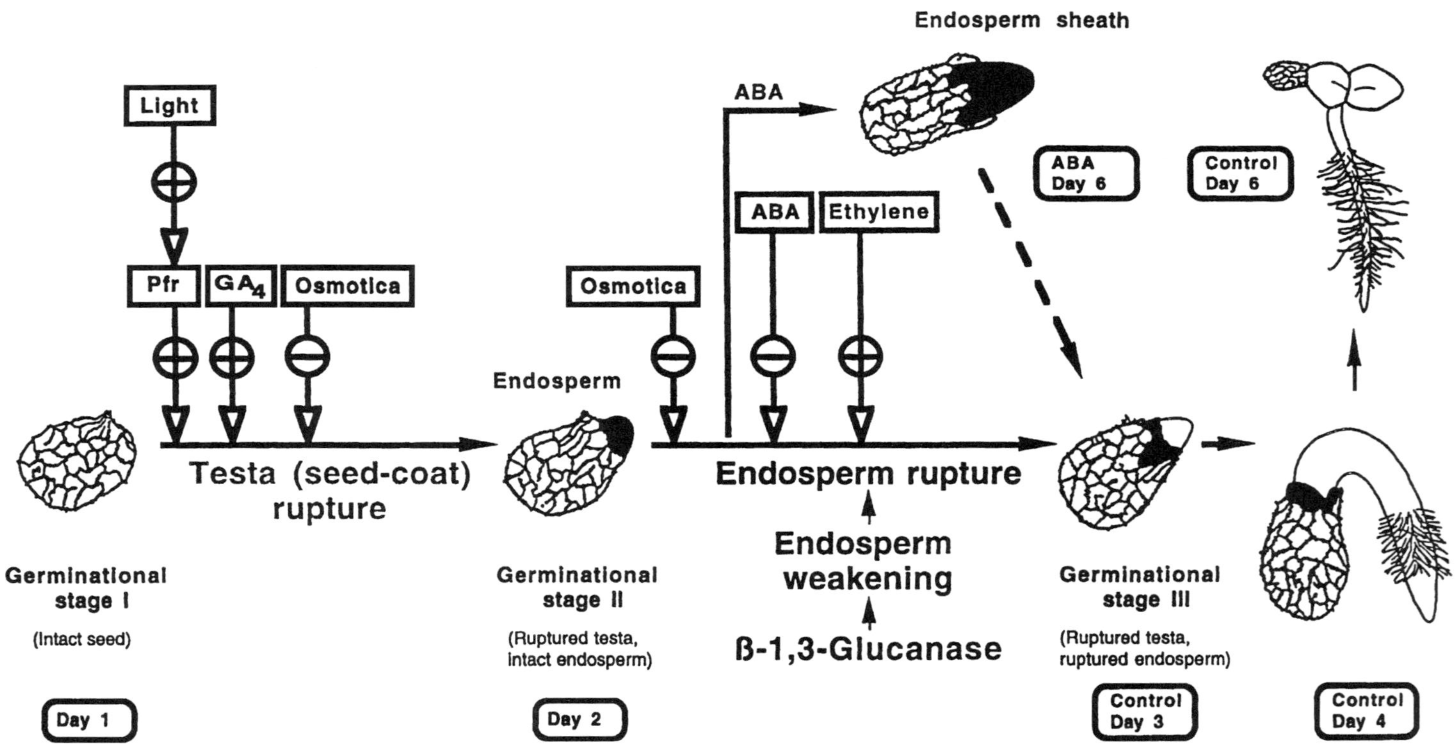

FIGURE 3.1 Working model for tobacco seed germination. Rupture of the testa and rupture of the endosperm are separate events in *Nicotiana tabacum*. Class I β-1,3-glucanase accumulates just prior to endosperm rupture and is proposed to promote radicle protrusion by weakening the endosperm. Plant hormones and environmental factors alter the germination process, and in strict correlation with this, either promote (+) or inhibit (–) class I β-1,3-glucanase induction. GA = gibberellin(s); ABA = abscisic acid; Pfr = Phytochrome. The model summarizes results from Leubner-Metzger et al.[19,65,71]

3.3.2 Pathogenesis-Related Functions

In 1971, Abeles et al.[75] suggested that the glucanohydrolases β-1,3-glucanase (βGlu) and chitinase might function as a defense against fungal pathogens. At about the same time, the PR proteins were first described as a novel set of abundant proteins accumulating in leaves of resistant tobacco cultivars reacting with hypersensitive response (HR) to infection with TMV.[3,76,77] Later, it was shown that the PR proteins include βGlu (the PR-2 family)[45] and Chn (the PR-3 family).[2,24,78]

There is now compelling evidence that βGlu and Chn, acting alone and particularly in combination, can help defend plants against fungal infection. It has been proposed that these glucanohydrolases act in at least two different ways: directly, by degrading the cell walls of the pathogen; and, indirectly by promoting the release of cell-wall derived materials that can act as elicitors of defense reactions.[4,79-83]

3.3.2.1 *In Vitro* Antifungal Activity

βGlu and Chn can hydrolyze β-1,3-glucans and chitin, respectively, which are major components of the cell walls of many pathogenic and potentially pathogenic fungi.[84] Although, in some cases, treatment with βGlu or Chn can inhibit fungal growth *in vitro*, more often combinations of the two enzymes are required for antifungal activity.[79,81,85]

Several studies have been made in which different βGlu and Chn isoforms were tested for *in vitro* antifungal activity.[86-93] Only class I vacuolar isoforms of tobacco βGlu and Chn were effective in promoting the lysis of hyphal tips and inhibiting the growth of *Fusarium solani.*[89] These effects were greatly enhanced by using combinations of βGlu I and class I Chn. In contrast, the class II βGlu PR-2a, PR-2b, and PR-2c did not exhibit antifungal activity either alone or in any combinations tested. Similar studies with tomato βGlu and Chn have shown that the vacuolar class I isoforms, but not the secreted class II isoforms, inhibit growth of *Alternaria solani, Trichoderma viride,* and *Phytophthora infestans,*[91,93] and that none of the combinations of βGlu and Chn tested inhibited growth of the tomato pathogen *Cladosporium fulvum.*[92]

3.3.2.2 Release of Fungal Elicitors

Plant βGlu can release oligosaccharides from cell walls of the pathogens, which can then act as elicitors of defense reactions.[81-83] This is well documented for interactions between soybean and the β-glucan elicitor from the pathogenic oomycete *Phytophthora megasperma* f. sp. *glycinea.*[94-107] Following fungal attack, soybean βGlu releases β-glucans from the fungal cell wall, which then induce accumulation of the phytoalexin glyceollin. The smallest β-glucan released with elicitor-activity was a β-1,3-β-1,6-heptaglucoside,[95] and the structural requirements for elicitor activity of these oligosaccharides have been investigated.[96]

Proteins which bind this oligosaccharide elicitor have been purified from soybean membranes.[104,105] Recently, Umemoto et al.[97] isolated a cDNA for a β-glucan elicitor binding protein (GEBP), which is localized in the plasma membrane of soybean root cells. Expression of the soybean GEBP gene has been shown to confer β-glucan binding activity to *Escherichia coli* and to tobacco cells cultured in suspension, suggesting that GEBP might be an elicitor receptor.

Soybean βGlu have been purified that are able to release active β-glucan elicitors from fungal cell walls.[98,99,106,107] The enzyme described by Takeuchi et al.[98] is 63% identical in amino acid sequence to the tobacco class III βGlu PR-Q′, but only 55% and 51% identical to the class I and class II enzymes of tobacco, respectively, suggesting that the soybean enzyme is a class III βGlu. Albersheim and Valent[94] have reported that the fungus *Colletotrichum lindemuthianum* secretes a protein that inhibits an endo-βGlu of its host, French bean. Recently, Ham et al.[99] presented evidence that fungal pathogens secrete proteins that can selectively inhibit plant βGlu. They purified two basic pathogenesis-related βGlu, $EnGL_{soy}$-A and $EnGL_{soy}$-B, from soybean seedlings as well as a β-1,3-glucanase inhibitor protein (GIP-1) from the culture fluid of *Phytophthora sojae* f. sp. *glycines* (formerly *Phytophthora megasperma* f. sp. *glycinea*). GIP-1 specifically inhibited the soybean $EnGL_{soy}$-A, but not $EnGL_{soy}$-B or several other βGlu, including tobacco class II PR-2c and enzymes secreted by the fungus. GIP-1 does not exhibit proteolytic activity but does appear to physically bind to $EnGL_{soy}$-A. The fungal pathogen can also secrete GIPs with other host βGlu as targets. Thus, fungal pathogenesis appears to involve a complex interplay between host βGlu and pathogen βGlu and βGlu inhibitors.

3.3.2.3 Enhanced Resistance to Fungal Pathogens Resulting from Transgene Expression

There is strong evidence that expression of *βGlu* transgenes alone or in combination with *Chn* transgenes regulated by the strong CaMV 35S RNA promoter can reduce the susceptibility of plants to infection by certain fungi. Transgenic tobacco plants expressing a soybean β-1,3-glucan-elicitor releasing βGlu or the tobacco class II βGlu PR-2b show reduced symptoms when infected with *Alternaria alternata* or the oomycetes *Phytophthora parasitica* var. *nicotianae* and *Peronospora tabacina.*[102,102a] β-1,3-Glucans are the major components of the cell walls of oomycetes, a group of fungi that do not contain chitin.[84]

In many cases, a pronounced synergistic effect is obtained when *βGlu* and *Chn* transgenes are expressed in combination. Tomato plants expressing tobacco class I *βGlu* and *Chn* transgenes show reduced susceptibility to infection by *Fusarium oxysporum* f.sp. *lycopersici*, whereas expression of either gene alone is not effective.[108] Melchers et al.[85] and Sela-Buurlage et al.[89] transformed tobacco plants with transgenes encoding modified class I tobacco βGlu and Chn that are secreted. They found that the extracellular wash fluid from the leaves of plants expressing both βGlu and Chn showed strong antifungal activity against *Fusarium solani,* whereas this effect was less for plants expressing either transgene alone. Tobacco plants expressing a bean class I *Chn* gene show decreased susceptibility to the root pathogen *Rhizoctonia solani.*[109] Resistance to infection was further enhanced by co-expression of this *Chn* gene with barley class II *βGlu* and *Chn* genes.[110] Alfalfa plants expressing alfalfa *Aglu1* acidic *βGlu* and rice *Rch10* basic *Chn* transgenes showed reduced disease symptoms when infected with the oomycete pathogen *Phytophthora megasperma* f. sp *medicaginis*, which does not contain chitin in its cell walls, whereas no reduction in symptoms was observed with several chitin-containing fungal pathogens.[111] Expression of the *Rch10* and *Aglu1* transgenes also substantially increased protection of tobacco against the chitin-containing fungus *Cercospora nicotianae,* the causal agent of the frogeye disease, relative to plants expressing either of the transgenes alone.[112] In contrast, susceptibility of *N. sylvestris* to *C. nicotianae* was not affected by high-level expression of tobacco class I Chn[113] or deficiencies in host βGlu I generated by antisense transformation.[114] The latter results suggest that host βGlu I may not be required for defense against this pathogen.

3.3.2.4 Decreased Susceptibility to Viral Disease of β-1,3-Glucanase-Deficient Plants

The induction of βGlu as part of the hypersensitive reaction is a stereotypic response, i.e., the pattern of induction is similar for viral, bacterial, and fungal pathogens.[115] Although antifungal βGlu I appears to be tailored for defense against fungi, recent studies of βGlu I-deficient mutants generated by antisense transformation suggest that these enzymes also play an important role in viral pathogenesis.

TMV infection of leaves of tobacco cultivars showing a local-lesion response induces the expression of all three βGlu classes.[27] Antisense transformants of Havana 425 tobacco and *Nicotiana sylvestris* transformed with sequences of *Gla*, the *N. sylvestris* homeolog of tobacco βGlu I, in reverse orientation regulated by the CaMV 35S RNA promoter show greatly reduced levels of βGlu I.[43,114,116,117] Expression of class I but not class II or class III βGlu is effectively and specifically blocked when these antisense lines are infected with TMV.[117] The antigen content for the βGlu I in lower leaves of healthy transgenic plants is reduced about 20-fold. They are fertile, develop normally under greenhouse conditions, and as in the wild-type plants, they accumulate the known pathogen-inducible class II and class III isoforms of βGlu when infected with necrotizing viruses.[43] A novel intracellular form of βGlu serologically distinct from any of the known tobacco βGlu is induced in βGlu I-deficient plants, but not in wild-type plants by virus infection. Thus, plants can compensate for a deficiency in enzyme activity by producing a functionally equivalent replacement — i.e., "ersatz" protein or proteins. The fact that compensation occurred specifically in response to virus infection suggests an important role of βGlu in pathogenesis.

Unexpectedly, the βGlu I-deficient mutants showed markedly reduced lesion size, lesion number, and virus yield in the local-lesion response of Havana 425 tobacco to TMV and *N. sylvestris* to tobacco necrosis virus.[117] In contrast to βGlu I, no change in resistance to TMV was reported for antisense and sense transformation with constructs for the class II βGlu PR-2b.[102a] The βGlu I-deficient *N. sylvestris* mutants[117] also showed decreased severity and spread of mosaic disease symptoms and reduced virus yield in the susceptible response to TMV. Moreover, the symptoms of disease in both plant species were positively correlated with βGlu I content in a series of independent transformants providing direct evidence for a function of these enzymes in viral pathogenesis.

Callose deposition is known to act as a physical barrier to the spread of virus.[118] Callose deposition in and around TMV-induced lesions is increased in βGlu I-deficient tobacco, suggesting that decreased susceptibility to virus resulted from increased callose deposition in response to infection. These findings are of particular interest because they suggest a novel means, based on antisense transformation with host genes, for protecting plants against viral infection. They also raise the intriguing possibility that viruses can use a defense response of the host against fungal infection — production of βGlu I — to promote their own replication and spread.

3.4 REGULATION OF β-1,3-GLUCANASE EXPRESSION

βGlu show developmental regulation and regulation in response to treatment with hormones or infection with pathogens. Early studies were done on the basis of measurements of enzyme activity before it was recognized that there are different classes of βGlu.[119-123] More recently, specific βGlu proteins have been measured immunologically and their mRNAs have been measured semiquantitatively by RNA-blot hybridization. In a limited number of cases, regulation of transcription has been studied using plants transformed with *Gus* reporter genes under the control of the promoter region of *βGlu* genes, namely: (1) the tobacco class I *N. tomentosiformis* homolog *Glb*,[18,124] and the tobacco class I *N. sylvestris* homologs *Gglb50*[125,153a] and *Gln2*;[126] (2) the less-related βGlu I gene *Gn1* of *N. plumbaginifolia*;[41,127] and (3) the tobacco class II βGlu genes *PR-2b* (*G19, PR-N*)[125] and *PR-2d*.[128,129]

In the following sections, we summarize the patterns of βGlu regulation, and review progress being made in identifying *cis*-acting promoter elements and *trans*-acting factors important in transcriptional regulation and signal transduction.

3.4.1 Developmental and Hormonal Regulation

3.4.1.1 Class I β-1,3-Glucanases

βGlu I accumulates at high concentrations in the roots and in lower leaves of mature, healthy tobacco plants.[120,130,131] The βGlu I content of leaves decreases toward the top of the plant. Within leaves, they are localized in the vacuole of epidermal cells.[34] βGlu I transcripts accumulate at low levels in developing floral tissues of tobacco[36,132] and potato[133] and accumulate at high levels during *de novo* flower formation of tobacco.[131] The pattern of expression in leaves and roots of βGlu I proteins and steady-state mRNA is very similar and is correlated with promoter activity of the approximately 1.5 to 1.7 kb 5′ flanking region.[120,124,125,130,132] Therefore, regulation of βGlu I in these organs appears to be primarily at the level of transcription. Similar conclusions may be drawn for the less related βGlu I gene *Gn1* of *N. plumbaginifolia*, which shows low promoter activity in upper leaves and high promoter activity in lower leaves and roots.[41]

The accumulation of βGlu I protein and mRNA is strongly down-regulated when discs of leaf tissue or callus cultures are incubated for less than 7 days on medium containing combinations of auxin and cytokinin at physiological concentrations.[120,130,134] This is a particularly rapid form of down-regulation detectable at the mRNA level one hour after suspension-cultured cells are treated with auxin and cytokinin.[135] Down-regulation by auxin and cytokinin also appears to be at least in part transcriptional since the decrease in steady-state RNA is correlated with decreased activity of the *Glb* promoter.[124,136]

Many plant species react to treatment with the stress hormone ethylene with induction of βGlu I activity, protein, and mRNA in leaves.[30,79,121,137-139] Although ethylene increases the βGlu I content of epidermal cells slightly, its inductive effect is most pronounced on mesophyll cells of the tobacco leaf.[34] Studies with inhibitors of ethylene production and ethylene action have shown that ethylene is required for the induction of βGlu I in cultured tobacco cells.[121] Ethylene treatment also dramatically increases the promoter activity of the tobacco βGlu I gene *Glb*.[124] Similarly, ethephon (2-chloroethylphosphonic acid), which releases ethylene, increases the promoter activity of the tobacco βGlu I genes *Gglb50*[125] and *Gln2*[126] in leaves of transgenic tobacco. When applied to plants, ethephon also gives rise to HCl and H_3PO_3.[140] Since acids can induce accumulation of PR-protein transcripts,[141] control experiments have been performed suggesting the induction of *Gglb50*[125] and *Gln2*[126] by ethephon is, in fact, due to ethylene. Judging from the approximately 20-fold increase in *Gus* reporter-gene activity following treatment with ethylene or ethephon, βGlu *I* promoters are highly induced by ethylene. Although activity of the homologous *N. plumbaginifolia Gn2* promoter is also highly induced by ethylene in transgenic tobacco leaves, activity of the less homologous *N. plumbaginifolia Gn1* promoter shows only a weak, approximately 2.5-fold induction.[41]

Regulation of βGlu I and class I Chn is often tightly coordinated.[4,26,30] In tobacco, their expression gradients in leaves and roots of the mature plant, their kinetics of down-regulation is cultured by auxin and cytokinin, and their responses in leaves to ethylene treatment and infection by pathogens are very similar. In contrast, during the germination of tobacco seeds, βGlu I, but not class I Chn, are transcriptionally induced in the micropylar endosperm.[19] ABA inhibits this seed-specific induction,[19] and also down-regulates βGlu I, but not class I Chn, at the transcriptional level in tobacco pith-cell suspensions and cultured leaf discs.[136]

3.4.1.2 Class II and Class III β-1,3-Glucanases

The class II βGlu PR-2a, PR 2b, and PR-2c are present in sepals but not in other floral organs.[128,142] In general, these acidic βGlu do not appear to accumulate in vegetative tissues of mature, healthy tobacco plants.[27,28] Reporter gene experiments suggest that the 1.7 kb promoter of the tobacco class II βGlu *PR-2d* gene is active in sepals, in the base of flowers, and in young seedlings, but not in leaves, roots, or the stem of mature tobacco plants.[128] In contrast, the 1.75 kb class II βGlu *PR-2b* promoter is active in leaves, stem, and root of mature tobacco plants, but at levels much lower than that of the tobacco βGlu I *Gglb50* promoter.[125]

Treatment of plants with ethephon results in no detectable induction, or very weak induction of class II and class III βGlu in leaves of tobacco and tomato.[26,137,139,143]

3.4.2 Pathogenesis-Related Regulation

In general, βGlu and Chn are induced in plants infected with viral, bacterial, and fungal pathogens. Similarly, elicitors including fungal glucans,[144,145] chitosan,[122] *N*-acetylchitooligosaccharides,[146] and glycoprotein[147] can induce the accumulation of the two enzymes.[79,81,82]

Tobacco cultivars carrying the dominant *N* gene from *N. glutinosa* show a local lesion, HR response to TMV rather than systemic symptoms of mosaic disease.[2,148,149] These plants also show decreased disease symptoms in response to secondary infection with certain other viral, bacterial, and fungal pathogens. This is an example of a type of induced long-lasting, broad-spectrum resistance called systemic acquired resistance (SAR).[150-153]

Class I βGlu proteins and their mRNAs are induced in TMV-infected leaves of tobacco as part of the local-lesion, HR response.[2,4,28,44,45,116] *Gus* reporter-gene experiments with the tobacco βGlu I *Gglb50* and *Glb* promoters have shown that βGlu I is transcriptionally induced by up to about 10-fold in TMV infected leaves showing HR.[124,125,153a] Histological studies indicate that the *Glb* promoter is active in a ring of cells around necrotic lesions induced by TMV infection, but not in cells immediately adjacent to the lesions or in the lesions themselves. *Gus* activity is also higher in areas with lesions, compared to lesion-free areas,[124] which is in agreement with the accumulation of basic PR-2 proteins in and around lesions described by Heitz et al.[154]

Activity of the weakly ethylene-inducible *Gn1* promoter of *N. plumbaginifolia* is strongly induced (about 21-fold) as part of the HR of tobacco leaves infected with the incompatible bacterium *Pseudomonas syringae* pv *syringae* and is localized around the necrotic lesions.[41,127] Induction of this promoter is much weaker in leaves infected with the compatible bacterium *Erwinia carotovora* subsp *carotovora* or a saprophytic strain of *Pseudomonas fluorescens.*[41] The *Gn1* promoter is also induced in a *Cf*-gene dependent manner in the interaction of transgenic tomato with incompatible and compatible races of the leaf mold pathogen *Cladosporium fulvum.*[155]

Either no or weak and erratic RNA-blot hybridization signals for βGlu I mRNAs have been detected in uninfected leaves of TMV-infected tobacco plants, indicating that βGlu I induction is a local response associated with HR.[26-28,142,153,156]

The PR-related class II and III βGlu are induced both locally in TMV-infected leaves and systemically in noninfected leaves of the same plant.[26-28,44,142,153,156-158] *Gus* reporter gene studies have shown that the promoters of the tobacco class II *PR-2b* and *PR-2d* are induced both locally around necrotic lesions on TMV-infected tobacco and systemically in noninfected leaves.[125,128] The close correlation between systemic induction of class II and class III βGlu has led to the use of these genes as markers for SAR.[150-153] For example, reporter gene constructs based on the *Arabidopsis* βGlu II *Bgl2* promoter have been used to isolate *Arabidopsis* SAR mutants such as *cpr1* (constitutive expresser of PR genes) and *npr1* (nonexpresser of PR genes).[150,151]

Systemic accumulation of salicylic acid (SA) is associated with the HR of tobacco, *Arabidopsis thaliana,* and certain other plants.[150,151] Treatment of mature, wild-type tobacco plants with SA strongly induces accumulation of mRNAs of PR-related class II and III βGlu and certain other PR proteins;[153,158a] and promoter activity of the class II *PR-2b*[125] and *PR-2d*[128] genes is induced in response to SA. While SA is probably not the long-distance systemic signal for SAR activation, it is required for transduction of this signal in leaves distal to the primary infection site.[150,151] Thus, transgenic tobacco plants expressing a bacterial salicylate hydroxylase gene (*NahG*) that fail to accumulate significant amounts of SA are unable to develop SAR and do not show increased expression of SAR markers including class II and III βGlu.

Transcripts of tobacco class I βGlu and Chn are either not induced or only weakly induced in response to SA.[44,132,153,158a,159] In contrast to the tobacco class I *Gglb50* and *Glb* promoters, the promoter of the less related *N. plumbaginifolia* class I *Gn1* promoter is strongly induced (about 14-fold) in transgenic tobacco plants treated with SA.[41]

3.4.3 *Cis*-Acting Elements

3.4.3.1 Class I β-1,3-Glucanase Genes

Expression studies of the *Gus* coding region fused to deletion series of 5′-flanking sequences have been used to identify regions of class I βGlu promoters important for transcriptional regulation. Elements for responsiveness to SA, ethylene, and TMV are present in the –0.45 to –1.5 kb region of the 1.5 kb tobacco *Gglb50* promoter.[125,153a] The homeologous 1.6 kb tobacco *Glb* promoter confers proper regulation to reporter gene expression in leaves, roots, seeds, and cultured cells.[18,19,65,71,124,136] A more detailed analysis of the *Glb* promoter indicates that the distal –1193 to –1452 region is required for high levels of expression in leaves and for responsiveness to ethylene and TMV infection.[124] The distal region contains a 61-bp enhancer of transcription in *N. plumbaginifolia* protoplasts.[138] A slightly modified 49-bp sequence from the highly homologous enhancer region of Bright Yellow tobacco *Gln2* gene[132] is an ethylene-responsive element (ERE) essential for ethylene responsiveness when combined with a minimal CaMV 35S promoter.[126] Enhancer activity and ethylene responsiveness depend on the integrity of two copies of the AGC-box, AGCCGCC, present in the promoters of several ethylene-responsive genes.[126,138,160,160a]

The same *Glb* deletion series has been used to analyze transcriptional regulation of βGlu I in the micropylar endosperm of germinating tobacco seeds.[71] The distal –1452 to –1193 region, which contains the ERE, is required for high-level, ethylene-sensitive expression; the regions –1452 to –1193 and –402 to 0 are important for down-regulation by ABA; and the region –402 to –211 is necessary and sufficient for micropylar–endosperm specific expression. The –402 *Glb* promoter is the shortest fragment giving developmental regulated expression in seeds[71] and leaves.[124] It is not, however, the minimal promoter *per se* since the shorter –211 fragment confers root-specific expression.[124]

Analysis of the *N. plumbaginifolia* class I *Gn1* gene has shown that a short –138 bp promoter is sufficient to confer full activity in transgenic tobacco leaves and is more active than the –736 and –2000 promoters.[127] The region –138 to –98 of the

Gn1 promoter is sufficient for high-level response to *Pseudomonas syringae* pv *syringae infection.*

3.4.3.2 Class II β-1,3-Glucanase Genes

Multiple regions of the approximately 1.7 kb tobacco class II *PR-2b* and *PR-2d* promoters contain elements for inducibility by SA and TMV.[125,128] For the *PR-2d* gene, this includes a major *cis*-acting element in the region –364 to –288, which confers to a core CaMV 35S promoter high-level expression in response to SA.[129]

3.4.4 Signal Transduction and *Trans*-Acting Factors

A putative ethylene receptor and several upstream components of the ethylene signaling pathway have been identified.[161-164] Far less is known about downstream components closer to the activation of transcription. Nuclear factors from tobacco leaves have been described that bind defined regions of the promoters of the class I *Gn1* of *N. plumbaginifolia,*[127] the class II *PR-2d* of tobacco,[129] and the class I *Glb* of tobacco.[138,153a]

Tobacco cDNA clones have been identified representing four novel DNA-binding proteins called *ethylene-responsive element binding proteins* (EREBPs) that specifically bind the ERE AGC box.[126] The mRNAs of EREBP-1 and EREBP-2 in the same class and EREBP-3 and EREBP-4, each in different classes, show distinctive expression patterns in leaves, roots, and cultured cells, which are correlated with the pattern of βGlu I expression. Accumulation of mRNAs for these EREBPs in tobacco leaves is induced by ethylene treatment. Thus, it is likely that the EREBPs are transcription factors important for ethylene-dependent, high-level transcription of βGlu I genes.

The deduced amino acid sequences of EREBPs are not similar to those of classical DNA binding proteins or transcription factors, i.e., they do not contain a basic leucine zipper or zinc finger motif.[126] The EREBP DNA binding domain is highly homologous in sequence to a domain present in the APETALA2 (AP2) protein, a regulator of meristem identity, floral organ specification, and seed coat development.[164a] Recently, homologs of the tobacco EREBPs have been isolated from *Arabidopsis thaliana* and tomato. The *Arabidopsis* EREBP, AtEBP, binds specifically to TAAGAGCCGCC, an AGC-box containing sequence and confers ethylene responsiveness to promoters of genes encoding PR proteins.[160] AtEBP interacts with *ocs* element binding factors (OBFs), belonging to a specific class of basic-region leucine zipper (bZIP) transcription factors. This suggests that cross-coupling between EREBP and bZIP transcription factors is important in regulating plant defense-related gene expression.

Further evidence linking EREBPs with the defense response has come from an analysis of the tomato Pto resistance gene against *Pseudomonas syringae* pv *tomato.* Three tomato genes, *Pti4, Pti5,* and *Pti6* have been identified that physically interact with the Pto kinase.[165] Each of these genes encodes a protein with characteristics that are typical of transcription factors and are similar to the tobacco EREBPs. These proteins specifically bind a DNA sequence present in the promoter region of a large number of PR genes. These findings are of particular interest because they establish a direct relationship between EREBPs, a disease-resistance gene, and the specific activation of plant defense genes.

In animals, plants, and fungi, cholera toxin (CTX) can activate signaling pathways dependent on heterotrimeric GTP binding proteins (G-proteins).[139] Tissues of transgenic tobacco plants expressing CTX show greatly reduced susceptibility to the bacterial pathogen *Pseudomonas tabaci,* accumulate high levels of SA, and constitutively express PR protein genes including PR-1 and the class II βGlu. In contrast, the class I βGlu are not induced and display normal developmental and ethylene-responsive regulation. In good agreement with these results, microinjection experiments demonstrate that CTX or GTP-γ-S induce the expression of a *Gus* reporter gene regulated by the PR-1 promoter, but not by the class I *Glb* promoter. Microinjection and grafting experiments strongly suggest that CTX-sensitive G-proteins are important in inducing the expression of a subset of PR genes and that these G-proteins act locally rather than systemically upstream of SA induction.

Multiple signal transduction pathways in tobacco and *Arabidopsis* and "cross-talk" between these pathways seem to be utilized to signal the induction of different subsets of PR genes in response to different pathogens. One type of pathway depends on ethylene and also leads to induction of the highly ethylene-responsive tobacco class I βGlu and Chn.[4,80,139,161,166,167] The SA-dependent pathways associated with SAR induction also activate expression of class II and III βGlu genes. These pathways appear to be ethylene-independent in tobacco and *Arabidopsis*.[139,141,150,151,168]

Vidal et al.[159] have recently identified an SA-independent pathway in the interaction between the soft-rot pathogen *Erwinia carotovora* subsp. *carotovora* and tobacco. Treatment of tobacco with this bacterium or a sterile culture filtrate (CF) containing elicitor(s) very rapidly induces local and systemic accumulation of transcripts for PR-proteins including βGlu I, but not of acidic PR-1a, which, in contrast to βGlu I, is induced by SA but not by ethylene.[2,169] SA is not the signal molecule leading to the early response of plants to *Erwinia*,[159] since induction of βGlu I transcripts in SA-deficient transgenic *NahG* tobacco plants and wild-type plants in response to CF is comparable. Therefore, the induction of βGlu- and other PR-protein genes by *Erwinia* and SA appear to involve two distinct pathways.

3.5 CONCLUDING REMARKS

Over the last 10 years, considerable progress has been made in understanding the structure and regulation of plant βGlu. It is now recognized that species of higher plants produce a broad range of βGlu differing in primary structure, cellular localization, and catalytic activity. The most striking structural variation is the C-terminal extension that distinguishes the vacuolar class I βGlu from the extracellular class II and III isoforms. This peptide is posttranslationally removed during intracellular transport and is likely to contain a vacuolar targeting signal.

A major problem is establishing the biological functions of these βGlu and understanding how intracellular localization and structure are related to these functions. The available evidence suggests that different classes of βGlu have different functions in plant-microbe interactions. For example, the class I βGlu, particularly in combination with class I Chn, inhibit the growth of certain fungi both *in vitro* and when over-expressed in transgenic plants. On the other hand, the extracellular βGlu have weak antifungal activity, but may be involved in releasing elicitors that activate host defense reactions. These βGlu are also induced systemically after infection. Their role, if any, in SAR is still an open question.

Several PR-like βGlu are localized in specific floral organs. Dissolution of the callosic wall of microspores, which can be broken down by class I βGlu, is required for pollen formation. Correlative evidence suggests that precisely regulated expression of specific anther βGlu is required for this process. Direct evidence for a causal role of these βGlu has yet to be established. The results of antisense experiments with the stylar βGlu are inconclusive, and the function of these proteins in the conducting tissues is not known. Class I βGlu show a novel pattern of expression and regulation during seed germination and may have a role in weakening the endosperm to allow the penetration of the radicle. Although there is a close correlation between βGlu I induction and endosperm rupture under a variety of physiological conditions, direct evidence for a causal role of these enzymes in germination is lacking.

Comparisons of the patterns of enzyme and steady-state mRNA accumulation and reporter-gene experiments indicate that transcription is an important site of developmental, hormonal, and pathogenesis-related regulation of βGlu genes. Nevertheless, these studies do not rule out additional regulation — at the level of translation or protein degradation, for example. Transcriptional regulation involves multiple signaling pathways linking different signal molecules to the same and different target βGlu genes. Promoter regions have been identified that are important for responses to ethylene, SA, elicitors, and infection with viral and bacterial pathogens. The most interesting of these is the ERE present in class I βGlu and Chn genes which is responsible for high-level expression in response to ethylene. Several components of the ethylene-signaling pathway have now been identified, and studies aimed at relating these components to the EREBP transcription factors will undoubtedly provide insight into the regulation and function of the βGlu.

REFERENCES

1. Van Loon, L. C., Pierpoint, W. S., Boller, T., and Conejero, T., Recommendations for naming plant pathogenesis-related proteins, *Plant Mol. Biol. Rep.,* 12, 245, 1994.
2. Stintzi, A., Heitz, T., Prasad, V., Wiedemann-Merdinoglu S., Kaufmann, S., Geoffroy, P., Legrand, M., and Fritig, B., Plant "pathogenesis-related" proteins and their role in defense against pathogens, *Biochimie*, 75, 687, 1993.
3. Antoniw, J. F., Ritter, C. E., Pierpoint, W. S., and van Loon, L. C., Comparison of three pathogenesis-related proteins from plants of two cultivars of tobacco infected with TMV, *J. Gen. Virol.,* 47, 79, 1980.
4. Meins, F., Jr., Neuhaus, J. M., Sperisen, C., and Ryals, J., The primary structure of plant pathogenesis-related glucanohydrolases and their genes, in *Genes Involved in Plant Defense,* Boller, T. and Meins, F., Jr., Eds., Vienna, Springer-Verlag, New York, 1992, 245.

5. Høj, P. B. and Fincher, G. B., Molecular evolution of plant β-glucan endohydrolases, *Plant J.,* 7, 367, 1995.
6. Simmons, C. R., The physiology and molecular biology of plant 1,3-β-D-glucanases and 1,3;1,4-β-D-glucanases, *Crit. Rev. Plant Sci.,* 13, 325, 1994.
7. Stone, B. A. and Clarke, A. E., *Chemistry and Biology of (1→3)-β-Glucans,* La Trobe University Press, Victoria, Australia, 1992.
8. Waterkeyn, L., Sur l'existence d'un "stade callosique" présenté par la paroi cellulaire, au cours de la cytokinèse, *C.R. Acad. Sci. Paris D,* 265, 1792, 1967.
9. Fulcher, R. G., McCully, M. E., Setterfield, G., and Sutherland, J., β-1,3-glucans may be associated with cell plate formation during cytokinesis, *Can. J. Bot.,* 54, 459, 1976.
10. Worrall, D., Hird, D. L., Hodge, R., Paul, W., Draper, J., and Scott, R., Premature dissolution of the microsporocyte callose wall causes male sterility in transgenic tobacco, *Plant Cell,* 4, 759, 1992.
11. Bucciaglia, P. A. and Smith, A. G., Cloning and characterization of *Tag1*, a tobacco anther β-1,3-glucanase expressed during tetrad dissolution, *Plant Mol. Biol.,* 24, 903, 1994.
12. Roggen, H. P. and Stanley, R. G., Cell wall hydrolyzing enzymes in wall formation as measured by pollen-tube extension, *Planta,* 84, 295, 1969.
13. Meikle, P. J., Bonig, I., Hoogenraad, N. J., Clarke, A. E., and Stone, B. A., The location of (1-3)-β-glucans in the walls of pollen tubes of *Nicotiana alata* using a (1-3)-β-glucan-specific monoclonal antibody, *Planta,* 185, 1, 1991.
14. Lotan, T., Ori, N., and Fluhr, R., Pathogenesis-related proteins are developmentally regulated in tobacco flowers, *Plant Cell,* 1, 881, 1989.
15. Ori, N., Sessa, G., Lotan, T., Himmelhoch, S., and Fluhr, R., A major stylar matrix polypeptide (Sp41) is a member of the pathogenesis-related proteins superclass, *EMBO J.,* 9, 3429, 1990.
16. Dong, J. Z. and Dunstan, D. I., Endochitinase and β-1,3-glucanase genes are developmentally regulated during somatic embryogenesis in *Picea glauca, Planta,* 201, 189, 1997.

16a. Helleboid, S., Bauw, G., Belingheri, L., Vasseur, J., and Hilbert, J. L., Extracellular β-1,3-glucanases are induced during early somatic embryogenesis in *Cichorium, Planta,* 205, 56, 1998.

17. Hinton, D. M. and Pressey, R., Glucanase in fruits and vegetables, *J. Amer. Soc. Hort. Sci.,* 105, 499, 1980.
18. Vögeli-Lange, R., Fründt, C., Hart, C. M., Beffa, R., Nagy, F., and Meins, F., Jr., Evidence for a role of β-1,3-glucanase in dicot seed germination, *Plant J.,* 5, 273, 1994.
19. Leubner-Metzger, G., Fründt, C., Vögeli-Lange, R., and Meins, F., Jr., Class I β-1,3-glucanase in the endosperm of tobacco during germination, *Plant Physiol.,* 109, 751, 1995.
20. Fincher, G. B. and Stone, B. A., Physiology and biochemistry of germination in barley, in *Barley: Chemistry and Technology,* MacGregor, A. W. and Bhatty, R. S., Eds., American Association of Cereal Chemists, St. Paul, 1993, 247.
21. Krabel, D., Eschrich, W., Wirth, S., and Wolf, G., Callase-(1,3-β-D-glucanase) activity during spring reactivation in deciduous trees, *Plant Sci.,* 93, 19, 1993.
22. Ernst, D., Bodemann, A., Schmelzer, E., Langebartels, C., and Sandermann, H. J., β-1,3-Glucanase mRNA is locally, but not systemically induced in *Nicotiana tabacum* L. cv. BEL W3 after ozone fumigation, *J. Plant Physiol.,* 148, 215, 1996.

23. Thalmair, M., Bauw, G., Thiel, S., Doehring, T., Langebartels, C., and Sandermann, H., Jr., Ozone and ultraviolet B effects on the defense-related proteins β-1,3-glucanase and chitinase in tobacco, *J. Plant Physiol.,* 148, 222, 1996.
24. Linthorst, H. J. M., Pathogenesis-related proteins of plants, *Crit. Rev. Plant Sci.,* 10, 123, 1991.
25. Hincha, D. K., Meins, F., Jr., and Schmitt, J. M., β-1,3-Glucanase is cryoprotective *in vitro* and is accumulated in leaves during cold acclimation, *Plant Physiol.,* 114, 1077, 1997.
26. Brederode, F., Linthorst, H., and Bol, J., Differential induction of acquired resistance and PR gene expression in tobacco by virus infection, ethephon treatment, UV light and wounding, *Plant Mol. Biol.,* 17, 1117, 1991.
27. Ward, E. R., Payne, G. B., Moyer, M. B., Williams, S. C., Dincher, S. S., Sharkey, K. C., Beck, J. J., Taylor, H. T., Ahl-Goy, P., Meins, F., Jr., and Ryals, J. A., Differential regulation of β-1,3-glucanase messenger RNAs in response to pathogen infection, *Plant Physiol.,* 96, 390, 1991.
28. Payne, G., Ward, E., Gaffney, T., Goy, P. A., Moyer, M., Harper, A., Meins, F., Jr., and Ryals, J., Evidence for a third structural class of β-1,3-glucanase in tobacco, *Plant Mol. Biol.,* 15, 797, 1990.
29. Oh, H. Y. and Yang, M. S., Nucleotide sequence of genomic DNA encoding the potato β-1,3-Glucanase, *Plant Physiol.,* 107, 1453, 1995.
30. Beerhues, L. and Kombrink, E., Primary structure and expression of mRNAs encoding basic chitinase and 1,3-β-glucanase in potato, *Plant Mol. Biol.,* 24, 353, 1994.
31. Domingo, C., Conejero, V., and Vera, P., Genes encoding acidic and basis class III β-1,3-glucanases are expressed in tomato plants upon viroid infection, *Plant Mol. Biol.,* 24, 725, 1994.
32. Van Kan, J. A. L., Joosten, M. H. A. J., Wagemakers, C. A. M., van den Berg-Velthius, G. C. M., and de Wit, P. J. G. M., Differential accumulation of mRNAs encoding extracellular and intracellular PR proteins in tomato induced by virulent and avirulent races of *Cladosporium fulvum, Plant Mol. Biol.,* 20, 513, 1992.
33. Van den Bulcke, M., Bauw, G., Castresana, C., van Montagu M., and Vandekerckhove, J., Characterization of vacuolar and extracellular β(1,3)-glucanases of tobacco: evidence for a strictly compartmentalized plant defense system, *Proc. Natl. Acad. Sci. U.S.A.,* 86, 2673, 1989.
34. Keefe, D., Hinz, U. and Meins, F., Jr., The effect of ethylene on the cell-type-specific and intracellular localization of β-1,3-glucanase and chitinase in tobacco leaves, *Planta,* 182, 43, 1990.
35. Sticher, L., Hinz, U., Meyer, A. D., and Meins, F., Jr., Intracellular transport and processing of a tobacco vacuolar β-1,3-glucanase, *Planta,* 188, 559, 1992.
36. Shinshi, H., Wenzler, H,. Neuhaus, J. M., Felix, G., Hofsteenge, J., and Meins, F., Jr., Evidence of amino and carboxyl-terminal processing of a plant defense-related enzyme primary structure of tobacco prepro-β-1,3-glucanase, *Proc. Natl. Acad. Sci. U.S.A.,* 85, 5541, 1988.
37. Neuhaus, J. M., Sticher, L., Meins, F., Jr., and Boller, T., A short C-terminal sequence is necessary and sufficient for the targeting of chitinases to the plant vacuole, *Proc. Natl. Acad. Sci. U.S.A.,* 88, 10362, 1991.
38. Bednarek, S. Y. and Raikhel, N. V., The barley lectin carboxyl-terminal propeptide is a vacuolar protein sorting determinant in plants, *Plant Cell,* 3, 1195, 1991.

39. Melchers, L. S., Sela-Buurlage, M. B., Vloemans, S. A., Woloshuk, C. P., Vanroekel, J. S. C., Pen, J., van del Elzen, P. J. M., and Cornelissen, B. J. C., Extracellular targeting of the vacuolar tobacco proteins — AP24, chitinase and β-1,3-glucanase — in transgenic plants, *Plant Mol. Biol.,* 21, 583, 1993.
39a. Kunze, I., Kunze, G., Bröker, M., Manteuffel, R., Meins, F., Jr., and Müntz, K., Evidence for secretion of vacuolar (α-mannosidase, class I chitinase, and class I β-1,3-glucanase in suspension cultures of tobacco cells, *Planta*, 205, 92, 1998.
40. Gheysen, G., Inzé, D., Soetaert, P., van Montagu, M., and Castresana, C., Sequence of a *Nicotiana plumbaginifolia* β-1,3-glucanase gene encoding a vacuolar isoform, *Nucleic Acids Res.,* 18, 6685, 1990.
41. Castresana, C., De Carvalho, F., Gheysen, G., Habets, M., Inzé, D., and van Montagu, M., Tissue-specific and pathogen-induced regulation of a *Nicotiana plumbaginifolia* β-1,3- glucanase gene, *Plant Cell,* 2, 1131, 1990.
42. Sperisen, C., Ryals, J., and Meins, F., Jr., Comparison of cloned genes provides evidence for intergenomic exchange of DNA in the evolution of a tobacco glucan endo-1,3-β-glucosidase gene family, *Proc. Natl. Acad. Sci. U.S.A.,* 88, 1820, 1991.
43. Beffa, R. S., Neuhaus, J. M., and Meins, F., Jr., Physiological compensation in antisense transformants: specific induction of an ersatz glucan endo-1,3-β-glucosidase in plants infected with necrotizing viruses, *Proc. Natl. Acad. Sci. U.S.A.,* 90, 8792, 1993.
44. Linthorst, H. J. M., Melchers, L. S., Mayer, A., van Roekel, J. S. C., Cornelissen, B. J. C., and Bol, J. F., Analysis of gene families encoding acidic and basic β-1,3-glucanases of tobacco, *Proc. Natl. Acad. Sci. U.S.A.,* 87, 8756, 1990.
45. Kauffmann, S., Legrand, M., Geoffroy, P., and Fritig, B., Biological function of "pathogenesis-related" proteins: four PR proteins of tobacco have β-1,3-glucanase activity, *EMBO J.,* 6, 3209, 1987.
46. Davies, G. and Henrissat, B., Structures and mechanisms of glycosyl hydrolases, *Structure*, 3, 853, 1995.
47. Varghese, J. N., Garrett, T. P. J., Colman, P. M., Chen, L., Høj, P. B., and Fincher, G. B., Three-dimensional structures of two plant β-glucan endohydrolases with distinct substrate specificities, *Proc. Natl. Acad. Sci. U.S.A.,* 91, 2785, 1994.
48. McCormick, S., Male gametophyte development, *Plant Cell,* 5, 1265, 1993.
49. Scott, R., Hodge, R., Paul, W., and Draper, J., The molecular biology of anther differentiation, *Plant Sci.,* 80, 167, 1991.
50. Steiglitz, H. and Stern H., Regulation of β-1,3-glucanase activity in developing anthers of *Lilium, Dev. Biol.,* 34, 169, 1973.
51. Frankel, R., Izhar, S., and Nitsan, J., Timing of callase activity and cytoplasmic male sterility in *Petunia, Biochem. Genet.,* 3, 451, 1969.
52. Steiglitz, H., Role of β-1,3-glucanase in postmeiotic microspore release, *Dev. Biol.,* 57, 87, 1977.
53. Izhar, S. and Frankel R., Mechanism of male sterility in *Petunia:* the relationship between pH, callase activity in the anthers, and the breakdown of the microsporogenesis, *Theor. Appl. Genet.,* 41, 104, 1971.
54. Warmke, H. E. and Overman, M. A., Cytoplasmic male sterility in sorghum. 1. Callose behavior in fertile and sterile anthers, *J. Hered.,* 63, 103, 1972.
55. Jin, W., Horner, H. T., and Palmer, R. G., Genetics and cytology of a new genic male-sterile soybean [*Glycine max* (L) Merr], *Sex. Plant Reprod.,* 10, 13, 1997.

56. Hird, D. L., Worrall, D., Hodge, R., Smartts, S., Paul, W., and Scott, R., The anther-specific protein encoded by the *Brassica napus* and *Arabidopsis thaliana* A6 gene displays similarity to β-1,3-glucanases, *Plant J.,* 4, 1023, 1993.
56a. Curtis, I. D., He, C., Scott, R., Power, J. B., and Davey, M. R., Genomic male sterility in lettuce, a baseline for the production of F_1 hybrids, *Plant Sci,* 113, 113, 1996.
57. Tsuchiya, T., Toriyama K., Yoshikawa, M., Ejiri, S. I., and Hinata, K., Tapetum-specific expression of the gene for an endo-β-1,3-glucanase causes male sterility in transgenic tobacco, *Plant Cell Physiol.,* 36, 487, 1995.
58. Gasser, C. S. and Robinsonbeers, K., Pistil development, *Plant Cell,* 5, 1231, 1993.
59. Mascarenhas, J. P., Molecular mechanisms of pollen tube growth and differentiation, *Plant Cell,* 5, 1303, 1993.
60. Sessa, G. and Fluhr, R., The expression of an abundant transmitting tract-specific endoglucanase (Sp41) is promoter-dependent and not essential for the reproductive physiology of tobacco, *Plant Mol. Biol.,* 29, 969, 1995.
61. Black, M., Liberating the radicle: a case for softening-up, *Seed Sci. Res.,* 6, 39, 1996.
62. Bewley, J. D., Seed germination and dormancy, *Plant Cell,* 9, 1055, 1997.
63. Avery, G. S. J., Structure and germination of tobacco seed and the developmental anatomy of the seedling plant, *Am. J. Bot.,* 20, 309, 1933.
64. Arcila, J. and Mohapatra, S. C., Development of tobacco seedling. 2. Morphogenesis during radicle protrusion, *Tob. Sci.,* 27, 35, 1983.
65. Leubner-Metzger, G., Fründt, C., and Meins, F., Jr., Effects of gibberellins, darkness and osmotica on endosperm rupture and class I β-1,3-glucanase induction in tobacco seed germination, *Planta,* 199, 282, 1996.
66. Ogawara, K., Interaction of light and darkness in the germination of the light-favored tobacco seeds, *Proc. Jpn. Acad.,* 30, 504, 1954.
67. Kasperbauer, M. J., Germination of tobacco seed — I. Inconsistency of light sensitivity, *Tob. Sci.,* 12, 20, 1968.
68. Bihlmeier, M., Der Einfluss der Vorquellung und der Samenschale auf die Keimung lichtgeförderter Samen, *Jahrb. wiss. Bot.,* 67, 702, 1927.
69. Böhmer, K., Die Bedeutung der Samenteile für die Lichtwirkung und die Wechselbeziehung von Licht und Sauerstoff bei der Keimung lichtempfindlicher Samen, *Jahrb. wiss. Bot.,* 68, 549, 1928.
70. Kincaid, R. R., The effects of certain environmental factors on the germination of Florida cigar-wrapper tobacco seeds, *Tech. Bull. Univ. Florida. Agr. Exp. Station,* 277, 5, 1935.
71. Leubner-Metzger, G., Petruzzelli, L., Waldvogel, R., Vögeli-Lange, R., and Meins, F., Jr., Ethylene responsive element binding protein (EREBP) expression and the transcriptional regulation of class I β-1,3-glucanase during tobacco seed germination, *Plant Mol. Biol.,* 38, 785, 1998.
72. Chen, L., Fincher, G. B., and Høj, P. B., Evolution of polysaccharide hydrolase substrate specificity — catalytic amino acids are conserved in barley 1,3-1,4- and 1,3-β-glucanase, *J. Biol. Chem.,* 268, 13318, 1993.
73. Fincher, G. B., Molecular and cellular biology associated with endosperm mobilization in germinating cereal grains, *Annu. Rev. Physiol. Plant Mol. Biol.,* 40, 305, 1989.
74. Cordero, M. J., Raventos, D., and San Segundo, B., Differential expression and induction of chitinases and β-1,3-glucanases in response to fungal infection during germination of maize seeds, *Mol. Plant-Microbe Interact.,* 7, 23, 1994.
75. Abeles, F. B., Bosshart, R. P., Forrence, L. E., and Habig, W. H., Preparation and purification of glucanase and chitinase form bean leaves, *Plant Physiol.,* 47, 129, 1971.

76. Van Loon, L. C. and van Kammen, A., Polyacrylamide disc electrophoresis of the soluble leaf proteins from *Nicotiana tabacum* var. "Samsun" and "Samsun NN." II. Changes in leaf protein constitution after infection with tobacco mosaic virus, *Virol.,* 40, 199, 1970.
77. Gianinazzi, S., Martin, C., and Vallee, J. C., Hypersensibilité aux virus, temperatures et proteines solubles chez le *Nicotiana* Xanthi-nc. Apparition de nouvelles macromolecules lors de la repression de la synthese virale, *C.R. Acad. Sci. Paris D.,* 270, 2382, 1970.
78. Cutt, J. R. and Klessig, D. F., Pathogenesis-related proteins, in *Genes Involved in Plant Defense,* Boller, T. and Meins, F., Jr., Eds., Springer-Verlag, Wien, New York, 1992, 209.
79. Boller, T., Ethylene and the regulation of antifungal hydrolases in plants, *Oxford Surveys Plant Mol. & Cell Biol.,* 5, 145, 1988.
80. Boller, T., Ethylene and plant-pathogen interactions, in *Current Topics in Plant Physiology: An American Society of Plant Physiologists Series,* Flores, H. E., Arteca, R. N., and Shannon, J. C., Eds., 1990, 138.
81. Boller, T., Antimicrobial functions of the plant hydrolases chitinase and β-1,3-glucanase, in *Developments in Plant Pathology,* Fritig, B. and Legrand, M., Eds., 1993, 391.
82. Boller T., Chemoperception of microbial signals in plant cells, *Annu. Rev. Plant Physiol. Plant Mol. Biol.,* 46, 189, 1995.
83. Bowles, J. D., Defense-related proteins in higher plants, *Annu. Rev. Biochem.,* 59, 873, 1990.
84. Wessels, J. G. H. and Sietsma, J. H., Fungal cell walls: a survey, in *Encyclopedia of Plant Physiology. Plant Carbohydrates II,* Vol. 13B, Tanner, W. and Loewus, F. A., Eds., Springer Verlag, Berlin, 1981, 352.
85. Melchers, L. S., Ponstein, A. S., Sela-Buurlage, M. B., Vloemans, S. A., and Cornelissen, B. J. C., In vitro anti-microbial activities of defense proteins and biotechnology, in *Mechanisms of Plant Defense Responses,* Fritig, B. and Legrand, M., Eds., Kluwer Academic Publishers, Dordrecht, 1993, 401.
86. Mauch, F., Mauch-Mani, B., and Boller, T., Antifungal hydrolases in pea tissue. II. Inhibition of fungal growth by combinations of chitinase and β-1,3-glucanase, *Plant Physiol.,* 88, 936, 1988.
87. Ludwig, A. and Boller, T., A method for the study of fungal growth inhibition by plant proteins, *FEMS Microbiol. Lett.,* 69, 61, 1990.
88. Leah, R., Tommerup, H., Svendsen, I., and Mundy, J., Biochemical and molecular characterization of three barley seed proteins with antifungal properties, *J. Biol. Chem.,* 266, 1564, 1991.
89. Sela-Buurlage, M. B., Ponstein, A. S., Bres-Vloemans, S. A., Melchers, L. S., van den Elzen, P. J. M., and Cornelissen, B. J. C., Only specific tobacco (*Nicotiana tabacum*) chitinases and β-1,3-glucanases exhibit antifungal activity, *Plant Physiol.,* 101, 857, 1993.
90. Kim, Y. J. and Hwang, B. K., Isolation of a basic 34 kiloDalton β-1,3-glucanase with inhibitory activity against *Phytophthora capsici* from pepper stems, *Physiol. Molec. Plant Pathol.,* 50, 103, 1997.
91. Lawrence, C. B., Joosten, M. H. A. J., and Tuzun, S., Differential induction of pathogenesis-related proteins in tomato by *Alternaria solani* and the association of a basic chitinase isozyme with resistance, *Physiol. Mol. Plant Pathol.,* 48, 361, 1996.

92. Joosten, M. H. A. J., Verbakel, H. M., Nettekoven, M. E. van Leeuwen, J., van der Vossen, R. T. M., and de Wit, P. J. G. M., The phytopathogenic fungus *Cladosporium fulvum* is not sensitive to the chitinase and β-1,3-glucanase defense proteins of its host, tomato, *Physiol. Mol. Plant Pathol.,* 46, 45, 1995.
93. Anfoka, G. and Buchenauer H., Systemic acquired resistance in tomato against *Phytophthora infestans* by pre-inoculation with tobacco necrosis virus, *Physiol. Molec. Plant Pathol.,* 50, 85, 1997.
94. Albersheim, P. and Valent, B. S., Host-pathogen interactions. VII. Plant pathogens secrete proteins which inhibit enzymes of the host capable of attacking the pathogen, *Plant Physiol.,* 53, 684, 1974.
95. Sharp, J. K., Valent, B., and Albersheim, P., Purification and partial characterization of a β-glucan fragment that elicits phytoalexin accumulation in soybean, *J. Biol. Chem.,* 259, 11312, 1984.
96. Okinaka, Y., Mimori, K., Takeo, K., Kitamura, S., Takeuchi, Y., Yamaoka, N., and Yoshikawa, M., A structural model for the mechanisms of elicitor release from fungal cell walls by plant β-1,3-endoglucanase, *Plant Physiol.,* 109, 839, 1995.
97. Umemoto, N., Kakitani, M., Iwamatsu, A.,Yoshikawa, M., Yamaoka, N., and Ishida, I., The structure and function of a soybean β-glucan-elicitor-binding protein, *Proc. Natl. Acad. Sci. U.S.A.,* 94,1029, 1997.
98. Takeuchi, Y., Yoshikawa, M., Takeba, G., Tanaka, K., Shibata, D., and Horino, O., Molecular cloning and ethylene induction of mRNA encoding a phytoalexin elicitor-releasing factor, β-1,3-endoglucanase, in soybean, *Plant Physiol.,* 93, 673, 1990.
99. Ham, K. S., Wu, S. C., Darvill, A. G., and Albersheim, P., Fungal pathogens secrete an inhibitor protein that distinguishes isoforms of plant pathogenesis-related endo-β-1,3-glucanases, *Plant J.,* 11, 169, 1997.
100. Schmidt, W. E. and Ebel, J., Specific binding of a fungal glucan phytoalexin elicitor to membrane fractions from soybean Glycine max, *Proc. Natl. Acad. Sci. U.S.A.,* 84, 4117, 1987.
101. Yoshikawa, M., Yamaoka, N., and Takeuchi, Y., Elicitors: their significance and primary modes of action in the induction of plant defense reactions, *Plant Cell Physiol.,* 34, 1163, 1993.
102. Yoshikawa, M., Tsuda, M., and Takeuchi, Y., Resistance to fungal diseases in transgenic tobacco plants expressing the phytoalexin elicitor-releasing factor, β-1,3-endoglucanase, from soybean, *Naturwiss*, 80, 417, 1993.
102a. Lusso, M. and Kuc, J., The effect of sense and antisense expression of the PR-N gene for β-1,3-glucanase on disease resistance of tobacco to fungi and viruses, *Physiol. Molec. Plant Pathol.,* 49, 267, 1966.
103. Cheong, J. J., Birberg, W., Fügedi, P., Pilotti, A., Garegg, P. J., Hong, N., Ogawa, T., and Hahn, M. G., Structure-activity relationships of oligo-β-glucoside elicitors of phytoalexin accumulation in soybean, *Plant Cell,* 3, 127, 1991.
104. Cheong, J. J., Alba, R., Côté, F., Enkerli, J., and Hahn, M. G., Solubilization of functional plasma membrane-localized hepta-β-glucoside elicitor-binding proteins from soybean, *Plant Physiol.,* 103, 1173, 1993.
105. Mithöfer, A., Lottspeich, F., and Ebel, J., One-step purification of the β-glucan elicitor-binding protein from soybean (*Glycine max* L.) roots and characterization of an anti-peptide antiserum, *FEBS Lett.,* 381, 203, 1996.
106. Ham, K. S., Kauffmann, S., Albersheim, P., and Darvill, A. G., Host-pathogen interactions XXXIX. A soybean pathogenesis-related protein with β-1,3-glucanase activity releases phytoalexin elicitor-active heat-stable fragments from fungal walls, *Molec. Plant-Microbe Interact.,* 4, 545, 1991.

107. Ham, K. S., Albersheim, P., and Darvill, A. G., Generation of β-glucan elicitors by plant enzymes and inhibition of the enzymes by a fungal protein, *Can. J. Bot.*, 73, S1100, 1995.
108. Jongedijk, E., Tigelaar, H., van Roekel, J. S. C., Bres-Vloemans, S. A., Dekker, I. van den Elzen, P. J. M., Cornelissen, B. J. C., and Melchers, L. S., Synergetic activity of chitinases and β-1,3-glucanases enhances fungal resistance in transgenic tomato plants, *Euphytica*, 85, 173, 1995.
109. Broglie, K., Chet, I., Holliday, M., Cressman, R., Biddle, P., Knowlton, S., Mauvais, C. J., and Broglie, R., Transgenic plants with enhanced resistance to the fungal pathogen *Rhizoctonia solani, Science,* 254, 1194, 1991.
110. Jach, G., Gornhardt, B., Mundy, J., Logemann, J., Pinsdorf, R., Leah, R., Schell, J., and Maas, C., Enhanced quantitative resistance against fungal disease by combinatorial expression of different barley antifungal proteins in transgenic tobacco, *Plant J.*, 8, 97, 1995.
111. Masoud, S. A., Zhu, Q., Lamb, C., and Dixon, R. A., Constitutive expression of an inducible β-1,3-glucanase in alfalfa reduces disease severity caused by the oomycete pathogen *Phytophthora megasperma* f. sp *medicaginis,* but does not reduce disease severity of chitin-containing fungi, *Transgenic Res.,* 5, 313, 1996.
112. Zhu, Q., Maher, E. A., Masoud, S., Dixon, R. A., and Lamb, C. J., Enhanced protection against fungal attack by constitutive co-expression of chitinase and glucanase genes in transgenic tobacco, *Bio/technology,* 12, 807, 1994.
113. Neuhaus, J. M., Ahl-Goy, P., Hinz, U., Flores, S., and Meins, F., Jr., High-level expression of a tobacco chitinase gene in *Nicotiana sylvestris.* Susceptibility of transgenic plants to *Cercospora nicotianae* infection, *Plant Mol. Biol.,* 16, 141, 1991.
114. Neuhaus, J. M., Flores, S., Keefe, D., Ahl, G. P., and Meins, F., Jr., The function of vacuolar β-1,3-glucanase investigated by antisense transformation. Susceptibility of transgenic *Nicotiana-sylvestris* plants to *Cercospora-nicotianae* infection, *Plant Mol. Biol.,* 19, 803, 1992.
115. Collinge, D. B. and Slusarenko, A. J., Plant gene expression in response to pathogens, *Plant Mol. Biol.,* 9, 389, 1987.
116. Beffa, R. and Meins, F., Jr., Pathogenesis-related functions of plant β-1,3-glucanases investigated by antisense transformation: a review, *Gene*, 179, 97, 1996.
117. Beffa, R. S., Hofer, R. M., Thomas, M., and Meins, F., Jr., Decreased susceptibility to viral disease of β-1,3-glucanase-deficient plants generated by antisense transformation, *Plant Cell,* 8, 1001, 1996.
118. Wu, J. H. and Dimitman, J. E., Leaf structure and callose formation as determinants of TMV movement in bean leaves as revealed by UV irradiation experiments, *Virology,* 40, 820, 1970.
119. Moore, A. E. and Stone, B. A., Effect of senescence and hormone treatment on the activity of a β-1,3-glucan hydrolase in *Nicotiana glutinosa* leaves, *Planta*, 104, 93, 1972.
120. Felix, G. and Meins, F., Jr., Developmental and hormonal regulation of β-1,3-glucanase in tobacco *Nicotiana tabacum* cultivar Havana 425, *Planta*, 167, 206, 1986.
121. Felix, G. and Meins, F., Jr., Ethylene regulation of β-1,3-glucanase in tobacco, *Planta*, 172, 386, 1987.
122. Mauch, F., Hadwiger, L. A., and Boller, T., Ethylene: symptom, not signal for the induction of chitinase and β-1,3-glucanase in pea pods by pathogens and elicitors, *Plant Physiol.*, 76, 607, 1984.
123. Clarke, A. E. and Stone, B. A., β-1,3-glucan hydrolases from the grape vine (*Vitis vinifera*) and other plants, *Phytochemistry,* 1, 175, 1962.

124. Vögeli-Lange, R., Fründt, C., Hart, C. M., Nagy, F., and Meins, F., Jr., Developmental, hormonal, and pathogenesis-related regulation of the tobacco class I β-1,3-glucanase B promoter, *Plant Mol. Biol.,* 25, 299, 1994.
125. Van de Rhee, M. D., Lemmers, R., and Bol, J. F., Analysis of regulatory elements involved in stress-induced and organ-specific expression of tobacco acidic and basic β-1,3-glucanase genes, *Plant Mol. Biol.,* 21, 451, 1993.
126. Ohme-Takagi, M. and Shinshi, H., Ethylene-inducible DNA binding proteins that interact with an ethylene-responsive element, *Plant Cell,* 7, 173, 1995.
127. Alonso, E., de Carvalho Niebel, F., Obregon, P., Gheysen, G., Inzé, D., van Montagu, M., and Castresana, C., Differential *in vitro* DNA binding activity to a promoter element of the *gn1* β-1,3-glucanase gene in hypersensitively reacting tobacco plants, *Plant J.,* 7, 309, 1995.
128. Henning, J., Dewey, R. E., Cutt, J. R., and Klessig, D. F., Pathogen, salicylic acid and developmental dependent expression of β-1,3-glucanase/GUS gene fusion in transgenic tobacco plants, *Plant J.,* 4, 481, 1993.
129. Shah, J. and Klessig, D. F., Identification of a salicylic acid-responsive element in the promoter of the tobacco pathogenesis-related β-1,3-glucanase gene, PR-2d, *Plant J.,* 10, 1089, 1996.
130. Shinshi, H., Mohnen, D., and Meins, F., Jr., Regulation of a plant pathogenesis-related enzyme inhibition of chitinase and chitinase messenger RNA accumulation in cultured tobacco tissues by auxin and cytokinin, *Proc. Natl. Acad. Sci. U.S.A.,* 84, 89, 1987.
131. Neale, A. D., Wahleithner, J. A., Lund, M., Bonnett, H. T., Kelly, A., Meeks Wagne, D. R., Peacock, W. J., and Dennis, E. S., Chitinase, β-1,3-glucanase, osmotin, and extensin are expressed in tobacco explants during flower formation, *Plant Cell,* 2, 673, 1990.
132. Ohme-Takagi, M. and Shinshi, H., Structure and expression of a tobacco β-1,3-glucanase gene, *Plant Mol. Biol.,* 15, 941, 1990.
133. Garcia-Garcia, F., Schmelzer, E., Hahlbrock, K., and Roxby, R., Differential expression of chitinase and β-1,3-glucanase genes in various tissues of potato plants, *Z. Naturforsch. C.,* 9, 195, 1994.
134. Mohnen, D., Shinshi, H., Felix, G., and Meins, F., Jr., Hormonal regulation of β-1,3-glucanase EC-3.2.1.39 messenger RNA levels in cultured tobacco *Nicotiana-tabacum* cultivar Havana-425 tissues, *EMBO J.,* 4, 1631, 1985.
135. Sperisen, C., Structure, evolution and regulation of β-1,3-glucanase genes of tobacco, Dissertation ETH No 10219, Swiss Federal Institute of Technology, Zürich, 1993.
136. Rezzonico, E., Flury, N., Meins, F., Jr., and Beffa, R., Transcriptional down-regulation by abscisic acid of pathogenesis-related β-1,3-glucanase genes in tobacco tissue and cell cultures, *Plant Physiol.,* 117, 585, 1998.
137. Memelink, J., Linthorst, H. J. M., Schilperoort, R. A., and Hoge, J. H. C., Tobacco genes encoding acidic and basic isoforms of pathogenesis-related proteins display different expression patterns, *Plant Mol. Biol.,* 14, 119, 1990.
138. Hart, C. M., Nagy, F., and Meins, F., Jr., A 61 bp enhancer element of the tobacco β-1,3-glucanase B gene interacts with one or more regulated nuclear proteins, *Plant Mol. Biol.,* 21, 121, 1993.
139. Beffa, R., Szell, M., Meuwly, P., Pay, A., Vögeli-Lange, R., Métraux, J. P., Neuhaus, G., Meins, F., Jr., and Nagy, F., Cholera toxin elevates pathogen resistance and induces pathogenesis-related gene expression in tobacco, *EMBO J.,* 14, 5753, 1995.
140. Yang, S. F., Ethylene evolution from 2-chloroethylphosphonic acid, *Plant Physiol.,* 44, 1203, 1969.

141. Lawton, K. A., Potter, S. L., Uknes, S., and Ryals, J., Acquired resistance signal transduction in *Arabidopsis* is ethylene independent, *Plant Cell,* 6, 581, 1994.
142. Côté, F., Cutt, J. R., Asselin, A., and Klessig, D. F., Pathogenesis-related acidic β-1,3-glucanase genes of tobacco are regulated by both stress and developmental signals, *Mol. Plant-Microbe Interact.,* 4, 173, 1991.
143. Van Kan, J. A. L., Cozijnsen, T., Danhash, N., and de Wit, P. J. G. M. Induction of tomato stress protein mRNAs by ethephon, 2,6- dichloroisonicotinic acid and salicylate, *Plant Mol. Biol.,* 27, 1205, 1995.
144. Kombrink, E., Schröder, M., and Hahlbrock, K., Several "pathogenesis-related" proteins in potato are 1,3-β-glucanases and chitinases, *Proc. Natl. Acad. Sci. U.S.A.,* 85, 782, 1988.
145. Kombrink, E. and Hahlbrock, K., Responses of cultured parsley cells to elicitors from pathogenic fungi. Timing and dose dependency of elicitor-induced reactions, *Plant Physiol.,* 81, 216, 1986.
146. Kaku, H., Shibuya, N., Xu, P. L., Aryan, A. P., and Fincher, G. B., N-acetylchitooligosaccharides elicit expression of a single (1→3)-β-glucanase gene in suspension-cultured cells from barley (*Hordeum vulgare*), *Physiol. Plant.,* 100, 111, 1997.
147. Münch-Garthoff, S., Neuhaus J. M., Boller, T., Kemmerling, B., and Kogel, K. H., Expression of β-1,3-glucanase and chitinase in healthy, stem-rust-affected and elicitor-treated near-isogenic wheat lines showing *Sr5*- or *Sr24*-specified race-specific rust resistance, *Planta,* 201, 235, 1997.
148. Dinesh-Kumar, S. P., Witham, S., Choi, D., Hehl, R., Corr, C., and Baker, B., Transposon tagging of tobacco mosaic virus resistance gene *N:* Its possible role in the TMV-*N*-mediated signal transduction pathway, *Proc. Natl. Acad. Sci. U.S.A.,* 92, 4175, 1995.
149. Witham, S., Dinesh-Kumar, S. P., Choi, D., Hehl, R., Corr, C., and Baker, B., The product of the tobacco mosaic virus resistance gene N: similarity to Toll and the interleukin-1 receptor, *Cell,* 78, 1101, 1994.
150. Delaney, T. P., Genetic dissection of acquired resistance to disease, *Plant Physiol.,* 113, 5, 1997.
151. Ryals, J. A., Neuenschwander, U. H., Willits, M. G., Molina, A., Steiner, H. Y., and Hunt, M. D., Systemic acquired resistance, *Plant Cell,* 8, 1809, 1996.
152. Friedrich, L., Lawton, K., Ruess, W., Masner, P., Specker, N., Gut Rella, M., Meier, B., Dincher, S., Staub, T., Uknes, S, Métraux, J. P., Kessmann, H., and Ryals, J., A benzothiadiazole derivative induces systemic acquired resistance in tobacco, *Plant J.,* 10, 61, 1996.
153. Ward, E. R., Uknes, S. J., Williams, S. C., Dincher, S. S., Wederhold, D. L., Alexander, D. C., Ahl-Goy, P., Métraux, J. P., and Ryals, J. A., Coordinate gene activity in response to agents that induce systemic acquired resistance, *Plant Cell,* 3, 1085, 1991.
153a. Livne, B., Factor, O., Zeitoune, S., Edelbaum, O., and Sela, I., TMV-induced expression of tobacco β-glucanase promoter activity is mediated by a single, inverted, GCC motif, *Plant Sci.,* 130, 159, 1997.
154. Heitz, T., Fritig, B., and Legrand, M., Local and systemic accumulation of pathogenesis-related proteins in tobacco plants infected with tobacco mosaic virus, *Mol. Plant-Microbe Interact.,* 7, 776, 1994.
155. Ashfield, T., Hammon Kosack, K. E., Harrison, K., and Jones, J. D. G., *Cf* gene-dependent induction of a β-1,3-glucanase promoter in tomato plants infected with *Cladosporium fulvum, Mol. Plant-Microbe Interact.,* 7, 645, 1994.

156. Bol, J. F., Buchel, A. S., Knoester, M., Baladin, T., van Loon, L. C., and Linthorst, H. J. M., Regulation of the expression of plant defense genes, *Plant Growth Reg.,* 18, 87, 1996.
157. Lusso, M. and Kuc, J., Evidence for transcriptional regulation of β-1,3-glucanase as it relates to induced systemic resistance of tobacco to blue mold, *Mol. Plant-Microbe Interact.,* 8, 473 1995.
158. Ye, X. S., Pan, S. Q., and Kuc, J., Specificity of induced systemic resistance as elicited by ethephon and tobacco mosaic virus in tobacco, *Plant Sci.,* 84, 1, 1992.
158a. Niki, T., Mitsuhara, I., Seo, S., Ohtsubo, N., and Ohasshi, Y., Antagonistic effect of salicylic acid and jasmonic acid on the expression of pathenogenesis-related (PR) protein genes in wounded mature tobacco leaves, *Plant Cell Physiol.,* 39, 500, 1998.
159. Vidal, S., Ponce de León, I,. Denecke, J., and Plava, E. T., Salicylic acid and the plant pathogen *Erwinia carotovora* induce defense genes via antagonistic pathways, *Plant J.,* 11, 115, 1997.
160. Büttner, M. and Singh, K. B., *Arabidopsis thaliana* ethylene-responsive element binding protein (AtEBP), a ethylene-inducible, GCC box DNA-binding protein interacts with an *ocs* element binding protein, *Proc. Natl. Acad. Sci. U.S.A.,* 94, 5961, 1997.
160a. Sato, F., Kitajima, S., Koyama, T., and Yamada, Y., Ethylene-induced gene expression of osmotin-like protein, a neutral isoform of tobacco PR-5, is mediated by the AGC-CGCC *cis*-sequence, *Plant Cell Physiol.,* 37, 249, 1996.
161. Ecker, J. R., The ethylene signal transduction pathway in plants, *Science*, 268, 667, 1995.
162. Fluhr, R. and Mattoo, A. K., Ethylene — biosynthesis and perception, *Crit. Rev. Plant Sci.,* 15, 479, 1996.
163. Kieber, J. J., The ethylene response pathway in *Arabidopsis, Annu. Rev. Plant Physiol. Plant Mol. Biol.,* 48, 277, 1997.
164. Chao, Q., Rothenberg, M., Solano, R., Roman, G., Terzagi, W., and Ecker, J. R., Activation of the ethylene gas response pathway in Arabidopsis by the nuclear protein ETHYLENE-INSENSITIVE3 and related proteins, *Cell,* 89, 1133, 1997.
164a. Weigel, D., The APETALA2 domain is related to a novel type of DNA binding domain, *Plant Cell,* 8, 388, 1995.
165. Zhou, J. M., Tang, X. Y., and Martin, G. B., The Pto kinase conferring resistance to tomato bacterial speck disease interacts with proteins that bind a *cis*-element of pathogenesis-related genes, *EMBO J.,* 16, 3207, 1997.
166. Boller, T., Ethylene in pathogenesis and disease resistance, in *The Plant Hormone Ethylene,* Mattoo, A. K. and Suttle, J. C., Eds., CRC Press, Boca Raton, FL, 1991, 293.
167. Penninckx, I. A. M. A., Eggermont. K., Terras., F. R. G., Thomma, B. P. H. J., de Samblanx, G. W., Buchala, A., Métraux, J. P., Manners, J. M., and Broekaert, W. F., Pathogen-induced systemic activation of a plant defensin gene in *Arabidopsis* follows a salicylic acid-independent pathway, *Plant Cell,* 8, 2309, 1996.
168. Silverman, P., Nuckles, E., Ye, X. S., Kuj, J., and Raskin, I., Salicylic acid, ethylene, and pathogen resistance in tobacco, *Mol. Plant-Microbe Interact.,* 6, 775, 1993.
169. Eyal, Y., Sagee, O., and Fluhr, R., Dark-induced accumulation of a basic pathogenesis-related (PR-1) transcript and a light requirement for its induction by ethylene, *Plant Mol. Biol.,* 19, 589, 1992.

4 Plant Chitinases (PR-3, PR-4, PR-8, PR-11)

Jean-Marc Neuhaus

CONTENTS

4.1 Introduction 77
4.2 Structure of the Proteins 79
4.2.1 PR-3, A Plant-Specific Chitinase Family (Family 19, Genes *Chia*) 79
4.2.2 Family 18, The Ubiquitous Chitinase Family 84
4.2.3 PR-8/Class III Chitinases 84
4.2.4 PR-11 Chitinases 86
4.2.5 Other Related Proteins 87
4.2.6 The PR-4 Family 87
4.3 Catalytic Mechanisms and Specificities 88
4.3.1 Catalytic Mechanism of Family 18 Chitinases 89
4.3.2 Catalytic Mechanism of Family 19 Chitinases 91
4.3.3 Catalytic Mechanism of PR-4 Chitinases 92
4.3.4 The Role of Chitin-Binding Domains 93
4.3.5 Intracellular Location 94
4.4 Structure and Regulation of the Genes 94
4.4.1 *Chib* (PR-8) and *Chic* (PR-11) Genes 94
4.4.2 *Chia* Genes (Family 19, PR-3) 94
4.4.3 *Chid* (PR-4) Gene 95
4.5 Functions of Plant Chitinases 95
4.5.1 Antifungal Activity *In Vitro* 95
4.5.2 Antifungal Activity *In Vivo* 96
4.5.3 Nodulation 96
4.5.4 Embryogenesis 97
4.5.5 Other Functions 97
4.5.6 A Note of Caution for Biotechnological Applications 97
4.6 Conclusion 98
References 98

4.1 INTRODUCTION

Chitinases (EC 3.2.1.14) are ubiquitous enzymes of bacteria, fungi, animals, and plants. They hydrolyze the β-1,4-linkage between *N*-acetylglucosamine residues of

0-8493-0697-3/99/$0.00+$.50

chitin, a structural polysaccharide of the cell wall of many fungi and of the exoskeleton of invertebrates. The development of research in plant defense mechanisms has led to a rapid and steady interest in chitinases, since they were the first pathogen-induced proteins whose function was identified. Their substrate is present in the cell wall of many fungi, as well as in insects and nematodes, which are major pathogens and pests of crop plants.

Chitinases have been purified and/or cloned from many plants. As of September 1997, 145 sequences from 33 species were available in the EMBL database. Several nomenclatures coexist for chitinases: as glycosyl hydrolases, as pathogenesis-related (PR) proteins, as chitinase classes, and as gene families (Table 4.1). A classification into three classes was introduced[1] and later extended to six classes,[2] based on sequence comparisons. In parallel, the identification of several PR proteins as chitinases[3] led to their inclusion into a new nomenclature of PR proteins.[4] PR-3 became the name for chitinases related to class I, while the unrelated class III chitinases were assigned the name PR-8, and a new type of chitinases, the name PR-11. PR-3 chitinases also correspond to the family 19 of glycosyl hydrolases, and both PR-8 and PR-11 belong to family 18.[5] Finally, a tobacco PR-4 protein, CBP20, was characterized as yet another chitinase, leading to the inclusion of all PR-4 proteins into the current chitinase gene nomenclature.[123] An extension of the gene nomenclature could also include chitin-binding proteins, e.g., Chil for chitin-binding lectins or Chir for chitin-binding receptors.

TABLE 4.1
Nomenclatures of Chitinases[1]

PR-Proteins	Family of glycosyl hydrolases	Gene name	Class	Chitin-binding domain	Catalytic domain	Number of known sequences[2]
PR-3	19	*Chia1*	I	1	I	50
		Chia2	IIb	—	I	11
		Chia2	IIa	—	II	16
		Chia4	IV	1	IV	23
		Chia5	V	2	V	1
		Chia6	VI	1/2+Pro		1
		Chia7	VII	—	IV	2
PR-8	18	*Chib1*	III	—		19
PR-11	18	*Chic1*	I	—		3
PR-4	Unassigned	*Chid1*	I	1		7
		Chid2	II	—		5

[1] The different nomenclatures are essentially equivalent, with the exception of class numbers. For example, class V has been used simultaneously for proteins from three different families. Thus, the family as well as the class should be indicated whenever confusion is possible: PR-3 class V or *Chia5* or class a5, PR-4 class II or *Chid2* or class d2. Classes I, II, and IV are well established for PR-3 chitinases, as well as class III for PR-8 chitinases.

[2] September 1997.

4.2 STRUCTURE OF THE PROTEINS

4.2.1 PR-3, A PLANT-SPECIFIC CHITINASE FAMILY (FAMILY 19, GENES *CHIA*)

The chitinases of this family share a common catalytic domain, the tertiary structure of which has been determined for a barley seed chitinase.[6,7] This structure is mostly a-helical and forms a globular domain with a groove where the catalytic residues (two glutamates) can be found (Figure 4.1). There is a certain resemblance to the fold of lysozymes, which has led to the suggestion of a common ancestry, despite the lack of sequence homology.[8] The only nonplant example of this family comes from the bacterium *Streptomyces griseus*.[9] Its singularity among bacterial chitinases

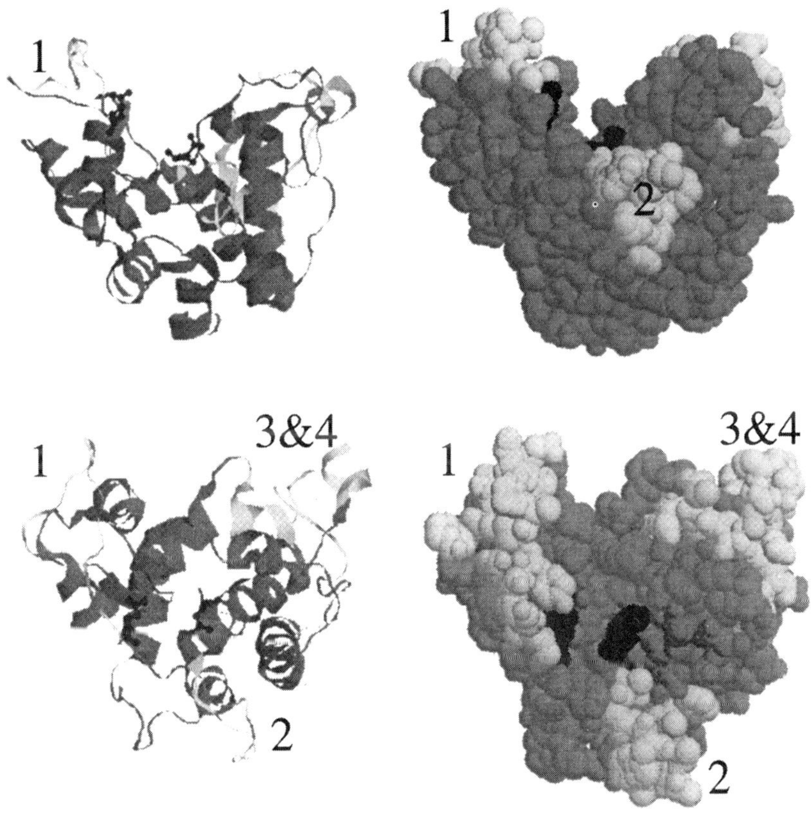

FIGURE 4.1 Catalytic domain of PR-3 chitinases. The structure of a barley class IIb chitinase[6] was represented as ribbons (left) and as space-filling model (right), using the program RASMOL. On top, the chitinase is shown from the front, emphasizing the cleft; on the bottom, the chitinase is shown in a perpendicular view into the cleft. The catalytic glutamates are shown in black. The loops 1 to 4, which are characteristic for the different classes (see also Figure 4.2), are shown in light gray and labeled.

and its close homology to the class IV chitinases from plants suggests a horizontal gene transfer from a plant to the bacterium, followed by the acquisition of bacterial domains and control elements.

The various classes of this family have first been defined according to sequence homology and the presence or absence of a chitin-binding domain (CBD) (Figure 4.2). Class II lacks the CBD of class I. Class IV chitinases differ from class I by several internal deletions within both chitin-binding and catalytic domains (loops 1, 3, and 4). Classes V and VI (one single sequence each) were defined based on the presence of two CBDs in tandem, or the presence of a truncated CBD and a long proline-rich spacer, respectively. Class VII, the latest addition, possesses a catalytic domain homologous to class IV, but lacks the CBD. In hindsight, another classification based on catalytic domains could have been more adequate, since they seem more conserved than the presence of a CBD, which appears to have been lost several times independently.

4.2.1.1 Class I Chitinases

The prototypes of this class are the tobacco basic chitinases.[10] They are encoded by four genes, two of which are strongly expressed. These chitinases are synthesized as precursors with an N-terminal signal sequence which directs them to the secretory pathway. Most class I chitinases also have a C-terminal propeptide, which is required for their targeting to the vacuole.[11] Within the mature protein, the N-terminal CBD is linked to the catalytic domain by a spacer, variable both in length and sequence, but usually rich in proline or glycine. In the case of the tobacco chitinases, the spacer was shown to be modified by hydroxylation of some of the prolines.[12] The function of this modification is unclear. It may be due to an accidental sequence similarity between the spacer and secreted hydroxyproline-rich glycoproteins. Another modification has been found in a single potato chitinase which has an N-glycosylation site and is indeed glycosylated.[13] Based on the comparison of the different classes, four dispensable loops can be distinguished (Figure 4.1). Loop 1 is located in the prolongation of the catalytic cleft. It is suggested that it provides a sugar-binding subsite. Loop 2 is held at the base by a disulfide bond and, in the crystal structure of barley class II chitinase, it extends away on one side of the catalytic cleft. The other two loops 3 and 4 fold together on an outer surface of the enzyme, away from the catalytic cleft. Loop 4 is actually the C-terminus of class I or II chitinases, following the last cysteine which forms a disulfide bond. Loop 3 holds loop 4 in place, such that the C-terminus of loop 4 is located on the surface of the protein.

4.2.1.2 Class II Chitinases

The tobacco pathogenesis-related proteins P and Q were identified as chitinases[3] and became the prototypes of class II. They lack a CBD and also have an internal deletion that removes loop 2 (Figure 4.2). Related chitinases have been found in monocots and dicots, and even in a pine. A major flour protein from barley was identified as another class II chitinase.[14] This is the only PR-3 chitinase that has been crystallized (Figure 4.1). It is most closely related in sequence to a class I

```
                   1                                                         60
Tobacco    a1;1 EQCGsQaGGarCasgLCCSkfGWCGnTndYCGpGnCQ-SQCpGgptppgg----------
Barley     a1;2 qQCGsQaGGatCpncLCCSrfGyCGsTsdYCGaG-CQ-SQCsGcgptppgpspg------
Tomato     a2;1 ------------------------------------------------------------
Barley     a2;1 ------------------------------------------------------------

Barley     a2;2 ------------------------------------------------------------
Tobacco    a2;1 ------------------------------------------------------------
P.strobus  a2;1 -----------imssLCCyvsaq--------------------------------------

Bean       a4;1 QNCG-------CaegLCCSqyGYCGTgedYCGtG-CQqGPCttaspppsn----------
Maize      a4;1 QNCG-------CqpnfCCSkfGYCGTTdaYCGdG-CQsGPCrsggggggggggggsgg
P.glauca   a4;1 QNCG-------CasgvCCSqyGYCGTTsaYCGkG-CksGPCyssgggspsaggg------
Rice       a7;1 ------------------------------------------------------------

Nettle*    a5;1 qrCGsqggggtCpalwCCSiwGwCGdsepYCGrt-Ce-nkCwsgersd-----------
Nettle     a5;1 hrCGaavgnppCgqdrCCSvhGwCGggndYCsgskCQ-yrCsssvrgprvalsgnstan-
Beet       a6;1 ---------------------vqCGrq-CnttdtnCl-sgCsvgr...90 Pro / 130 aa
   SS-bonds

                   61                                                        120
Tobacco    a1;1 GdlgSiISsSmFDQMLkHRNDnaCqgKG-FYsYnAFInAarsFPgFGTsGDttaRKREiA
Barley     a1;2 GgvsSiISrdlFeQfLlHRdr--CqdaagFYTYDAFlaAaatFPaFGTTGsteTRKqEvA
Tomato     a2;1 qniSSLISknLFeriLvHRNDraCgaKG-FYTYeAFItAtktFaaFGTTGDtnTRnkEIA
Barley     a2;1 -svSSivSraqFdrmLLHRNDgaCqaKG-FYTYdAFvaAaaaFPgFGtTGsadaqKREvA

Barley     a2;2 QgvGSVITrsvyasMLpnRdnslCParG-FYTYDAFIaAaNtFPGFGTTGsaddiKrElA
Tobacco    a2;1 QgiGSIVTndLFneMLknRNdgrCPanG-FYTYDAFIaAaNSFPGFGTsGDDTaRrKEiA
P.strobus  a2;1 QgvaSIisedvFhqfLkhRNddaCsakG-FYTYsAFIaAaNSFPdFGniGDqdsRKrElA

Bean       a4;1 nvnadIlTadFl-NGIIdQadsgCaGKn-FYTRdAFLsAlnsYtdFGrvGSeddsKREIA
Maize      a4;1 anvanvVTdaFF-NGIknQagsgCeGKn-FYTRsAFLsAvnaYPgFahgGtevegKREIA
P.glauca   a4;1 -svggIisqsFF-NGlaggagssCeGKg-FYTynAFiaAanafsgFGttGSndvkKRElA
Rice       a7;1 -svesvVTeaFF-NGIknQapngCaGKs-FYTRqsFLnAarsYsgFandrtnddsKREIA
S.griseus       -gegpggnngFv-vseaqfnqmfpnrna-FYTykgltdAlsaYPaFaktGSdevkKREaA

Nettle     a5;1 signvvvteplFdqmfshrkd--Cpsqg-FYsyhsFlvAaetfPeFGntGndeirKrEiA
Beet       a6;1 thltdiiseemFnefllnriqprCpgrw-FYTyqAFitAaesfPaFGtiGdvatrKrEvA
   SS-bonds

                   121     *                           *                     180
Tobacco    a1;1 AFfaQTSHETTGGWaTAPDGPYaWGYCwlrEqgsp--gdYCtP--SgQWPCapGr-KYfG
Barley     a1;2 AFfgQTSHETTGGWaTAPDGPYsWGYCyrrElgsp--pdYCqP--SsQWPCvqdr-qYYG
Tomato     a2;1 AFLaQTSHETTGGWaTAPDGPYsWGYCynrEQGsP--gDYCas--sQQWPCaPGK-kYfG
Barley     a2;1 AFLaQTSHETTGGWaTAPDGafaWGYCFKqErGas--sDYCtp--saQWPCaPGK-rYyG

Barley     a2;2 AFFGQTSHETTGGtrgAaD-qFqwGYCFkeEIska-------------------LSp pYYG
Tobacco    a2;1 AFFGQTSHETTGGslSA--epFtGGYCFVRQndQ--------------------Sd-rYYG
P.strobus  a2;1 AFFGQTSHETTGGwntApDGpyawGYCFkdQvnst--------------------d-rYrG

Bean       a4;1 AaFaHfTHET-------------GhFCYIEEIdGask-dYCDeesiaQYPCsssK-gYhG
Maize      a4;1 AFFaHvTHET-------------GhFCYIsEIn--ksnaYCDa-snrQwPCaagq-kYyG
P.glauca   a4;1 AFFanvmHET-------------GglCYInEknppmk--YCqs--sstwPCtsGK-sYhG
Rice       a7;1 AFFaHvTHET-------------GhmCYInEInGanm-dYCDk-snkQwPCqpGK-kYyG
S.griseus       AFlanvsHET-------------GglfYIkEvneanyphYCDt--tqsYgCpagqaaYyG

Nettle     a5;1 AFlahisqaTsGersdven-phawglChintttvten-dfCt---ssdwPCaagk-kYsp
Beet       a6;1 AFfgqtsHETsG-eptAqhGpftwgyCfieEigagplsqYCap--svewPCirgr-fYyg
   SS-bonds
   Loops                <-  loop 1 ->                <-      loop 2     ->
   Intron 1             Δ
```

FIGURE 4.2 *Continued on next page.*

```
                 181 !     !!                                                      240
Tobacco    a1;1 RGPIQIShNYNYGPcGRaIGvDLLNNPDLVATDPvISFKsAlWFWMTPQSPK-PScHDVI
Barley     a1;2 RGPImlSwNYNYGPaGRaIGvDLLNNPDLVATDatvSFrtAlWFWMTPQanK-PSsHaVI
Tomato     a2;1 RGPIQISyNYNYGaaGsaIGvnLLNNPDLVAnDaVvSFKTALWFWMTaQqPK-PSaHDVI
Barley     a2;1 RGPIQlShNYNYGPaGraIGvDLLaNPDLVAtDatvgFKTAiWFWMTaQpPK-PSsHaVI

Barley     a2;2 RGPIQLTgrSNYdlaGraIGkDLVsNPDLVsTDavvSFrTAmWFWMTaQgNK-PSsHnVa
Tobacco    a2;1 RGPIQLTnQnNYekaGnaIrqDLVnNPDLVATDatiSFKTAIWFWMTpQdNK-PSsHDVI
P.strobus  a2;1 RGPIQLTgdyNYkaaGnalGyDLinNPDLVATDatvSFKTAvWFWMTaQspK-PScHDVI

Bean       a4;1 RGPiQLSWNfNYGpaGsannFDGLgaPetVsnDvVvsFKTALWyWMqhVrp---------
Maize      a4;1 RGPLQiSWNYNYGpaGrdiGFnGLadPnrVAqDaViaFKTALWFWMnnVhg---------
P.glauca   a4;1 RGPLQLSWNYNYGavGkSiGFDGLnnPekVgkDpttssKTAvWFWMknrnc---------
Rice       a7;1 RGPLQiSWNfNYGpaGkniGFDGLrdPdkVAqDptisFKTALWFWMnnVhq---------
S.griseus       RGPiQLSWNfNYkaaGdalGinlLanPylVeqDpavawKTglWyWnsqngp---------

Nettle     a5;1 RGPiQlthnfNYglaGqaiGeDliqnPdlVekDpiisFKtAlWFWMsqhdnK-PScHDiv
Beet       a6;1 RGPvQltwnfNYgkqvkhlGlDllfnPdiVAhDpvisFetAiWFWMtpegnK-PSsHeVI
P.taeda    0;1  .....................nsaadhaeylsenatlaFaaAiWrWMtpmkvKqPSaHqVm
   Loops                                                              <- loop 3
   Intron 2            Δ


                 241                                                               300
Tobacco    a1;1 iGrWqPSsADraanRlPGfGVITNIINGGLECGrGtDsRVqD-RIGFYrRYCsiLGVs--
Barley     a1;2 TGqWtPtaADtaaGRvPGYGVITNIINGGLECGrGaDsRVaD-RIGFYqRYCniLGVg--

Tomato     a2;1 TGrWsPSvADsapGRVPGFGVITNIINGGmECnsGSNalmdn-RIGFYrRYCqILGVd--
Barley     a2;1 aGqWsPSgADraaGRVPGFGVITNIINGGIECGhGqdsrVaD-RIGFYkRYCdILGVg--

Barley     a2;2 lrrWTPtaADTaagRVPGYGVITNIINGGlECGmGrNdanvD-RIGYYtRYCGMLgta--
Tobacco    a2;1 iGsWTPSaADqsaNRaPGcGVITNIINGGiECGvGpNaaVED-RIGYYRRYCGMLNVa--
P.strobus  a2;1 lGrlTPSvtDTaagRVaGYGmlTdIINGGpECGtGtisdvqqgRIGfYqRYCkMLgVd--

Bean       a4;1 -------------VinQGFGaTIRaINGalECdGaNpttVqa-RVnYYteYCrQLGVa--
Maize      a4;1 -------------VmpQGFGaTIRaINGalECNGnNpaqmNa-RVgYYkqYCqQLrVd--
P.glauca   a4;1 ---------hsaitsgkGlGgTIkaINs-mECNGgNsgeVNs-RVnYYkkiCsQLGVd--
Rice       a7;1 -------------VmsQGFGaTIRaINGalECNGkNpgaVNa-RVnYYkdYCrQfGVs--
S.griseus       ----gtmtphnaiVnnaGFGeTIRsINGalECNGgNpaqVqs-RinkftqftqiLGtt--

Nettle     a5;1 ln------AnsaanripnkGvigniIsrafghd--dfavrss-sigfYkrYCdmLgVs--
Beet       a6;1 tGqWtPtpADiarnrlpGyGlitnifNGalECgthgpdnrgenRiqfYqrYCdlLdVs--
P.taeda    0;1  vGkWvPtknDtealrlpGfGmtinilkadaECgtdsddkqmntRiahYldfldhmdVgre
   SS-bonds                                              |_______________________
   Loops        loop 3      ->

               301                   325
Tobacco   a1;1  -pGdNLDCgNQRsFgn|gll--vdtm*
Barley    a1;2  -YGgNLDCYNQRPFve glliqrvte*
Tomato    a2;1  -pGnNLDCaNQRPFg*
Barley    a2;1  -yGnNLDCysQRPFa*

Barley      a2;2  -tGgNLDCYtQRNFas*
Tobacco     a2;1  -pGdNLDCYnQRNFaqg*
P. strobus  a2;1  -vGsNLDyknQkpygt*

Bean        a4;1 -tGdNLtC*
Maize       a4;1 -PGpNLiC*
P. glauca   a4;1 -PGaNvaC*
Rice        a7;1 -PGgNLyC*
S. griseus       -tGpNLsC*

Nettle      a5;1 -yGhdLkywfdntpssefqriqmrvaa*
Beet        a6;1 -yGdNLdCyrqtpfdw-glkklqgareswsss*
P. taeda    a0;1 naGdNvdCseqkvlnpfllrlhlaripflqinedkfvmrlcgifsyispdc...
P. taeda    a0;1              ...ymysdifavqvhtdcnistmklffpgwclfedqksdiaqqg*
   SS-bonds      __________
   Loops                   <loop 4><CTPP ->              ►
```

chitinase from the same plant. It was the first example of a heterogeneous collection of class II chitinases with less homology to the pathogen-induced class II chitinases than to class I. Sequence comparisons suggest that the CBD has been lost several times independently by various class II chitinases. The pathogen-induced class II chitinases form a distinct subclass of higher homology and the diagnostic internal deletion of loop 1. To distinguish them from the other class II chitinases mentioned, it is suggested to call them class IIa; the others could be called class IIb.

4.2.1.3 Class IV Chitinases

The bean chitinase originally called PR-4[15] is the prototype of this class. Its low homology to the former two classes prevented an identification by DNA hybridization and also explains the lack of serological relatedness. No member of this class has yet been identified in tobacco, the reference plant for PR proteins, but if looked for, there is little doubt that genes would be found, since this class has been found in monocots and dicots, and a related gene has recently been found in spruce. The diagnostic features of this class are several deletions (Figure 4.2). A loop is missing in the CBD, without affecting its sugar-binding properties. Within the catalytic domain, there are three deletions. The loss of loop 1, located in the prolongation of the catalytic cleft, probably means the loss of a sugar-binding subsite. This could allow these chitinases to hydrolyze a glycosidic bond closer to the surface of a pathogen. The coordinated loss of loops 3 and 4 reduces the volume of the protein on an outer surface away from the catalytic cleft. The consequences of this change are unknown.

4.2.1.4 Class V Chitinases

This class is represented so far by a single protein. When the cDNA for the precursor of stinging nettle (*Urtica dioica*) lectin was cloned, it was a big surprise to discover that the precursor was synthesized as a chitinase homolog with two CBDs.[16] Expression of the catalytic domain in *E. coli* led to a small increase in chitinase activity. Sequence comparison indicates, however, that this protein is unlikely to have any catalytic activity, since both catalytic residues are mutated. No chitinase activity could be detected upon transient expression in plant protoplasts (B. Iseli and J.-M. Neuhaus, unpublished observation).

4.2.1.5 Class VI Chitinases

The only class VI chitinase known so far was isolated from sugar beet.[17] Its heavily truncated CBD lacks four out of eight cysteines and can at best form a single correct

FIGURE 4.2 *Continued. (Opposite page)* Examples for chitinase classes of the PR-3 family (hydrolase family 19, gene name *Chia*). The only nonplant chitinase from this family, from the bacterium *Streptomyces griseus,* is also shown. Residue conservation within a class is indicated in bold (>75% of known sequences) and in capitals (>50% of known sequences). Gaps are indicated with —. Disulfide bonds, the loops 1 to 4, the position of introns (Δ), and the C-terminal propeptides (CTPP) are indicated below the sequences. The CTPP processing site of tobacco *Chia1;1* is indicated (|). The catalytic glutamates (*) and other residues shown by chemical modification or mutagenesis to be involved in substrate binding (!) are indicated above the sequences. Both N-terminal chitin-binding domains of the nettle *Chia5;1* are shown.

disulfide bond instead of four. The absence of any aromatic residue further reduces the possibility of binding sugars. On the other hand, this chitinase possesses by far the longest spacer sequence known, a stretch of more than 135 amino acids, of which 90 are prolines. This suggests a role as a cell wall protein, which are typically rich in proline and hydroxyproline. The catalytic domain includes the catalytic residues and the four loops described above and is expected to be active.

4.2.1.6 Class VII Chitinases

A rice chitinase without CBD but with a catalytic domain highly homologous to the domain of class IV chitinases was recently cloned (Y. Itoh, Acc. AB003194). The Chitinase Nomenclature Commission has decided to give this type of chitinase a new class name, by analogy with class II, which also lacks the CBD compared to the related class I.

4.2.1.7 Unclassified Chitinases

Several partial sequences have been obtained from various plants by PCR, using primers derived from highly conserved motifs. Without knowledge of the presence of diagnostic domains and/or loops, they cannot be classified.

The cDNA clone from the loblolly pine (*Pinus taeda*) is a special case because it includes several short open-reading frames before the last longer frame, which encodes a heavily truncated catalytic domain.[18] It lacks the whole part containing the catalytic residues and the loops 1 and 2 but contains the highly conserved WxWM motif and the last disulfide bond, the loops 3 and 4. This truncated domain is followed by an additional sequence containing four cysteines and is without homology to known proteins. The authors have excluded the possibility that they isolated an artifactual double clone, but the role of such a complex mRNA is unclear.

4.2.2 Family 18, The Ubiquitous Chitinase Family

This family has been found in both bacteria and eukaryotes, in fungi, animals, and plants. It includes endochitinases, exochitinases, and *N*-acetylglucosaminidases (EC 3.2.1.96). The members of this family have a common tertiary structure, an $(\alpha/\beta)_8$ barrel. The most conserved sequence elements are two short segments corresponding to the two β-strands containing the catalytic residues.

Several bacterial and eukaryotic chitinases also have additional domains. The yeast (*Saccharomyces cerevisiae*) chitinase possesses a Ser/Thr-rich linker and a C-terminal CBD (unrelated to the plant CBD), which is needed to retain this secreted enzyme on the yeast's own cell wall.[19] Another yeast, *Kluyveromyces lactis,* produces a killer toxin, one subunit of which includes a chitinase domain and a CBD which is related to the plant CBD of family 19 chitinases.[20] Several bacterial chitinases also have an N- or C-terminal CBD.

Plants have two subfamilies with very little sequence homology apart from the two mentioned strands: the PR-8 (or class III) and the PR-11 chitinases.

4.2.3 PR-8/Class III chitinases

These chitinases were first described as lysozymes.[21] A cucumber chitinase was purified and cloned,[22] soon followed by many others (Figure 4.3). A major protein of *Hevea brasiliensis* latex, hevamine, was identified as a dual lysozyme and chiti-

```
                       1                                                          60
Tobacco      b1;1      AGdIvIYWGQNGNEGSLadTCaTnNYaiVNIAFLvvFGnGQn-PvLNLAGHCdPna-gaC
Hevea        b1;1      .GGIAIYWGQNGNEGtLtqTCsTrkYsyVNIAFLnkFGnGQT-PqiNLAGHCnPaa-ggC
Sesbania     b1;2      easIgvnWGQNkrEGSLssTCdsGNYdtVhlgyLnvFGcGri-PsgNfgGHCggyr-NpC
ConB         b1;1      steIAvYWGQ-redGlLrdTCkTnNYkiVfIsFLdkFGceirkPeLeLeGvCgPsvgNpC
Cucumber     b1;1      AaGIAIYWGQNGNEGSLasTCaTGNYefVNIAFLssFGsGQa-PvLNLAGHCnPdn-NgC
Cucumber     b1;2      AaGIAIYWGQNGNEGSLasTCaTGNYefVNIAFLssFGsGQT-PvLNLAGHCnPdn-NgC
Cucumber     b1;3      AaGIgIYWGQNGNEGSLasTCaTGNYefVNIAFLssFGgGQT-PvLNLAGHCnPdn-NgC
Maize        b1;1      AGnIAvYWGQNGNEGSLadaCnsGlYayVNIAFLttFGnGQT-PvLNLAGHCdPgs-gsC
   SS bonds

                       61                !!!!!!!!!                                120
Tobacco      b1;1      TglSndIraCQnqGIKVmLSLGGaGsYfLsSaDDArnVAnYLWNNyLGGQSntRPLGDA
Hevea        b1;1      TivSngIrsCQiqGIKVmLSLGGiGsYtLaSqaDAknVAdYLWNNFLGGkSsSRPLGDA
Sesbania     b1;2      TilepqIqhCQqkGIKlfLSLGGpyGdYSLcSrrDAkqVAnYLyNNFLsGQy--gPLGsv
ConB         b1;1      sflesqIkeCQrmGvKVfLaLGGpkGtYSacSaDyAkdlAeYLhtyFLserre-gPLGkv
Cucumber     b1;1      aflSdeInsCksqnvKVlLSiGGaGsYSLsSaDDAkqVAnfiWNsyLGGQSdSRPLGaA
Cucumber     b1;2      afvSdeInsCQsqnvKVlLSiGGvGrYSLsSannAkqVAgfLWNNyLGGQSdSRPLGDA
Cucumber     b1;3      TilSneInsCQsqnvKVlLSiGGtGsYSLySaDDAkeVAnfiWNsyLGGQSdSRPLGDA
Maize        b1;1      TgqSsdIqtCQslGIKVlLSiGGasGsYgLsStDDAnsVAdYLWdNFLGGsgsSRPLGaA
   SS bonds
   Intron 1                              Δ

                        !!!!!!*!*                                                 180
Tobacco      b1;1      VLDGIDFDIE-gGttqhWDeLAktLsq--fsqq--rkVYLtAAPQCPfPDtwLngAlsTG
Hevea        b1;1      VLDGIDFDIE-hGStlyWDDLAryLsa--yskqg-kkVYLtAAPQCPfPDryLgtAlnTG
Sesbania     b1;2      tLDGIDleIk-gGSnryWDDLAneLas--lkshny-qfsLsAvPQCamPDyyLDrAIgTG
ConB         b1;1      aLDGIhFDIqkpvdelnWDnLleeLyqikdvyq--stflLsAAPgClsPDeyLDnAIqTr
Cucumber     b1;1      VLDGvDFDIE-sGSgqfWDvLAqeLkn--fgq-----ViLsAAPQCPfPDaqLDaAIrTG
Cucumber     b1;2      VLDGvDFvIg-fGSgqfWDvLAreLks--fgq-----ViLsAAPQCPiPDahLDaAIkTG
Cucumber     b1;3      VLDGvDFDIE-fGSdqfWDvLAqeLks--fgq-----ViLsAAPQCPiPDahLDaAIrTG
Maize        b1;1      VLDGIDFDIE-nGqsahyDDLAnaLkg--kgs-----VlLtAAPQCPyPDasLgpAlqTG
   SS bond
   Intron 2                                        Δ

                       181                                                        240
Tobacco      b1;1      LFDYVWVQFYNNPPCQYSgGsad-NLknyWNQWn-aiqa--gkiFLGLPAaqgAAgSG-f
Hevea        b1;1      LFDYVWVQFYNNPPCQYSsGNin-NiinSWNrWTtsina--gkiFLGLPAapeAAgSG-y
Sesbania     b1;2      vFDdilVQFYNsPtCQYSrGNte-rLldSWNgWasvgeafnstvymGLaASpemspfGgy
ConB         b1;1      hFDYifVrFYNdrsCQYStGNiq-rirnaWlsWTksvyprdknlFLeLPASqatApgGgy
Cucumber     b1;1      LFDsVWVQFYNNPPCmY-adNad-NLlsSWNQWTafpt---sklymGLPAareAApSGgf
Cucumber     b1;2      LFDsVWVQFYNNPPCmf-adNad-NLlsSWNQWaaypi---sklymGLPAapeAApSGgf
Cucumber     b1;3      LFDsVWVQFYNNPsCmY-adNtd-dilsSWNQWaaypi---lklymGLPAapeAApSGgf
Maize        b1;1      qFDnVWiQFYNNPgCaYanGddt-NLvnaWNtWTssita--gsfyLGvPASpqAAgSGts
   SS bond

                       241                                                        300
Tobacco      b1;1      IPsDVLvSQVLPlI-ngSpKYGGVMLWSkfyD--nGYSsaIKanV*
Hevea        b1;1      vPpDVLiSriLPeI-KkSpKYGGVMLWSkfyDdkgnYSssIldSV...
Sesbania     b1;2      IqprVLtSeViPfv-KngpnYGGiMLWSRyyDiqnqYSdkIKpylsny---vwqsvteai
ConB         b1;1      IPpsaLigQVLPyIpdlqtyaaGiaLWnRqaDketGYStnIiryln*
Cucumber     b1;1      IPaDVLiSQVLPtI-KaSsnYGGVMLWSkafD--nGYSdsIKgSig*
Cucumber     b1;2      IPaDVLiSQVLPtI-KtSsnYGGVMLWSkafD--nGYSdaIKgrillkkssycygvrrlt
Cucumber     b1;3      IPvDeLiSeVLPtI-KaysnYGGVMLWSkafD--nGYSdaIKdSiyqlkgss*
Maize        b1;1      tPa--LtgtViPaI-rgignYGGiMvWdRfnDvqnnYSsqvKgSV*
                                                                     < CTPP      ->
                       301
Sesbania     b1;2      seaasaalhrilh-kpyvattpti*
Cucumber     b1;2      mathlplsleieqvlv*
                       CTPP                  ->
```

FIGURE 4.3 Examples of chitinases of the PR-8 family (hydrolase family 18, gene name *Chib*). Residue conservation is indicated in bold (>75% of known sequences) and in capitals (>50% of known sequences). Gaps are indicated with —. Disulfide bonds, intron positions (Δ), and C-terminal propeptides (CTPP) are indicated below the sequences. The catalytic aspartate and glutamate (*) and the two β-strands (!) conserved in family 18 chitinases (PR-8 and PR-11) are indicated above the sequences.

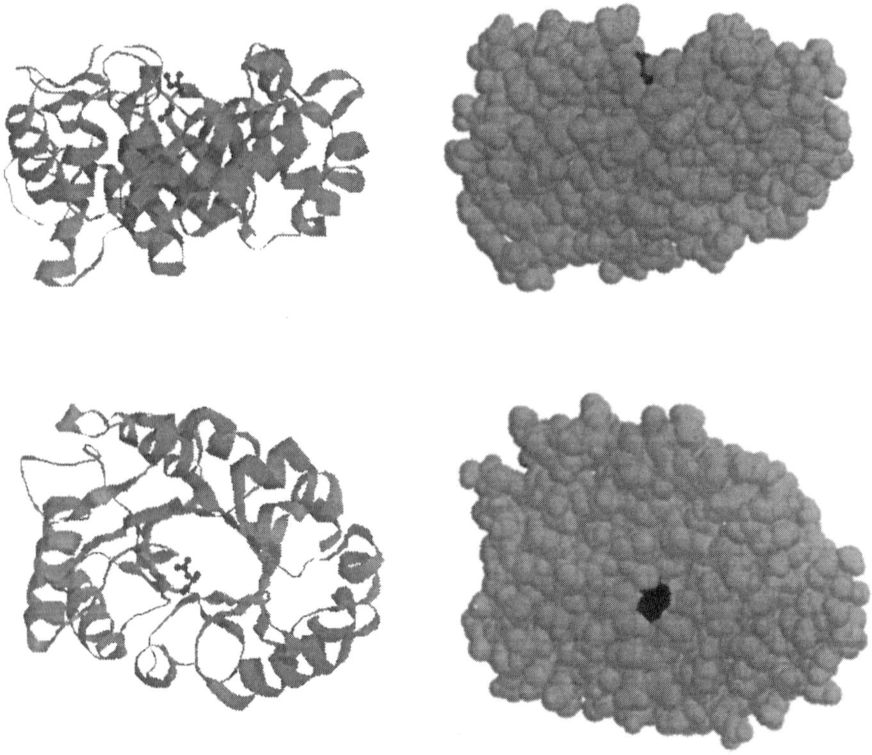

FIGURE 4.4 Catalytic domain of PR-8 chitinases. The structure of a *Hevea* class III[24] is represented as ribbons (left) and as space-filling model (right) using the program RASMOL. On top, the chitinase is shown from the front, emphasizing the cleft; on the bottom the chitinase is shown in a perpendicular view into the cleft. The catalytic glutamate and aspartate are shown in black.

nase.[23] It was crystallized and the structure could be determined.[24] This structure was similar to the simultaneously published structure of a bacterial family 18 chitinase.[25] Despite the low sequence similarity, they both had the same $(\alpha/\beta)_8$ barrel with a groove at the C-terminal end of the β-strands, with the two most conserved sequence elements lining the groove and presenting the catalytic residues (Figure 4.4). All plant chitinases of this class have very similar sequences, but differ widely in isoelectric points. The cucumber chitinase is acidic, as is one of the two tobacco chitinases, but the "acidic" chitinase from *Arabidopsis*[26] is basic, as is the other tobacco chitinase. One member of this family, concanavalin B, seems to have lost its catalytic activity and to have turned into a storage protein.[27]

4.2.4 PR-11 Chitinases

A new type of chitinase from tobacco was purified using a zinc affinity column and subsequently cloned.[28] It was simultaneously cloned by another group,[29] using an anonymous probe isolated long ago[30] from a library of pathogen-induced cDNA sequences. The field of chitinases might have developed differently had this probe

```
                  1                                                          60
Tobacco      c1;1 QNV-KGGYWFKDSGLALNNIDSTLFTHLFCAFADLNPQLNQLIISPENQDSFRQFTSTVQ
Arabidopsis  c1;1 QTVVKASYWFPASEFPVTDIDSSLFTHLFCAFADLNSQTNQVTVSSANQPKFSTFTQTVQ

                  61     !!!!!!!!!                                !!!!!!*!* 120
Tobacco      c1;1 RKNPSVKTFLSIAGGRANSTAYGIMARQPNSRKSFIDSSIRLARQLGFHGLDLDWEYPLS
Arabidopsis  c1;1 RrNPSVKTlLSIGGGIAdkTAYa.MAsnPt.RKi

                  121                                                       180
Tobacco      c1;1 AADMTNLGTLLNEWRTAINTEARNSGRAALLLTAAVSNSPRVNGLNYPVESLARNLDWIN
    Intron 1              Δ

                  181                                                       240
Tobacco      c1;1 LMAYDFYGPNWSPSQTNSHAQLFDPVNHVSGSDGINAWIQAGVPTKKLVLGIPFYGYAWR

                  241                                                       300
Tobacco      c1;1 LVNANIHGLRAPAAGKSNVGAVDDGSMTYNRIRDYIVESRATTVYNATIVGDYCYSGSNW

                  301                                              173
Tobacco      c1;1 ISYDDTQTVRNKVNYVKGRGLLGYFAWHVAGDQNWGLSRTASQTWGVSSQEMK
    Intron 2                                              Δ
```

FIGURE 4.5 Chitinases of the PR-11 family (hydrolase family 18, gene name *Chic*). One of two tobacco sequences and an *Arabidopsis* EST sequence are shown. Gaps are indicated with —. The position of introns (Δ) is indicated below the sequences. The catalytic aspartate and glutamate (*) and the two β-strands (!) conserved in family 18 chitinases (PR-8 and PR-11) are indicated above the sequences.

been chosen for detailed scrutiny eleven years ago. This chitinase now stands as an erratic block among the numerous examples of other plant chitinase families. The only other related sequence is an uncharacterized EST sequence from *Arabidopsis thaliana* (Figure 4.5). The next most related sequences belong to bacterial chitinases, e.g., from *Bacillus circulans* (31% identity) and *Serratia marcescens* (26%). Lower homology could be found to invertebrate and vertebrate chitinases. While all these proteins were classified as family 18 of the glycosyl hydrolases, there is very poor sequence similarity to the PR-8 family. The only conserved sequence elements are the two neighboring β-strands, the second of which contains the catalytic residues.

4.2.5 Other Related Proteins

There is another group of plant proteins with sequence and structure homology to the family 18 chitinases. They are storage proteins of legumes, known as narbonins (from *Vicia narbonensis*).[31] The cDNA sequences predict proteins without signal sequence, and are thus cytosolic. They also have lost the catalytic residues.

4.2.6 The PR-4 Family

4.2.6.1 Class I: Hevein and the Win Proteins

Hevein is a small antifungal protein found in high concentration in the latex fluid of the rubber tree (*Hevea brasiliensis*).[32] The three-dimensional structure of hevein[33] is similar to each of the four domains of wheat-germ agglutinin and other cereal lectins.[34] cDNAs encoding wound-induced proteins (hence called Win) were cloned from potato. The proteins were found to include a CBD related to hevein, the cereal chitin-binding lectins, and the CBD of class I chitinases.[35] The cDNA sequence for

```
                    1                                                          60
Tobacco      d1;1   QQCGRQRGGALCsGNLCCSQFGWCGSTPEYCSPSQGCQSQCsgggggggggaggggAqN
Hevea        d1;1   eQCGRQaGGkLCpnNLCCSQwGWCGSTdEYCSPdhnCQSnCkdsgegvgg.....gSAsN
Tobacco      d2;1   ........................................................qSAtN
Barley       d2;1   ........................................................eqAnd
   SS-bonds

                    61                                                        120
Tobacco      d1;1   VRATYHiYNPQNvgWDLy..AVSAYCsTWDgnKPLaWRrKYGWTAFCGPvGPRGrDSCGK
Hevea        d1;1   VlATYHlYNsQdhgWDLn..AaSAYCsTWDAnKPysWRsKYGWTAFCGPvGAhGQsSCGK
Tobacco      d2;1   VRsTYHlYNPQNinWDLr..AaSAfCATWDAdKPLaWRqKYGWTAFCGPaGPRGQDSCGr
Barley       d2;1   VRATYHyYrPaqinWDLgapAVSAYCATWDAsKPLsWRsKYGWTAFCGPaGPRGQaACGK
   SS-bonds

                    121                                                       180
Tobacco      d1;1   CLRVTNTGTGAQTTVRIVDQCSNGGLDLDVN.VFrQlDTDGrGnQrGHLiVNYeFVNCgD
Hevea        d1;1   CLsVTNTGTGAkTTVRIVDQCSNGGLDLDVN.VFrQlDTDGkGYerGHitVNYQFVdCgD
Tobacco      d2;1   CLRVTNTGTGtQTTVRIVDQCSNGGLDLDVN.VFnQlDTnGvGYQQGHLtVNYeFVNCnD
Barley       d2;1   CLRVTNpaTGAQiTaRIVDQCaNGGLDLDwdtVFtkIDTnGiGYQQGHLnVNYQFVdCrD
   SS-bonds

   Intron 1          Δ

                    181          196
Tobacco      d1;1   NmNvLvSpvDke*
Hevea        d1;1   sfNpLfSvmkssvine*
                    <-    CTPP    ->
```

FIGURE 4.6 Examples for chitinase classes of the PR-4 family (gene name *Chid*). Residue conservation is indicated in bold (>75% of known sequences) and in capitals (>50% of known sequences). Gaps are indicated with —. Disulfide bonds and the position of the intron (Δ) are indicated below the sequences.

the hevein precursor revealed a high homology to the potato Win proteins (Figure 4.6). In fact, the rubber tree latex also contains the unprocessed precursor as well as the processed second domain.[36]

So far, a single protein of this family, the tobacco CBP20, was found to have chitinase activity.[37] This was the reason for including the PR-4 family in a chitinase nomenclature.

4.2.6.2 Class II: Defense-related Proteins

A protein named PR-4 was among the classical pathogenesis-related proteins of tobacco. Once cloned from tobacco and tomato, PR-4 homologs were found to be homologous to the Win proteins, but lacked the CBD.[38,39] Their functions and possible catalytic activities are unknown. The three-dimensional structure of a barley PR-4 protein has been determined.[40] It is a compact domain without obvious catalytic cleft or catalytic site (Figure 4.7).

4.3 CATALYTIC MECHANISMS AND SPECIFICITIES

Chitinases are endo-β-1,4-glucosaminidases. They hydrolyze the β-glycosidic bond at the reducing end of glucosaminides, which can be parts of various polymers, such as chitin, chitosan, or peptidoglycan (Figure 4.8). The activity of chitinases has been assessed with a variety of substrates and test conditions. This is due to the insoluble nature of chitin, a β-1,4-polymer of *N*-acetylglucosamine that is found in a variety

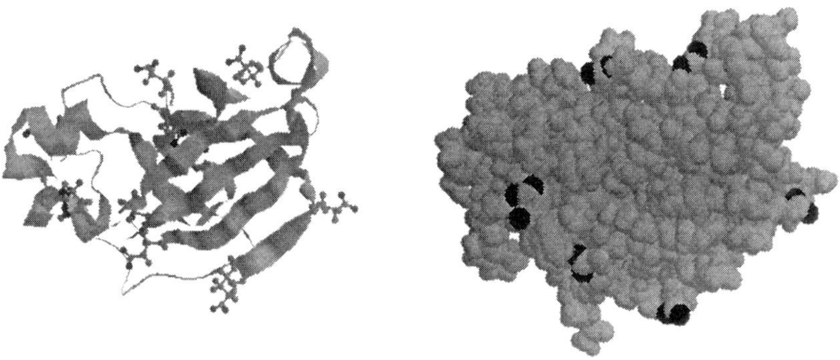

FIGURE 4.7 Catalytic domain of PR-4 chitinases. The structure of a barley Win protein[40] is represented as ribbons (left) and as space-filling model (right) using the program RASMOL. The glutamates and aspartates, one of which could be involved in catalysis, are shown in black.

of environments and arrangements. Enzymatic tests are usually based on an artificial substrate obtained by acetylation of chitosan, a soluble, partially deacetylated derivative of arthropod shells.[41,42] This creates a colloidal suspension with different properties for each batch. The soluble reaction products are measured either by radioactivity or by photometry. The hydrolysis reaction is not linear in time or with enzyme concentration, which imposes an extrapolation to low enzyme concentrations. Other artificial substrates can be used, with different results: glycol-chitin, chitosan, defined chitooligosaccharides, nodulation (Nod) factors (which are modified chitooligosaccharides), 4-methyl-umbelliferyl-chitotriose (MUCT), or *Micrococcus lysodeikticus* cells. The last two substrates are better known as lysozyme substrates, and some plant chitinases have indeed first been described as lysozymes.[21,43] Glycol-chitin and MUCT can be used to detect chitinases in gels after electrophoresis under nondenaturing conditions.[44] A manuscript comparing the specific activities of nine tobacco chitinases from all four families and five different classes against several substrates is in preparation.[124]

4.3.1 Catalytic Mechanism of Family 18 Chitinases

Glycan hydrolases can use two different mechanisms of catalysis, retaining or inverting the anomeric configuration at the newly formed reducing end.[5] Recent studies reveal that chitinases use either hydrolytic mechanism. Growing information about 3-D structures is available. Several bacterial chitinases and endo-β-N-acetylglucosaminidases, as well as the plant class III chitinase hevamine, all belonging to family 18 of glycosyl hydrolases, use a retaining mechanism producing the β-anomer, like hen egg white lysozyme. This catalytic mechanism requires, in general, the presence of a protonated and a deprotonated acidic residue as found in lysozymes, for example. The plant chitinases have only a single acidic residue (Glu127 of hevamine) at the active site (Figure 4.4). The reaction mechanism has been proposed to involve substrate-assisted catalysis, with the N-acetyl group of the –1 sugar forming an oxazolium intermediate with the glycosidic hydroxyl. This model is

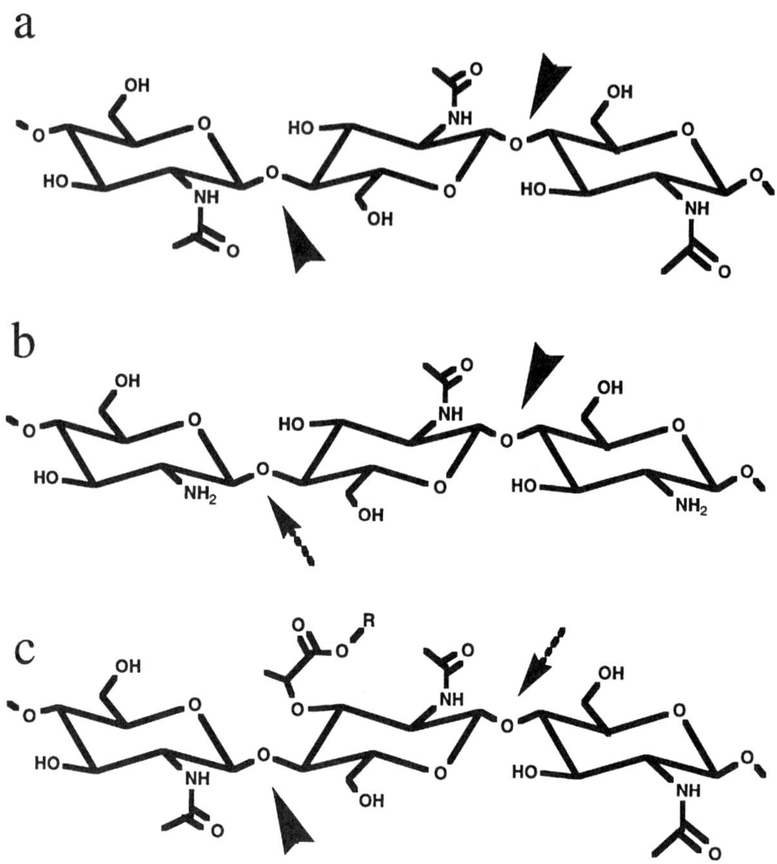

FIGURE 4.8 Chitinase substrates and the hydrolyzable bonds. (a) chitin (b) chitosan (c) peptidoglycan. Chitinases hydrolyze the glycosidic bond of N-acetylglucosamine residues (arrowheads) within a β-1,4-linked polysaccharide. The figures emphasize the alternating orientation of sugar residues. Chitinases do not hydrolyze the glycosidic bond of glucosamine residues of chitosan indicated by an arrow in (b) or the glycosidic bond of N-acetylmuramic acid residues in peptidoglycan[46] indicated by an arrow in (c). These bonds are hydrolyzed by chitosanases and lysozymes, respectively.

supported by the crystal structure of both substrate-bound and inhibitor-bound hevamine. The inhibitor allosamidine is a structural analog of the proposed reaction intermediate. It can be expected that all chitinases that share this mechanism will be sensitive to allosamidine derivatives. A neighboring aspartate (Asp125 of hevamine) is hydrogen-bonded to the catalytic glutamate and contributes to widen the catalytic pH range. The *Arabidopsis* chitinase has an asparagine at position 125 and is expected to have a much narrower pH range close to the pK of the Glu127.[45] Concanavalin B is closely related in sequence and folding to the chitinases, but its Glu127 is replaced by glutamine. It has no detectable activity. The same is probably true for one chitinase gene each from cucumber and *Sesbania rostrata*. Since the substrate-binding cleft is intact, these proteins can be expected to bind to chitin.

Some exclusively basic chitinases of family 18 also hydrolyze *M. lysodeikticus* cells or peptidoglycan and have thus been called chitinase/lysozymes. The peptidoglycan hydrolysis is restricted to a narrower pH range (around 4.5 to 5) than the chitin hydrolysis. The glycan backbone of peptidoglycan is composed of β-1,4 linked alternating *N*-acetylglucosamine and *N*-acetylmuramic acid. Lysozymes hydrolyze the glycosidic bond at the reducing end of *N*-acetylmuramic acid. In contrast, product analysis after hevamine digestion revealed that the cleavage occurred after the N-acetylglucosamine. Thus, these chitinases should not be called lysozymes,[46] but rather pseudolysozymes.

Since the PR-11 (Chic1) chitinases also belong to family 18 and are thus expected to have a fold very similar to hevamine, they are also likely to use the same retaining mechanism. Indeed, the two conserved sequence elements between PR-8 and PR-11 consist of two neighboring parallel β-strands, the second of which carries the same acidic residues Glu and Asp that have been characterized for hevamine and the bacterial chitinases.

4.3.2 Catalytic Mechanism of Family 19 Chitinases

These chitinases use the other possible mechanism of hydrolysis, which reverses the anomeric configuration at the cleavage site, as shown for a bean class I chitinase[47] as well as for a (presumably) class IV yam chitinase.[48] This mechanism displaces the aglycon by a molecule of water and requires the presence of a general acid to attack the glycosidic bond on one side and a general base to activate water on the other side of the bond. These two residues have been tentatively identified in the crystal structure of the class II barley seed chitinase C.[6,7] The highly conserved Glu67 and Glu89 face each other on both sides of a cleft that is suggested to bind the substrate, similar to the cleft known for family 18 chitinases and lysozymes. The two Glu are at a distance that is exactly as expected to fit the substrate and one molecule of water. The involvement of these two residues has been confirmed by site-directed mutagenesis in two different chitinases, the barley chitinase, for which the structure is known,[49] and the tobacco class I chitinase A.[50] In both cases, replacement of either Glu by Ala or Gln, the amide derivative of Glu, led to a massive loss of activity. While it dropped below the detection limit for the E67Q mutant, the residual activity of the E89Q mutant of barley chitinase still represented 0.25% of the wild-type activity. The catalytic domain of the tobacco chitinase was converted into a chitin-binding lectin by mutation of either glutamate.[50]

A few family 19 chitinases lack one or both glutamates. In the nettle lectin precursor, both catalytic residues are mutated, the first to Ala and the second to Thr, suggesting that it cannot have any activity. Indeed, it was not possible to detect any increase in chitinase activity in tobacco protoplasts overexpressing the precursor, while the catalytic domain could be detected immunologically[50] (B. Iseli and J.-M. Neuhaus, unpublished observations; Mirjam Does, pers. comm.). A more interesting case is the replacement of the second Glu by Gln in the pathogenesis-related class

IIa chitinases of tobacco (PR-P and -Q), potato, tomato, and *Pinus strobus.* These tobacco class IIa chitinases have a lower activity on colloidal chitin than the class I chitinases, but only by a factor of six.[124] There may be compensating mutations to improve the activity, or one could postulate a spontaneous deamidation of the glutamine by the time the protein has been purified. Other residues lining up along the catalytic cleft of a barley chitinase have been mutated to Ala: Gln118, Tyr123, and Asn124, of which only the last caused a strong reduction in activity.[49] Chemical modification of two different chitinases also destroyed their catalytic activity. Both treatments modified aromatic residues exposed within the catalytic cleft (see Figure 4.1), a Tyr in a maize chitinase[51] and a Trp in a rye chitinase,[52] sterically preventing substrate binding.

The catalytic cleft of barley chitinase is deeper than the cleft of hevamine (compare Figures 4.1 and 4.4). This may indicate a stronger requirement for accessibility of the substrate strand at the surface of a solid substrate, e.g., a fungal cell wall. In addition, two loops bordering the cleft are dispensable, as shown by the loss of loop 1 in class IV and VII chitinases and of loop 2 in class IIa. Loop 1 extends the cleft at one end and may increase the affinity for the substrate by providing an additional sugar-binding subsite. It may also increase the minimal distance of the hydrolyzable bond from the surface of an insoluble substrate.

Loop 2 is located on one side of the cleft and is constrained by a conserved disulfide bond. Similar surface exposed loops have been described in unrelated proteins and named "cystine nooses."[53] The loop 2 of chitinases has the potential of closing the cleft into a tunnel entrapping the substrate. This phenomenon has been described for other glycosyl hydrolases, conferring them the property of processivity: after hydrolysis, one half of the substrate is released easily, but the other half is held in the tunnel and can only diffuse longitudinally. Diffusion in one direction will automatically bring another hydrolyzable bond into position, with the release of a disaccharide (because of the alternating orientation of the *N*-acetylglucosamine units in chitin). This can occur a few times before the chitinase releases the chitin strand. Indeed, the class I enzymes rapidly produce chitobiose when digesting high-molecular-weight chitin.[124]

Loops 3 and 4 are folded together on the surface of barley chitinase, away from the catalytic cleft and are thus probably not involved in catalysis. The C-terminal targeting propeptide of vacuolar isoforms of class I is displayed on this side following loop 4.

4.3.3 Catalytic Mechanism of PR-4 Chitinases

Chitinase activity has been described for a single member of this family (tobacco CBP20).[37] To date, the mechanism is unknown. The 3-D structure of another member of this family, barley barwin, has been determined,[40] but there is no obvious catalytic cleft, nor is there any obvious pair of acidic residues likely to be involved in catalysis. Elucidation of the mechanism will require the analysis of reaction products as well as a systematic mutagenesis of surface-exposed acidic residues. The absence of a cleft suggests that the substrate is not bound strongly by the catalytic domain of CBP20, but rather by its CBD.

4.3.4 The Role of Chitin-Binding Domains

Chitinases with a CBD preceding or following the catalytic domain have been found in bacteria, yeasts, animals, and plants. Family 19 chitinases do not absolutely need a CBD for their catalytic activity, as is obvious from the many class II chitinases characterized to date. What, then, could be the role of an additional domain? Deletion of the CBD of the tobacco class I chitinase A modified its catalytic properties, increasing both apparent K_M and V_{max}. Analogous to the function of the unrelated CBD in the yeast family 18 chitinase, this was interpreted as follows: the CBD is strongly attaching the catalytic domain on the surface of a polymeric substrate, e.g., the cell wall of a pathogen, allowing the hydrolysis of many neighboring chitin strands. Without a CBD, the catalytic domain will diffuse away from the cell wall after a single cut (or a few cuts if there is processivity). This is the reason for the different behavior of class I and II chitinases on a chitin column. Class I chitinases bind to the column, while class II chitinases flow through. Indeed the tobacco chitinase with a deleted CBD could be separated from an intact chitinase.[54] On the other hand, a mutation causing the loss of catalytic activity increased the affinity for the chitin column and prevented this separation.[50] The CBD in PR-4 chitinases has probably the same function of fixing its catalytic domain on a chitinous structure.

The CBD is connected to the catalytic domain by a spacer of variable length and sequence, often rich in glycine, proline, and threonine. The spacer of tobacco chitinase A is modified and contains several hydroxyprolines.[12] Similarly, a sugar beet class IV chitinase contains hydroxyproline and is also glycosylated with xylose.[55] The spacer of a class IV maize chitinase contains sixteen glycines in tandem interrupted by a single serine, suggesting extreme flexibility. The most amazing spacer was found in the class VI sugar beet chitinase with 90 prolines out of 135 amino acids. This seems in fact an additional domain resembling cell wall proteins.[17]

In several chitinases, both PR-3 and PR-4, an *in vivo* proteolytic cleavage releases the CBD. In this case, the two domains may have separate functions. Hevein is such a released CBD and is *per se* an efficient antifungal agent.[32] The catalytic domain is also abundant in latex as well as the unprocessed precursor.[36] The stinging nettle antifungal lectin/agglutinin is composed of two CBDs in tandem and is derived from a precursor which appears to be a degenerate chitinase.[16] The processing occurs also in transgenic tobacco, where the lectin is found in the vacuole (M. Does, pers. comm.). Tobacco cells grown in a high-salt medium produce large amounts of four vacuolar chitinases, two of which are the known class I chitinases, while the other two lack the CBD. They could have been produced by proteolytic processing of the former two, though the ratio of the four proteins is surprisingly constant, and the N-terminal sequences of the shorter chitinases were not absolutely identical to internal sequences of the longer ones.[56] During infection of french bean by *Fusarium solani,* a class IV chitinase is processed, and the CBD is released. Whether this is a means for the fungus to detoxify the chitinase or for the plant to create two different antifungal compounds is unclear.[57]

4.3.5 Intracellular Location

A rule of thumb for PR proteins is that there are acidic and basic isoforms of each family and that the basic isoforms are both vacuolar and antifungal, while the acidic isoforms are extracellular and show little antifungal activity. Despite exceptions, this rule indicates that a basic pI may be more adequate for the vacuole and an acidic pI more adequate for the cell wall. It certainly also hints at different functions for both compartments in the defense mechanisms of plants.[58]

The intracellular localization of class I chitinases depends on the presence or absence of a vacuolar targeting propeptide at the C-terminus of the precursor, as has been thoroughly studied for a tobacco class I chitinase.[11,59] A similar C-terminal propeptide can be found in most other class I, V, and VI chitinases. In the other families, there are also forms with a C-terminal extension which may also be vacuolar targeting propeptides. The hevein precursor is synthesized with a 16 amino acid-long propeptide, and hevein is found in lutoids, the vacuolar component of the latex. Other PR-4 proteins are also predicted to have C-terminal propeptides. Hevamine, a PR-8 chitinase, is also found in the lutoids, and its cDNA sequence indicates the presence of a C-terminal propeptide (E. Bokma and J. Beintema, pers. comm.).

4.4 STRUCTURE AND REGULATION OF THE GENES

4.4.1 *Chib* (PR-8) and *Chic* (PR-11) Genes

Several cDNAs but only a few genes have been cloned from this family. Within the PR-8 *Arabidopsis* gene, there are two introns flanking the coding region for the conserved two β-strands and the catalytic residues (Figure 4.3). The three genes of cucumber have no introns. They have been isolated on a single genomic clone, with 2.5 and 2.4 kb separating the stop codon of one gene and the start codon of the next. The tobacco PR-11 chitinases also have two introns, the first of which is located approximately at the same place as the second intron of the *Arabidopsis* PR-8 chitinase, while the second intron is close to the end of the coding sequence (Figure 4.5). Most chitinases of these two families are induced when plants are infected by a pathogen.[60-69] The relative abundance of the various families in the infected tissue varies from plant to plant. The most abundant chitinase upon infection is a PR-8 in cucumber, while in tobacco only small amounts are produced and the PR-3 chitinases dominate. The promoter of the *Arabidopsis* chitinase has been fused to a reporter gene to analyze the expression pattern in whole plants.[62]

4.4.2 *Chia* Genes (Family 19, PR-3)

Several *Chia* genes have been cloned. They mostly contain two introns (Figure 4.2). The first is located just before the coding sequence for loop 1. Deletion of this loop in class IV and VII chitinases (or acquisition in a common ancestor of the other classes) is thus most likely due to the displacement of the 3′ splicing acceptor site. Some chitinase genes lack introns, e.g., in wheat[70,71] or potato.[72] The five *Chia4* genes from *Arabidopsis* that have been identified within 13 kb of two overlapping BACs only have the first intron and are separated by 1 to 3.3 kb of noncoding sequence. After the cloning of a cluster of three *Chib1* genes from cucumber and of a cluster of two *Chid1* genes from potato, finding a cluster of five Chia genes in *Arabidopsis,* a plant with a very compact genome, forecasts the existence of gene clusters in many other plants.

While most of these chitinases are also induced by pathogens, there is differential regulation of different genes within a single plant. In tobacco, for example, the class I genes are constitutively expressed in roots. The same chitinases also accumulate in the epidermis of leaves in an age-dependent manner.[73,74] They are induced by ethylene[75] and repressed by auxin and cytokinin.[10] They are also induced locally by pathogen infection.[74,76] The class IIa chitinases are also induced by pathogens,[3] but they are also induced systemically.[77] Chitinases are also abundant in flowers.[78,79] In potato, a special class IIb chitinase is the most abundant protein in the stylar transmitting tissue.[80] Chitinases are abundant proteins in seeds of several cereals and of chestnut.[14,81-86]

Several chitinases are induced by wounding,[87-89] especially in poplar, where acidic class I chitinases are induced in a systemic manner.[90]

Several promoters have been fused to reporter genes and introduced in plants.[91-93] The tobacco *Chia1*; 1 promoter has been analyzed by a set of deletions fused to the reporter *GUS*, whereby an ethylene-responsive element was identified within 22 bp, an element to which nuclear factors were shown to bind.[92,94,95]

4.4.3 *Chid* (PR-4) Gene

Two genes have been cloned from potato on a single genomic fragment, separated by 2.2 kb of noncoding sequence. The coding regions are interrupted by a single intron as indicated in Figure 4.6. Both genes were induced by wounding, but with different tissue specificities — win1 RNA accumulating in leaves and stem, but not in roots or tubers, while win2 RNA also accumulated in roots and tubers. Expression of the win2 promotor was also tested fused to the reporter GUS. Wound induction was first local, but then spread away from the wound along the vascular tissue.[35,96]

4.5 FUNCTIONS OF PLANT CHITINASES

4.5.1 Antifungal Activity *In Vitro*

Chitinases were rapidly suspected to affect growth of fungi in whose walls chitin is a major structural polysaccharide. Growth inhibition by a class I chitinase was shown for *Trichoderma viride,* a saprophytic fungus.[97] The enzyme caused thinning of the growing tip of the fungus *Trichoderma longibrachiatum,* followed by a balloon-like swelling and eventual bursting of the hypha.[98] However, these two fungi are saprophytes, and only a few other fungi are sensitive to chitinases alone, while many more fungi are sensitive to a combination of chitinase and a β-1,3-glucanase[99,100] or another protein or compound which alters the membrane structure or permeability.[101] Even then, not every chitinase is effective: the class IIa chitinases of tobacco are not antifungal, but the class I are. Their antifungal activity does not depend on the presence of the chitin-binding domain, since the effect is only slightly reduced by the deletion of the CBD.[54] Both the class I and IIb chitinases from barley seed are antifungal, further confirming the lesser role of the CBD in this respect. The lack of antifungal activity of the tobacco class IIa chitinase could be due to the absence of loop 2 and the concomitant lack of processivity, or to the active site glutamine instead of glutamate.

Chitinases of all four families were found to have antifungal activity *in vitro*.[29,37] Furthermore, the single CBD of hevein and the double CBD of stinging nettle lectin, both derived from chitinase-related proteins, were also antifungal *in vitro*.[32,102] The same is true for small proteins from *Amaranthus caudatus* and *Mirabilis jalapa* synthesized directly as single CBDs,[103] but not for the larger chitin-binding lectins of cereals, which are dimers of chains of four CBDs each.[104] On the other hand, the larger lectins of cereals and stinging nettle are insecticidal, while hevein is not. Thus chitinases may harbor two different antifungal domains with different modes of action and which may also be active after proteolytic processing.

4.5.2 Antifungal Activity *In Vivo*

Transgenic plants have been produced to express chitinases in a constitutive manner. These plants were tested with several pathogenic fungi with negative results.[105] However, bean class I chitinase protected tobacco against *Rhizoctonia solani,* the agent that causes damping-off.[106,107] Later, tobacco class I chitinase was found to reduce root colonization by the same fungus, a reduction also observed when the CBD was deleted.[108] As in *in vitro*, many more fungi were affected *in vivo* by the simultaneous coexpression of chitinase and another antifungal protein, such as β-1,3-glucanase,[109] ribosome-inactivating protein, osmotin, etc. As mentioned above, antifungal proteins appear to be concentrated in the vacuoles, while their target organisms are growing extracellularly. What is the role of the extracellular PR proteins if they are not antifungal? Why are they systemically induced? One possible role could be to release from the cell wall of the pathogens small molecules such as oligosaccharides, which could then act as elicitors.[58]

4.5.3 Nodulation

The identification of nodulation factors as modified chitin oligomers[110] led to an immediate interest in chitinases and chitin-binding proteins as receptors or modulators of the nodulation in legumes. Chitinases were cloned from many legumes, and several chitinases are expressed in root hairs or nodules.[111,112] Nod factors are indeed substrates for chitinases; they even allow a precise determination of the hydrolysis products because of the specific chemical modifications (acylation, sulfation, etc.) harbored by the two extremities of the chitooligosaccharide backbone.[113,114] In addition, it is possible to assess the influence of the length of the oligosaccharide and of its modifications on the activity of chitinases purified from various sources. This allowed the identification of chitinases as modulators of the host-range of distinct rhizobial strains through differential sensitivities.[114] A receptor for Nod factors or other chitin-derived signals would be expected to be membrane-bound and to have an extracellular chitin-binding domain. Such a receptor-like protein was recently cloned. It harbors an extracellular PR-11 chitinase-like domain, a membrane segment and a cytosolic kinase domain.[115] The catalytic residues are mutated, and indeed the chitinase-like domain expressed in *E. coli* lacked any activity but was able to bind to regenerated chitin. It remains to be seen what its binding specificity is and what signal transduction pathway leads to what reaction of the cell.

4.5.4 Embryogenesis

A totally different and unexpected function for a chitinase is that of a differentiation factor in embryogenesis. A mutant cell line of carrot unable to form embryos was rescued by culture supernatant of a wild-type cell line. It was a big surprise when a class IV chitinase was identified as the rescuing factor, since no endogenous chitin or related substrates were known in plants.[116-118] A Nod-like factor could also be implicated in the embryogenesis, since Nod factors could rescue the mutant cell line.[119] Finally, a potential substrate was found: arabinogalactan proteins (AGPs) from seeds were found to contain chitinase-sensitive oligosaccharides (S. de Vries, pers. comm.). AGPs stimulated the formation of somatic embryos from protoplasts, while the carrot chitinase had the opposite effect. Addition of AGPs that had been repurified after chitinase treatment increased the rate of embryo formation even further.

4.5.5 Other Functions

Some chitinases also have additional functions unrelated to their catalytic activity. Along with other PR proteins, chitinases are antifreeze proteins in winter rye.[120] The partial protein sequence of an inhibitor of insect α-amylases isolated from the tropical cereal Job's Tears (*Coix lacrima-jobi*) also revealed a PR-3 class IV chitinase.[84] This amylase inhibitor–chitinase appears to form dimers stabilized by disulfide bonds. Both these accessory functions are compatible, since they imply certain surface features of the protein and need not affect the catalytic site. The distinction between defense protein and storage protein may also be difficult for some abundant chitinases in seeds or vegetative tissues.

4.5.6 A Note of Caution for Biotechnological Applications

While the overproduction of chitinases in crop plants can, and in some cases, does increase the resistance to pathogens, they can also have negative effects. In grape juice, the presence of a class IV chitinase and of a thaumatin-like PR-5 protein causes the appearance of precipitates in the wine and thus lowers the quality of the product[121] not an acceptable side effect of the use of transgenic plants.

More serious is the existence of several allergenic chitinases. Among the many allergens of rubber, several chitinases are prominent: the hevein precursor and its two derived fragments hevein and the PR-4 domain, the PR-8 chitinase hevamine, and a PR-3 chitinase.[122] A chitinase is also a major allergen of avocado.[125] Allergenicity is not completely unexpected since chitinases share a major feature of allergenic proteins, the resistance to proteolytic degradation under acidic conditions, which allows a high proportion of ingested proteins to reach the intestinal tract intact, where they may pass into the bloodstream. Protease resistance is probably necessary for a successful PR protein, but it is important to choose carefully the source of the chitinase genes to introduce into food crops, especially into those that are to be eaten uncooked. A cautious strategy would be to choose a chitinase already present in food plants, since their allergenic potential has already been tested in large-scale and long-term trials.

4.6 CONCLUSION

The chitinases have become very popular proteins to study in plants because of their obvious potential in defense reactions against various pathogens. There is still much to study about the enzymology of the various families and classes, their specificities and their main natural substrates. At a time when identification of pathogen-induced chitinases could have reached a certain saturation, new and very different roles of chitinases attracted new attention. Their possible implication in nodulation attracted another major group of researchers, and the involvement in embryogenesis promises to appeal to a still wider audience. The number of publications devoted to chitinases is thus likely to increase further in the years to come.

REFERENCES

1. Shinshi, H., Neuhaus, J.-M., Ryals, J., and Meins, F., Jr., Structure of a tobacco endochitinase gene: evidence that different chitinase genes can arise by transposition of sequences encoding a cysteine-rich domain, *Plant Mol. Biol.,* 14, 357, 1990.
2. Meins, F., Fritig, B., Linthorst, H. J. M., Mikkelsen, J. D., Neuhaus, J. M., and Ryals, J., Plant chitinase genes, *Plant Mol. Biol. Rep.,* 12, S22, 1994.
3. Legrand, M. and Fritig, B., Biological function of pathogenesis-related proteins: four tobacco pathogenesis-related proteins are chitinases, *Proc. Natl. Acad. Sci. U.S.A.,* 84, 6750, 1987.
4. van Loon, L. C., Pierpoint, W. S., Boller, T., and Conejero, V., Recommendations for naming plant pathogenesis-related proteins, *Plant Mol. Biol. Rep.,* 12, 245, 1994.
5. Henrissat, B., A classification of glycosyl hydrolases based on amino acid sequence similarities, *Biochemical J.,* 280, 309, 1991.
6. Hart, P. J., Monzingo, A. F., Ready, M. P., Earnst, S. R., and Robertus, J. D., Crystal structure of an endochitinase from *Hordeum vulgare* L. seeds, *J. Mol. Biol.,* 229, 189, 1993.
7. Hart, P. J., Pfluger, H. D., Monzingo, A. F., Hollis, T., and Robertus, J. D., The refined crystal structure of an endochitinase from *Hordeum vulgare* L. seeds at 1.8 angstrom resolution, *J. Mol. Biol.,* 248, 402, 1995.
8. Holm, L. and Sander, C., Structural similarity of plant chitinase and lysozymes from animals and phage — an evolutionary connection, *FEBS Lett.,* 340, 129, 1994.
9. Ohno, T., Armand, S., Hata, T., Nikaidou, N., Henrissat, B., Mitsutomi, M., and Watanabe, T., A modular family 19 chitinase found in the prokaryotic organism *Streptomyces griseus* HUT 6037, *J. Bacteriol.,* 5065, 1996.

10. Shinshi, H. and Meins, F., Jr., Regulation of a plant pathogenesis-related enzyme: inhibition of chitinase and chitinase mRNA accumulation in cultured tobacco tissues by auxin and cytokinin, *Proc. Natl. Acad. Sci. U.S.A.,* 84, 89, 1987.
11. Neuhaus, J-M., Sticher, L., Meins, F., Jr., and Boller, T., A short C-terminal sequence is necessary and sufficient for the targeting of chitinases to the plant vacuole, *Proc. Natl. Acad. Sci.,* 88, 10362, 1991.
12. Sticher, L., Hofsteenge, J., Milani, A., Neuhaus, J. M., and Meins, F., Jr., Vacuolar chitinases of tobacco: a new class of hydroxyproline-containing proteins, *Science,* 257, 655, 1992.
13. Beerhues, L. and Kombrink, E., Primary structure and expression of mRNAs encoding basic chitinase and 1,3-β-glucanase in potato, *Plant Mol. Biol.,* 24, 353, 1994.
14. Leah, R., Mikkelsen, J. D., and Svendsen, I. B., Identification of a 28,000 Dalton endochitinase in barley endosperm, *Carlsberg Res. Commun.,* 52, 31, 1987.
15. de Tapia, M., Dietrich, A., and Burkard, G., In vitro synthesis and processing of a bean pathogenesis-related (PR4) protein, *Eur. J. Biochem.,* 166, 559, 1987.
16. Lerner, D. R. and Raikhel, N. V., The gene for stinging nettle lectin (*Urtica dioica* agglutinin) encodes both a lectin and a chitinase, *J. Biol. Chem.,* 267, 11085, 1992.
17. Berglund, L., Brunstedt, J., Nielsen, K. K., Chen, Z. C., Mikkelsen, J. D., and Marcker, K. A., A proline-rich chitinase from *Beta vulgaris, Plant Mol. Biol.,* 27, 211, 1995.
18. Chang, S. J., Puryear, J., Funkhouser, E. A., Newton, R. J., and Calmey, J., Cloning of a cDNA for a chitinase homologue which lacks chitin-binding sites and is down-regulated by water stress and wounding, *Plant Mol. Biol.,* 31, 693, 1996.
19. Kuranda, M. J. and Robbins, P. W., Chitinase is required for cell separation during growth of *Saccharomyces cerevisiae, J. Biol. Chem.,* 266, 19758, 1991.
20. Butler, A. R., O'Donnell, R. W., Martin, V. J., Gooday, G. W., and Stark, M. J. R., *Kluyveromyces lactis* toxin has an essential chitinase activity, *Eur. J. Biochem.,* 199, 483, 1991.
21. Bernasconi, P., Locher, R., Pilet, P. E., Jollés, J., and Jollés, P., Purification and N-terminal amino-acid sequence of a basic lysozyme from *Parthenocissus quinquifolia* cultured *in vitro, Biochim. Biophys. Acta,* 915, 254, 1987.
22. Métraux, J. P., Burkhart, W., Moyer, M., Dincher, S., Middlesteadt, W., Williams, S., Payne, S., Carnes, M., and Ryals, J., Isolation of a complementary DNA encoding a chitinase with structural homology to a bifunctional lysozyme/chitinase, *Proc. Natl. Acad. Sci. U.S.A.,* 86, 896, 1989.
23. Jekel, P. A., Hartmann, J. B. H., and Beintema, J. J., The primary structure of hevamine, an enzyme with lysozyme/chitinase activity from *Hevea brasiliensis* latex, *Eur. J. Biochem.,* 200, 123, 1991.
24. Terwisscha van Scheltinga, A. C., Kalk, K. H., Beintema, J. J., and Dijkstra, B. S., Crystal structures of hevamine, a plant defence protein with chitinase and lysozyme activity, and its complex with an inhibitor, *Structure,* 2, 1181, 1994.
25. Perrakis, A., Tews, I., Dauter, Z., Oppenheim, A. O., Chet, I., Wilson, K. S., and Vorgias, C. E., Crystal structure of a bacterial chitinase at 2.3 Å resolution, *Structure,* 2,1169, 1994.
26. Samac, D. A., Hironaka, C.M., Yallaly, P.E., and Shah, D. M., Isolation and characterization of the genes encoding basic and acidic chitinase in *Arabidopsis thaliana, Plant Physiol.,* 93, 907, 1990.
27. Hennig, M., Jansonius, J. N., Terwisscha van Scheltinga, A. C., Dijkstra, B. S., and Schlesier, B., Crystal structure of concanavalin B at 1.65 Å resolution, an "inactived" chitinase from seeds of *Canavalia ensiformis, J. Mol. Biol.,* 254, 237, 1995.

28. Heitz, T., Segond, S., Kauffmann, S., Geoffroy, P., Prasad, V., Brunner, F., Fritig, B., and Legrand, M., Molecular characterization of a novel tobacco pathogenesis-related (PR) protein: a new plant chitinase/lysozyme, *Mol. Gen. Genet.,* 245, 246, 1994.
29. Melchers, L. S., Apotheker-de Groot, M., van der Knaap, J., Ponstein, A. S., Sela-Buurlage, M. B., Bol, J. F., Cornelissen, B. J. Z., van den Elzen, P. J. M., and Linthorst, H. J. M., A new class of tobacco chitinases homologous to bacterial exo-chitinases displays antifungal activity, *Plant J.,* 5, 469, 1994.
30. Hooft van Huijsduijnen, R. A. M., Van Loon, L. C., and Bol, J. F., cDNA cloning of six mRNAs induced by TMV infection of tobacco and a characterization of their translation products, *EMBO J.,* 5, 2057, 1986.
31. Coulson, A. F. W., A proposed structure for "Family 18" chitinases — a possible function for narbonin, *FEBS Lett.,* 354, 41, 1994.
32. van Parijs, J., Broekaert, W. F., and Peumans, W. J., Hevein: an antifungal protein from rubber-tree (*Hevea brasiliensis*) latex, *Planta,* 183, 258, 1991.
33. Andersen, N. H., Cao, B., Rodriguez-Romero, A., and Arreguin, B., Hevein: NMR assignment and assessment of solution-state folding for the agglutinin-toxin domain, *Biochem.,* 32, 1407, 1993.
34. Wright, C. S. and Olafsdottir, S., Structural differences in the two major wheat germ agglutinin isolectins, *J. Biol. Chem.,* 261, 7191, 1986.
35. Stanford, A., Bevan, M., and Northcote, D., Differential expression within a family of novel wound-induced genes in potato, *Mol. Gen. Genet.,* 215, 200, 1989.
36. Soedjanaatmadja, U. M. S., Subroto, T., and Beintema, J. J., Processed products of the hevein precursor in the latex of the rubber tree (*Hevea brasiliensis*), *FEBS Lett.,* 363, 211, 1995.
37. Ponstein, A. S., Bles-Vloemans, S. A., Sela-Buurlage, M. B., van den Elzen, P. J. M., Melchers, L. S., and Cornelissen, B. J. C., A novel pathogen- and wound-inducible tobacco protein with antifungal activity, *Plant Physiol.,* 104, 109, 1994.
38. Linthorst, H. J. M., Danhash, N., Brederode, F. T., van Kan, J. A. L., de Wit, P. J. G. M., and Bol, J. F., Tobacco and tomato PR proteins homologous to Win and pro-hevein lack the "hevein" domain, Mol. Plant-*Microbe Interact.,* 4, 586, 1991.
39. Friedrich, L., Moyer, M., Ward, E., and Ryals, J., Pathogenesis-related protein-4 is structurally homologous to the carboxy-terminal domains of hevein, Win-1 and Win-2, *Mol. Gen. Genet.,* 230, 113, 1991.
40. Ludvigsen, S. and Poulsen, F. M., Three-dimensional structure in solution of barwin, a protein from barley seed, *Biochemistry,* 31, 8783, 1992.
41. Molano, J., Duran, A., and Cabib, E., A rapid and sensitive assay for chitinase using tritiated chitin, *Anal. Biochem.,* 83, 648, 1977.
42. Wirth, S. J. and Wolf, G. A., Dye-labelled substrates for the assay and detection of chitinase and lysozyme activity, *J. Microbiol. Meth.,* 12, 197, 1990.
43. Dahlquist, F. W., Borders, C. L., Jacobson, G., and Raftery, M. A., The stereospecificity of human, hen, and papaya lysozymes, *Biochem.,* 8, 694, 1969.
44. Trudel, J. and Asselin, A., Detection of chitinase activity after polyacrylamide gel electrophoresis, *Anal. Biochem.,* 178, 362, 1989.
45. Terwisscha van Scheltinga, A. C., Structure and mechanism of the plant chitinase hevamine, Ph.D. thesis, Rijksuniversiteit, Gronigen, 162, 1997.
46. Bokma, E., Van Koningsveld, G. A., Jeronimus-Stratingh, M., and Beintema, J. J., Hevamine, a chitinase from the rubber tree *Hevea brasiliensis,* cleaves peptidoglycan between the C-1 of N-acetylglucosamine and C-4 of N-acetylmuramic acid and therefore is not a lysozyme, *FEBS Lett.,* 411, 161, 1997.

47. Iseli, B., Armand, S., Boller, T., Neuhaus, J. M., and Henrissat, B., Plant chitinases use two different hydrolytic mechanisms, *FEBS Lett.*, 382, 186, 1996.
48. Fukamizo, T., Koga, D., and Goto, S., Comparative biochemistry of chitinases — anomeric form of the reaction products, *Biosci. Biotechnol. Biochem.*, 59, 311, 1995.
49. Andersen, M. D., Jensen, A., Robertus, J. D., Leah, R., and Skriver, J. K., Heterologous expression and characterization of wild-type and mutant forms of a 26 kda endochitinase from barley (*Hordeum vulgare* L), *Biochemical J.*, 322 (Part 3) 815, 1997.
50. Iseli-Gamboni, B., Boller, T., and Neuhaus, J-M., Mutation of either of two essential glutamates converts the catalytic domain of tobacco class I chitinase into a chitin-binding domain, *Plant Sci.*, 134, 45, 1998.
51. Verburg, J. G., Smith, C. E., Lisek, C. A., and Huynh, Q. K., Identification of an essential tyrosine residue in the catalytic site of a chitinase isolated from *Zea mays* that is selectively modified during inactivation with 1-ethyl-3-(3-dimethylaminopropyl)-carbodiimide, *J. Biol. Chem.*, 267, 3886, 1992.
52. Yamagami, T. and Funatsu, G., Identification of the tryptophan residue located at the substrate-binding site of rye seed chitinase-C, *Biosci. Biotech. Biochem.*, 59, 1076, 1995.
53. Lapthorn, A. L., Janes, R. W., Isaacs, N. W., and Wallace, B. A., Cystine nooses and protein specificity, *Nature Structural Biology,* 2, 266, 1995.
54. Iseli, B., Boller, T., and Neuhaus, J-M., The N-terminal cysteine-rich domain of tobacco class I chitinase is essential for chitin binding but not for catalytic or antifungal activity, *Plant Physiol.*, 103, 221, 1993.
55. Nielsen, K. K., Bojsen, K., Roepstorff, P., and Mikkelsen, J. D., A hydroxyproline-containing class IV chitinase of sugar beet is glycosylated with xylose, *Plant Mol. Biol.*, 25, 241, 1994.
56. Yun. D. J., Durzo, M. P., Abad, L., Takeda, S., Salzman, R., Chen, Z. T., Lee, H., Hasegawa, P. M., and Bressan, R. A., Novel osmotically induced antifungal chitinases and bacterial expression of an active recombinant isoform, *Plant Physiol.*, 111, 1219, 1996.
57. Lange, J., Mohr, U., Wiemken, A., Boller, T., and Vögeli-Lange, R., Proteolytic processing of class IV chitinase in the compatible interaction of bean roots with *Fusarium solani, Plant Physiol.*, 111, 1135, 1996.
58. Mauch, F. and Staehelin, L. A., Functional implications of the subcellular localization of ethylene-induced chitinase and β-1,3-glucanase in bean leaves, *Plant Cell,* 1, 447, 1989.
59. Neuhaus, J.-M., Pietrzak, M., and Boller, T., Mutation analysis of the C-terminal vacuolar targeting peptide of tobacco chitinase: low specificity of the sorting system, and gradual transition between intracellular retention and secretion into the extracellular space, *Plant J.*, 5, 45, 1994.
60. Métraux, J. P. and Boller, T., Local and systemic induction of chitinase in cucumber plants in response to viral, bacterial and fungal infections, *Physiol. Molec. Plant Pathol.*, 28, 161, 1986.
61. Awade, A., de Tapia, M., and Burkard, B., Biological function of bean pathogenesis-related (PR3 and PR4) proteins, *Plant Sci.*, 63, 121, 1989.
62. Samac, D. A. and Shah, D. M., Developmental and pathogen-induced activation of the Arabidopsis acidic chitinase promoter, *Plant Cell,* 3, 1063, 1991.
63. Martin, M. N, The latex of *Hevea brasiliensis* contains high levels of both chitinases and chitinases/lysozymes, *Plant Physiol.*, 95, 469, 1991.

64. Lawton, K., Ward, E., Payne, G., Moyer, M., and Ryals, J., Acidic and basic class III chitinase mRNA accumulation in response to TMV infection of tobacco, *Plant Mol Biol.,* 19, 735 1992.
65. Nielsen, K. K., Mikkelsen, J. D., Kragh, K. M., and Bojsen, K., An acidic class-III chitinase in sugar beet — induction by *Cercospora beticola,* characterization, and expression in transgenic tobacco plants, *Mol. Plant-Microbe Interact.,* 6, 495, 1993.
66. Ishige, F., Mori, H., Yamazaki, K., and Imaseki, H., Cloning of a complementary DNA that encodes an acidic chitinase which is induced by ethylene and expression of the corresponding gene, *Plant Cell Physiol.,* 34, 103, 1993.
67. Kragh, K. M., Jacobsen, S., Mikkelsen, J. D., and Nielsen, K. A., Tissue specificity and induction of class I, II and III chitinases in barley (*Hordeum vulgare*), *Physiol. Plant.,* 89, 490, 1993.
68. Domingo, C., Conejero, V., and Vera P., Genes encoding acidic and basic class III β-1,3-glucanases are expressed in tomato plants upon viroid infection, *Plant Mol. Biol.,* 24, 725, 1994.
69. Lawton, K. A., Beck, J., Potter, S., Ward, E., and Ryals, J., Regulation of cucumber class III chitinase gene expression, *Mol. Plant-Microbe Interact.,* 7, 48, 1994.
70. Liao, Y. C., Kreuzaler, F., Fischer, R., Reisener, H. J., and Tiburzy, R., Characterization of a wheat class Ib chitinase gene differentially induced in isogenic lines by infection with *Puccinia graminis, Plant Sci.,* 103, 177, 1994.
71. Münch-Garthoff, S., Neuhaus, J. -M., Boller, T., Kemmerling, B., and Kogel, K. H., Expression of β-1,3-glucanase and chitinase in healthy, stem rust-affected and elicitor-treated near-isogenic wheat lines showing Sr5 or Sr24-specified race-specific rust resistance, *Planta*, 201, 235, 1997.
72. Gaynor, J. J. and Unkenholz, K. M., Sequence analysis of a genomic clone encoding an endochitinase from *Solanum tuberosum, Nucl. Acid Res.,* 17, 5855, 1989.
73. van Buuren, M., Neuhaus J.-M., Shinshi, H., Ryals, J., and Meins, F., The structure and regulation of homeologous tobacco endochitinase genes of *Nicotiana sylvestris* and *N. tomentosiformis* origin, *Mol. Gen. Genet.,* 232, 460, 1992.
74. Vögeli-Lange, R., Hansen-Gehri, A., Boller, T., and Meins, F. J., Induction of the defense-related glucanohydrolases, β-1,3-glucanase and chitinase by tobacco mosaic virus infection of tobacco leaves, *Plant Sci.,* 54, 171, 1988.
75. Keefe, D., Hinz, U., and Meins, F., Jr., The effect of ethylene on the cell-type-specific and intracellular localization of β-1,3-glucanase and chitinase in tobacco leaves, *Planta*, 182, 43, 1990.
76. Hooft van Huijsduijnen, R. A. M., Cornelissen, B. J. C., and Bol, J. F., Virus-induced synthesis of mRNAs for precursors of pathogenesis-related proteins in tobacco, *EMBO J.,* 4, 2167, 1985.
77. van Loon, L. C. and Gerritsen, Y. A. M., Localization of pathogenesis-related proteins in infected and non-infected leaves of Samsun NN tobacco during the hypersensitive reaction to tobacco mosaic virus, *Plant Sci.,* 63, 131, 1989.
78. Lotan, T., Ori, N., and Fluhr, R., Pathogenesis-related proteins are developmentally regulated in tobacco flowers, *Plant Cell,* 1, 881, 1989.
79. Neale, A. D., Wahleithner, J. A., Lund, M., Bonnett, H. T., Kelly, A., Meeks-Wagner, D. R., Peacock, W. J., and Dennis, E. S., Chitinase, β-1,3-glucanase, osmotin, and extensin are expressed in tobacco explants during flower formation, *Plant Cell,* 2, 673, 1990.
80. Wemmer, T., Kaufmann, H., Kirch, H. H., Schneider, K., Lottspeich, F., and Thompson, R. D., The most abundant soluble basic protein of the stylar transmitting tract in potato (*Solanum tuberosum* L.) is an endochitinase, *Planta*, 194, 264, 1994.

81. Collada, C., Casado, R., and Aragoncillo, C., Endochitinases from *Castanea crenata* cotyledons, *J. Agr. Food Chem.*, 41, 1716, 1993.
82. Huynh, Q. K., Hironaka, C. M,. Levine, E. B., Smith, C. E., Borgmeyer, J. R., and Shah, D. M., Antifungal proteins from plants — purification, molecular cloning, and antifungal properties of chitinases from maize seed, *J. Biol. Chem.*, 267, 6635, 1992.
83. Lin, Z. F., Wu, D. Q., Luo, A. N., and Zhang, W. Q., Chitinases from seeds of *Zea mays* and *Coix lachryma-jobi* L. Purification and some properties, *Process Biochem.*, 27, 83, 1992.
84. Ary, M. B., Richardson, M., and Shewry, P. R., Purification and characterization of an insect alpha-amylase inhibitor/endochitinase from seeds of Job's Tears (*Coix lachryma-jobis*), *Biochim. Biophys. Acta,* 999, 260, 1989.
85. Swegle, M., Kramer, K. J., and Muthukrishnan, S., Properties of barley seed chitinases and release of embryo-associated isoforms during early stages of imbibition, *Plant Physiol.,* 99, 1009, 1992.
86. Yamagami, T. and Funatsu, G., Purification and some properties of 3 chitinases from the seeds of rye (*Secale cereale*), *Biosci. Biotechnol. Biochem.,* 57, 643, 1993.
87. Hedrick, S. A., Boller, T., and Lamb, C.J., Chitinase cDNA cloning and mRNA induction by fungal elicitor, wounding and infection, *Plant Physiol.,* 86, 182, 1988.
88. Esaka, M., Toyota, A., and Kitabayashi, M., Marked induction of basic chitinase in pumpkin in response to wounding, *Phytochemistry,* 34, 631, 1993.
89. Hamel, F. and Bellemare, G., Characterization of a class I chitinase gene and of wound-inducible, root and flower-specific chitinase expression in *Brassica napus, Biochimica et Biophysica Acta,* 1263, 212, 1995.
90. Parsons, T. J. and Gordon, M. P., Systemic accumulation of specific mRNAs in response to wounding in poplar, *Proc. Natl. Acad. Sci. U.S.A.,* 86, 7895, 1989.
91. Zhu, Q., Doerner, P. W., and Lamb, C. J., Stress induction and developmental regulation of a rice chitinase promoter in transgenic tobacco, *Plant J.,* 3, 203, 1993.
92. Shinshi, H., Usami, S., and Ohme-Takagi, M., Identification of an ethylene-responsive region in the promoter of a tobacco class I chitinase gene, *Plant Mol. Biol.,* 27, 923, 1995.
93. Xu, Y., Zhu, Q., Panbangred, W., Shirasu, K., and Lamb, C., Regulation, expression and function of a new basic chitinase gene in rice (*Oryza sativa* L), *Plant Mol. Biol.,* 30, 387, 1996.
94. Fukuda, Y. and Shinshi, H., Characterization of a novel cis acting element that is responsive to a fungal elicitor in the promoter of a tobacco class I chitinase gene, *Plant Mol. Biol.,* 24, 485, 1994.
95. Fukuda, Y., Interaction of tobacco nuclear protein with an elicitor-responsive element in the promotor of a basic class I chitinase gene, *Plant Mol. Biol.,* 34, 81, 1997.
96. Stanford, A. C., Northcote, D. H., and Bevan, M. W., Spatial and temporal patterns of transcription of a wound-induced gene in potato, *EMBO J.,* 9, 593, 1990.
97. Schlumbaum, A., Mauch, F., Vögeli, U., and Boller, T., Plant chitinases are potent inhibitors of fungal growth, *Nature*, 324, 365, 1986.
98. Arlorio, M., Ludwig, A., Boller, T., and Bonfante, P., Inhibition of fungal growth by plant chitinases and β-1,3-glucanases — a morphological study, *Protoplasma,* 171, 34, 1992.
99. Mauch, F., Mauch-Mani, B., and Boller, T., Antifungal hydrolases in pea tissue. 2. Inhibition of fungal growth by combinations of chitinase and β-1,3-glucanase, *Plant Physiol.,* 88, 936, 1988.

100. Sela-Buurlage, M. B., Ponstein, A. S., Bres-Vloemans, S. A., Melchers, L. S., van den Elzen, P. J. M., and Cornelissen, B. J. C., Only specific tobacco (*Nicotiana tabacum*) chitinases and β-1,3-glucanases exhibit antifungal activity, *Plant Physiol.,* 101, 857, 1993.
101. Lorito, M., Woo, S. L., Dambrosio, M., Harman, G. E., Hayer, C. K., Kubicek, Cp. P., and Scala, F., Synergistic interaction between cell wall degrading enzymes and membrane affecting compounds, *Mol. Plant-Microbe Interact.,* 9, 206, 1996.
102. Broekaert, W. F., van Parijs, J., Leyns F., Joos, H., and Peumans, W. J., A chitin-binding lectin from stinging nettle rhizomes with antifungal properties, *Science,* 245, 1100, 1989.
103. de Bolle, M. F. C., Osborn, R. W., Goderis, I. J., Noe, L., Acland, D., Hart, C. A., Torrekens, S., van Leuven, F., and Broekaert, W. F., Antimicrobial peptides from *Mirabilis jalapa* and *Amaranthus caudatus* — expression, processing, localization and biological activity in transgenic tobacco, *Plant Mol. Biol.,* 31, 993, 1996.
104. Chrispeels, M. J., Alt, M., and Raikhel, N. V., Lectins, lectin genes, and their role in plant defense, *Plant Cell,* 3, 1, 1991.
105. Neuhaus, J.-M., Ahl-Goy, P., Hinz, U., Flores, S., and Meins, F., Jr., High-level expression of a tobacco chitinase gene in *Nicotiana sylvestris* — susceptibility of transgenic plants to *Cercospora nicotianae* infection, *Plant Mol. Biol.,* 16, 141, 1991.
106. Broglie, K., Chet, I., Holliday, M., Cressman, R., Biddle, P., Knowlton, S., Mauvais, C. J., and Broglie, R., Transgenic plants with enhanced resistance to the fungal pathogen *Rhizoctonia solani, Science,* 254, 1194, 1991.
107. Benhamou, N., Broglie, K., Broglie, R., and Chet, I., Antifungal effect of bean endochitinase on *Rhizoctonia solani* — ultrastructural changes and cytochemical aspects of chitin breakdown, *Can. J. Microbiol.,* 39, 318, 1993.
108. Vierheilig, H., Alt, M., Neuhaus, J.-M., Boller, T., and Wiemken, A., Effect of chitinase overexpression on the colonization of roots by pathogenic and symbiotic fungi, *Mol. Plant-Microbe Interact.,* 6, 261, 1993.
109. Melchers, L. S., Sela-Buurlage, M. B., Vloemans, S. A., Woloshuk, C. P., van Roekel, J. S. C., Pen, J., van den Elzen, P. J. M., and Cornelissen, B. J. C., Extracellular targeting of the vacuolar tobacco proteins AP24, chitinase and β-1,3-glucanase in transgenic plant, *Plant Mol. Biol.,* 21, 583, 1993.
110. Dénarié, J. and Cullimore, J., Lipo-oligosaccharide nodulation factors: a new class of signaling molecules mediating recognition and morphogenesis, *Cell,* 74, 951, 1993.
111. Goormachtig, S., Valerio-Lepiniec, M., Szczyglowski, K., van Montagu, M., Holsters, J., and Debreijn, F. J., Use of differential display to identify novel *Sesbania rostrata* genes enhanced by *Azorhizobium caulinodans* infection1, *Mol. Plant — Microbe Interact.,* 8, 816, 1995.
112. Lan, V. T. T., Neuhaus, J.-M., Boller, T., Broughton, W. J., and Krause, A., Molecular cloning of *Vigna* cDNAs encoding class I (Accession No. X88800) and class IV (Accession No. X88803), *Plant Physiol.,* 110, 335, 1996.
113. Schultze, M., Kondorosi, E., Kondorosi, A., Staehelin, C., Mellor, R. B., and Boller, T., The sulfate group on the reducing end protects Nod signals of *R. meliloti* against hydrolysis by *Medicago* chitinases, in *New Horizons in Nitrogen Fixation,* Palacios, R., Mora, J., and Newton, W. E., Eds., Kluwer, Dordrecht, the Netherlands: 1993, 159.
114. Staehelin, C., Schultze, M., Kondorosi, E., Mellor, R. B., Boller, T., and Kondorosi, A., Structural modifications in *Rhizobium meliloti* Nod factors influence their stability against hydrolysis by root chitinases, *Plant J.,* 5, 319, 1994.

115. Kim, Y. S., Yoon, K., Liu, J. R., and Lee, H. S., Characterization of CHRK1, a receptor-like kinase containing a chitinase-related domain in its N-terminus, *5th International Congress of Plant Molecular Biology,* Singapore, 1997, 789.
116. De Jong, A. J., Cordewener, J., Lo Schiavo, F., Terzi, M., Vandekerckhove, J., van Kammen, A., and De Vries, S. C., A carrot somatic embryo mutant is rescued by chitinase, *Plant Cell,* 4, 425, 1992.
117. De Jong, A. J., Hendriks, T., Meijer, E. A., Penning, M., Lo Schiavo, F., Terzi, M. van Kammen, A., and De Vries, S. C., Transient reduction in secreted 32 kD chitinase prevents somatic embryogenesis in the carrot (*Daucus carota* L) variant ts11, *Developmental Genetics,* 16, 332, 1995.
118. Kragh, K. M., Hendriks, T., De Jong, A. J., Lo Schiavo, F., Buchema, N., Hojrup, P., Mikkelsen, J. D., and De Vries, S. C., Characterization of chitinases able to rescue somatic embryos of the temperature-sensitive carrot variant ts11, *Plant Mol. Biol.,* 31, 631, 1996.
119. De Jong, A. J., Heidstra, R., Spaink, H. P., Hartog, M. V., Meijer, E. A., Hendriks, T., Lo Schiavo, F., Terzi, M., Bisseling, T., van Kammen, A., and De Vries, S. C., *Rhizobium* lipooligosaccharides rescue a carrot somatic embryo mutant, *Plant Cell,* 5, 615, 1993.
120. Hon, W. C., Griffith, M., Mlynarz, A., Kwok, Y. C., and Yang, D. S. C., Antifreeze proteins in winter rye are similar to pathogenesis-related proteins, *Plant Physiol.,* 109, 879, 1995.
121. Waters, E. J., Shirley, N. J., and Williams, P. J., Nuisance proteins of wine are grape pathogenesis-related proteins, *J. Agric. Food Chem.,* 44, 3, 1996.
122. Posch, A., Chen, Z., Wheeler, C., Dunn, J., Faulf-Heimsoth, J., and Baur, X., Characterization and identification of latex allergens by two-dimensional electrophoresis and protein microsequencing, *J. Allergy Clin. Immunol.,* 99, 385, 1997.
123. Neuhaus, J.-M., Fritig, B., Linthorst, H.J.M., Meins, F., Jr., Mikkelson, J.D., and Ryals, J., A revised nomenclature for chitinase genes, *Plant Mol. Biol. Rep.,* 14, 102, 1996.
124. Brunner, F., Stintzi, A., Fritig, B., and Legrand, M., Substrate specificities of tobacco chitinases, *Plant J.,* 14, 225, 1998.
125. Sowka, S., Hsieh, L.S., Krebitz, M., Akasawa, A., Martin, B.M., Starrett, D., Peterbauer, C.K., Scheiner, O., and Breiteneder, H., Identification and cloning of Prs a 1, a 32-kda endochitinase and major allergen of avocado, and its expression in the yeast *Pichia pastoris, J. Biol. Chem.,* 273, 28091, 1998.

5 The PR-5 Family: Thaumatin-Like Proteins

Rethinasamy Velazhahan, Swapan K. Datta, and Subbaratnam Muthukrishnan

CONTENTS

5.1 Introduction 107
5.2 Occurrence 108
5.3 Physicochemical Properties of TLPs 109
5.4 Biological Properties of TLPs 113
5.4.1 Taste 113
5.4.2 Antifungal Activity 114
5.4.3 TLPs as Anti-Freeze Proteins 115
5.4.4 TLPs as Inhibitors? 115
5.5 Regulation of TLP Expression 116
5.5.1 Microbial Infection 116
5.5.2 Osmotic Stress 117
5.5.3 Abscisic Acid and Ethylene 117
5.5.4 Salicylate, Methyl Jasmonate, and Elicitors 118
5.5.5 Wounding 118
5.6 cDNAs and Genes for TLPs 119
5.6.1 cDNAs 119
5.6.2 Genes 119
5.7 Transgenic Plants Over-Expressing PR-5 Proteins Have Enhanced Disease Resistance 120
5.8 Extracellular Targeting of PR-5 Proteins 122
5.9 Do Other Defense Proteins Have PR-5 Protein Domains? 123
5.10 Osmotin Interferes with a Signal Transduction Pathway 123
5.11 Conclusion 124
References 124

5.1 INTRODUCTION

As outlined in Chapter 1, pathogenesis-related proteins (PR proteins) were initially identified in tobacco plants reacting hypersensitively to tobacco mosaic virus (TMV)

0-8493-0697-3/99/$0.00+$.50

infection. The PR-5 protein(s) was detected as the slowest moving band among several low-molecular-weight proteins resolved by polyacrylamide gel electrophoretic analyses of acidic extracts of tobacco leaves (Samsun NN or Xanthi-nc) infected with TMV.[1,2,3] It was only when these PR-5 proteins were sequenced that it became clear that they were related to the intensely sweet tasting protein, thaumatin, from the West African shrub, *Thaumatococcus daniellii,* described by Dr. Daniell in 1852.[4] Because of their sequence similarity to thaumatin, the PR-5 group proteins have been called thaumatin-like proteins (TLPs) even though none of them have been found to have a sweet taste. Similarly, two other proteins, osmotin and NP24, which accumulated in cultured tobacco and tomato cells, respectively, when subjected to osmotic stress, were also found to be related to thaumatin in amino acid sequence. When it was observed that these proteins were also induced upon infection by pathogens, it became clear that osmotins also belong to the PR-5 group of proteins.[5,6] In addition, several seed-associated proteins from cereals which had sequence similarity to thaumatin were shown to possess antifungal activity. These proteins are clearly seed homologs of TLPs even though it must be understood that the seed TLPs are not *bona fide* PR proteins because their accumulation does not depend on exposure to pathogens. The availability of sensitive methods of detection (Western and northern blots) have led to the identification of TLPs whose expression is developmentally regulated. In this review, we will include the seed homologues and other developmentally regulated TLPs (such as those expressed in floral organs, seeds, and fruits) in order to present an overall view of the metabolism of all TLPs. An excellent review of these three groups of proteins that belong to the TLP family was presented by Dudler et al.[7] In this chapter, we have attempted to summarize the data from several laboratories dealing with the isolation and characterization of TLPs, their cDNAs and genes, and recent attempts to enhance plant resistance to pathogens by over-expression of TLPs in dicotyledonous and monocotyledonous plants.

5.2 OCCURRENCE

TLPs are not normally detected in leaves of young healthy plants, but they rapidly accumulate to high levels in response to biotic or abiotic stress. However, older leaves of potato were shown to accumulate TLPs in the absence of deliberate stress, though the possibility that they might have been subjected to some sort of stress when grown in a greenhouse or growth chamber was not ruled out.[8] Low levels of osmotin were detected in greenhouse-grown tomato leaves, and very low levels were detected in stems, but the levels in roots were much higher when equal amounts of protein from these tissues from seedlings were analyzed.[9] Possibly, the expression of TLPs in roots may be due to exposure to soil-associated microbes. Similar results were seen in tobacco seedlings. LaRosa et al.[10] analyzed a variety of tissues from normal (unstressed) tobacco plants for osmotin expression and found that the highest levels of osmotin and its mRNA were found in roots, epidermal peel, corolla, and immature flower bud. The roots had the highest levels of this protein. Koiwa et al.[11] used an isoelectricfocusing procedure to resolve three different forms of tobacco TLP. They reported that while no TLPs were detected in tobacco leaves, roots accumulated large amounts of a neutral TLP which was distinct from osmotin (which is more basic). Using histochemical analysis, they concluded that this neutral TLP accumulated constitutively in the cortex and metaxylem cells of roots. Cultured tobacco cells had detectable levels of two basic TLPs (osmotin) and two neutral TLPs (called OLP for osmotin-like protein) and secreted an acidic form of TLP (PR-S) into the medium. TLPs are expressed in flower buds of Brassica,[12] tobacco flower tissues,[13] overripe fruits of *Diaspora texana,*[14] and overripe cherries[15] where they are apparently developmentally regulated. Seeds of several cereals contain large amounts of TLPs.[16,17] Thaumatin itself is a fruit protein and accumulates in the arils of fruits of the West African rain forest shrub, *Thaumatococcus daniellii* Benth.

TLPs are also induced in tobacco roots and shoots and cotyledons in response to elicitors.[18,19] In addition, TLPs are induced in leaves of a large number of plants belonging to diverse genera when they are infected by pathogens or subjected to various stresses. The induction of TLPs upon pathogen infection will be covered in detail later in this chapter. It would appear that potentially all plant tissues are capable of synthesizing TLPs, depending on a variety of factors such as age, exposure to elicitors and microbes, and developmental cues.

5.3 PHYSICOCHEMICAL PROPERTIES OF TLPs

TLPs have been purified from several plant sources, most notably from leaves of infected tobacco, tomato, and soybean, grains of barley, maize and wheat, fruits of cherries and *Diospyros texana,* and floral buds of *Brassica napus.* Partially purified TLPs have also been obtained from sorghum and oat seeds. See Table 5.1 for a summary of the properties of purified TLPs or those predicted from cDNA clones.

The TLPs are generally highly soluble proteins that accumulate to high levels (1 to 12%) in selective tissues or subcellular compartments under appropriate conditions or are secreted into the medium or extracellular space. They remain soluble even under acidic conditions and are quite resistant to proteolysis. These properties led to the easy visualization of tobacco TLPs in acidic extracts of infected leaves upon electrophoresis in polyacrylamide gels followed by staining with Coomassie Brilliant Blue in earlier studies. The molecular weights of mature TLPs fall into two size ranges. One group of proteins has a size range of 22 to 26 kDa (201 to 229 amino acids), while the members of the other group have sizes around 16 kDa (148 to 151 amino acids). The second group of TLPs has an internal deletion of 58 amino acids.

The TLPs have pI values ranging from very acidic to very basic (range 3.4 to 12; see Table 5.1). Thus, there are acidic, neutral, and basic TLPs. All three have been identified in tobacco.[11] Generally, the extracellular TLPs tend to be acidic, while the vacuolar (or other vesicular) TLPs tend to be basic. It is not clear whether there is any biological significance to this observation. There is no evidence for glycosylation of any of the TLPs, including thaumatin.

The TLPs are generally resistant to proteases and pH- or heat-induced denaturation, and this may be due to the presence of 16 cysteines, which have been shown to be involved in the formation of eight disulfide bonds in thaumatin and zeamatin, a TLP isolated from maize seeds.[16] The three-dimensional structure of thaumatin I

TABLE 5.1
Properties of TLPs

Source	Name	M. wt. (kDa)	# amino acids	pI	DNA (D) Protein (P)	RNA size (kb)	Cellular location	Fungi inhibited	Induction/ regulation by	References (#)
Thaumatococcus daniellii	Thaumatin	22	207	12	D, P		Cytoplasmic vesicles	*C. albicans*	Developmental	Iyengar et al. (71)
Tobacco leaves	PR-R minor	24.6 (pre)	226	6.2–6.8	D, P	1.0–1.1	Extracellular	*C. beticola*	TMV	Cornelissen et al. (72) Pierpoint et al. (1) Vigers et al. (27)
Tobacco leaves	PR-R major	24.7 (pre)	226	6.2–6.8	D, P		Extracellular		TMV	Payne et al. (73) Pierpoint at al. (1)
Tobacco cells	Osmotin	26.4 (pre)	243	8.2	D, P	1.2	Vacuolar inclusions	*P. infestans* *C. albicans* *N. crassa* *T. reesei*	NaCl	Singh et al. (47) Singh et al. (48)
Tobacco cells	pTOL1	29 (pre)	251	neutral	D, P	1.0	Intracellular		Stress	Takeda et al. (55)
Tomato	pNP24	24 (mature)	226	basic	D, P	1.0	Soluble fraction	*P. infestans*	NaCl	King et al. (74)
Arabidopsis INA-treated	PR-5	22.7 mature	215	4.5	D	—	Apoplast		INA, SA	Uknes et al. (41)
Arabidopsis floral meristem	ATLP-1	26 (pre)	243	9.6	D		—	—	Developmental	Hu & Reddy (75)
Arabidopsis	ATLP-3	25.7(pre)	245	4.8	D	—	—	*A. solani* *T. reesei* *C. albicans* *V. albo-atrum* *V. dahliae*	SA, pathogens	Hu & Reddy (42)
Cherry fruit	pCHER29/6	25.8 (pre)	245	4.18	D	—	—	None	Developmental	Fuls-Lycaon et al. (15)
Tomato flowers	tomf216	27.2 (pre)	252	7.8	D	—	—	—	Developmental	Chen et al. (76)

Wheat (powdery mildew)	pWIR2	17.5 (pre)	173	—	D	—	—	—	Pathogen	Rebmann et al. (40)
Wheat seeds	Trimatin	22 (mature)			P		Developmental	*C. albicans* *T. reesei* *N. crassa*		Vigers et al. (17)
Rice	pPIR2	18 (pre)	177	4.75	D	0.7	—	—	P. Syringae	Reimmann et al. (44)
Rice	D34	24.4 (pre)	231	6.56	D	1.1	—	—	Pathogen	Velazhahan et al. (45)
Rice	C22	23.9 (pre)	220	4.46	D	1.1	—	—	Pathogen	Velazhahan et al. (45)
Oat (infected)	pCRL101	17.4 (pre)	169	6.2	D	0.9	—	—	Pathogen	Lin et al. (43)
Oat (infected)	pCRL102	17.4 (pre)	169	6.2	D	0.9	—	—	Pathogen	Lin et al. (43)
Oat (infected)	pCRL103	17.3 (pre)	169	6.2	D	0.9	—	—	Pathogen	Lin et al. (43)
Oat (infected)	pCRL104	17.6 (pre)	169	4.1	D	0.9	—	—	Pathogen	Lin et al. (43)
Maize seeds	Zeamatin	22 (mature)	206	—	P	—		*A. solani* *F. oxysporum* *N. crassa* *T. reesei* *C. albicans*		Richardson et al. (32) Roberts & Selitrennikoff (16) Huynh et al. (33)
Maize seeds	Zlp	22 (pre)	227	9.1	D, P	1.0	Intracellular wash fluid	*C. albicans* *T. reesei* *N. crassa* *A. solani*	Developmental	Malehorn et al. (77)
Atriplex nummularia	pA8	23.8 (pre)	222	8.31	D	1.2, 1.0	—	—	ABA	Casas et al. (78)
Atriplex nummularia	pA9	23.8 (pre)	224	6.88	D	0.9	Cells, medium	—	NaCl	Casas et al. (78)
Barley plants	Hv-1	15.6	173	3.4	D, P	0.75–0.85	—		Pathogen	Bryngelson & Green (37) Hahn et al. (79)

TABLE 5.1 *(continued)*
Properties of TLPs

Source	Name	M. wt. (kDa)	# amino acids	pI	DNA (D) Protein (P)	RNA size (kb)	Cellular location	Fungi inhibited	Induction/ regulation by	References (#)
Barley grain	BP-R, BP-S	23 23						*T. viride* *C. albicans* *F. oxysporum*		Hejgaard et al. (28)
Potato leaves	Protein C	21 (mature)			P	—	Intercellular fluid	—		Pierpoint et al. (80)
Grape	VVTL1	21 (mature)	197	4.6	D, P				Developmental	Tattersall et al. (81)
Soybean leaves	P21	21.5	202	4.6	P			—	—	Graham et al. (82)

Abbreviations: TMV=tobacco mosaic virus; INA=2,6-dichloroisonicotinic acid; SA=salicylic acid; ABA=abscisic acid.

has been determined using X-ray crystallography.[20,21] Two building motifs are utilized in thaumatin structure: a folded β-sheet (which forms an identifiable domain I) and the β-ribbons and small loops stabilized by disulfide bonds which form two separate domains, II and III. The positions of the eight disulfide bonds in thaumatin have been assigned. It has been proposed that domains II and III may be important for binding to the sweetness receptors on membranes of taste buds. All but one of the disulfide bridges are in domains I and II. Disruption of the disulfide bonds results in loss of the unique tertiary structure of the thaumatin molecule and loss of sweetness.[22] Treatment with SDS or guanidine or a less polar solvent mixture (water/dioxan) results in loss of sweetness, indicating the importance of the three-dimensional structure for this property. The locations of the cysteines are highly conserved between thaumatin and the members of the class of larger TLPs. Interestingly, the lower-molecular-weight TLPs are missing most of domain II and several half cysteines which stabilize this domain. Functional differences between these two groups of thaumatins would be very revealing in assigning specific roles to domains II and III if these domains turn out to be general features of all TLPs.

The crystal structure of zeamatin has also been determined.[23] Zeamatin, whose tertiary structure closely resembles that of thaumatin, has a β-sandwich with 11 β-strands from which an arm-like structure extends to form a large cleft lined by several acidic residues. This arm has a short α-helix and three disulfide bonds. The amino acid sequences constituting the acidic cleft are highly conserved among TLPs. The exterior surface of zeamatin is lined by highly basic residues. The crystal structure of zeamatin indicates that this protein associates to form a homodimer, but does not suggest the formation of a cyclic aggregate that is characteristic of channel forming proteins. Batalia et al.[23] suggest that electrostatic interactions between zeamatin and some membrane channel proteins are responsible for destabilization of pressure gradients that maintain the hyphal tip in the extended state. Studies on the nature of the target proteins to which TLPs bind may shed additional light on the mechanism of TLP action.

5.4 BIOLOGICAL PROPERTIES OF TLPs

5.4.1 Taste

The sweetness of thaumatin to primates (of the catarrhine group; old world monkeys, apes, and humans) is probably used for the dispersal of seeds in nature. Thaumatin binds to receptors in taste cell membranes and brings about stimulation of the chorda tympani nerve of rhesus monkey at micromolar concentrations, indicating tight binding of thaumatin to the taste cell receptors.[24] Even though thaumatins are very sweet-tasting proteins, none of the other TLPs tested have a sweet taste. Apparently, the three-dimensional architecture of thaumatin is utilized in TLPs to bind to other receptors such as those found in fungal cell membranes. It is interesting to note that zeamatin sequence is significantly different in two of the five homologous tripeptide stretches that are conserved between the two sweet proteins, thaumatin and monellin.[25]

5.4.2 Antifungal Activity

The finding that zeamatin, a protein isolated from maize seeds based on its antifungal activity against *Candida albicans, Neurospora crassa,* and *Trichoderma reesei,*[16] has an amino acid sequence closely related to the two TLP from tobacco, namely osmotin and PR-R, as well as to thaumatin[17] has led to the notion that TLPs represent antifungal defenses of plants. Zeamatin caused leakage of cytoplasmic material from the fungi and caused hyphal rupture. Further studies with the TLPs, AP24 (same as osmotin II), and NP24 purified from TMV-inoculated tobacco and tomato, respectively, demonstrated that these proteins caused sporangial lysis of *Phytophthora infestans.*[26] Vigers et al.[27] compared the antifungal activities of purified zeamatin, osmotin (basic vacuolar form) and PR-S (acidic extracellular form), from tobacco against the plant pathogen *Cercospora beticola.* PR-S was most effective against this pathogen in the hyphal extension inhibition assay on agar plates (50% inhibition at 2 μg/disc compared to 25% for osmotin and zeamatin). The inhibition was seen as early as one day after exposure to PR-S, but for the other two proteins significant inhibition was seen after only 3 to 4 days. On the other hand, when growth of *T. viride, C. albicans,* and *N. crassa* was studied in a liquid culture in a microtiter plate, zeamatin and osmotin completely inhibited the growth of these fungi at concentrations in the range of 3 to 10 μg/ml. PR-S did not inhibit growth at all, even up to 100 μg/ml. Thus, different TLPs exhibit differences in their ability to inhibit specific genera of fungi. Thaumatin was shown to have antifungal activity against *C. albicans* (at 30 μg/disc) in the presence of 0.2 μg/ml nikkomycin (a competitive inhibitor of chitin synthase), even though, by itself, thaumatin had no inhibitory effect against this yeast. It appears that high concentrations of TLPs can actively lyse fungal membranes, but at low concentrations they may cause membrane permeability changes that cause leakage of cell constituents and increase the uptake of other antifungal compounds. Similar results were reported for two TLPs from barley grain (R and S) by Hejgaard et al.[28] These two proteins inhibited the growth of *Trichoderma viride* and *Candida albicans,* but only in the presence of nikkomycin.

Antifungal activities of TLPs from a variety of sources were summarized by Stintzi et al.[29] Table 5.1 indicates the antifungal activities of TLPs that have been tested using *in vitro* assays. In some cases, antifungal activities have been inferred from the enhanced disease resistance of transgenic plants expressing TLPs (see section 5.7). The TLPs shown to have antifungal activity included osmotin and PR-S from tobacco, AP-24 from tomato, zeamatin from corn, trimatin from wheat, and protein R and S from barley. The susceptible fungi included *Phytophthora infestans, Candida albicans, Neurospora crassa, Trichoderma reesei, Trichoderma viride, Verticillium albo-atrum, Verticillium dahliae, Cercospora beticola, Fusarium oxysporum,* and *Alternaria solani* and included fungi belonging to the classes *Oomycetes, Hyphomycetes,* and *Ascomycetes.* The effects were both at spore germination level and inhibition of hyphal extension. More recently, antifungal activities have been demonstrated for a 27 kDa TLP from the fruits of *Diaspora texana*[14] against *Phytophthora infestans* and for a 27 kDa TLP from flower buds of *Brassica campestris* against *Neurospora crassa.*[12] Abad et al.[30] carried out an extensive study of the antifungal activity of tobacco osmotin by surveying 31 isolates representing 18 fungal genera at three different concentrations of osmotin. They found that *Trichoderma, Bipolaris, Fusarium, Verticillium, Cercospora,* and *Phytophthora* species were inhibited to varying degrees by osmotin. But *Aspergillus, Macrophomina,* and *Rhizoctonia* species were not affected by high concentrations of osmotin. The effects of osmotin included spore lysis, inhibition of spore germination and reduced germling viability, and dissipation of membrane pH gradient. Interestingly, the inhibitory activity of osmotin was counteracted by inorganic salts, but not by choline chloride. Osmotin is also not active against plasmolyzed cells. Therefore, it has been proposed that osmotin acts through specific receptors on the plasma membrane but requires juxtaposition of cell wall and the plasma membrane for it to be effective.

5.4.3 TLPs as Anti-Freeze Proteins

An unexpected role for a rye PR-5 protein as an anti-freeze protein has been proposed.[31] Apoplasts of cold-acclimated winter rye contain several distinct proteins. The accumulation of these proteins in the apoplast upon exposure to cold temperatures is correlated with the increase in freezing tolerance. Two of these proteins with molecular weights of 25 and 16 kDa were shown by N-terminal protein sequencing to be highly similar to TLPs from barley and rice. Anti-freeze activity of the apoplastic proteins (which included chitinases, β-1,3-glucanases, and TLPs) was demonstrated by observing the shape of the ice crystals formed when these apoplastic extracts were subjected to freezing. The formation of multifaceted bipyramidal crystals, diagnostic of anti-freeze activity was observed only in apoplastic extracts from cold-acclimated plants. In contrast, apoplastic extracts of control plants formed flat hexagonal ice crystals with straight faces, indicating no inhibition of ice formation.

5.4.4 TLPs as Inhibitors?

An exciting idea that TLPs may have a role as anti-feedant proteins targeted against insect proteases and α-amylases was proposed based on the isolation of a maize protein with high sequence similarity to thaumatin with inhibitory activity against α-amylase of the flour beetle, *Tribolium castaneum* and against bovine trypsin.[32] However, none of the other TLPs tested so far, including zeamatin, barley permatin, barley TLPs, R and S, and tobacco TLPs, showed any inhibitory activity against α-amylase or trypsin.[17,28] Further it was shown that the maize TLP reported by Richardson et al.[32] had the same sequence as zeamatin for the N-terminal 61 amino acids. Finally, an antifungal protein isolated from maize seeds[33] had the same sequence as the TLP reported by Richardson et al.[32] but had no activity against α-amylase or trypsin. It appears that contamination of the latter protein with a bifunctional inhibitor of α-amylase or trypsin from maize may have contributed to the misidentification of the maize TLP as a bifunctional inhibitor of α-amylase or trypsin. Thus, it should be concluded that there is no convincing evidence that TLPs function as bifunctional inhibitors of proteases and α-amylases.

5.5 REGULATION OF TLP EXPRESSION

5.5.1 MICROBIAL INFECTION

The induction of TLPs in response to microbial infection has been studied most extensively in tobacco plants infiltrated with TMV where TLPs were first reported (see Chapter 1). While uninfected leaves do not contain any detectable TLPs, upon viral infection the leaves accumulate large amounts of TLPs within 6 to 8 days.[1,34] Bean leaves infected with tobacco ring spot virus, southern bean mosaic virus, bean pod mottle virus, and TMV also induce a 21 kDa TLP.[35] TLPs were also induced in 20 plant species, which included six dicotyledonous families after infection with tobacco ring spot virus.[36] In tomato, a protein of 23 kDa named P23, accumulates after infection with citrus exocortis viroid. P23 is highly homologous to tomato osmotin (NP24).[6] An 832 bp osmotin-like cDNA (TPM 1) has been isolated from a cDNA library constructed from tomato planta macho viroid (TPMV)-infected tomato.[37] Nucleotide sequence of TPM 1 shows high homology (nearly 90%) to tomato osmotin (NP24) cDNA. The predicted mature N-terminus of TPM 1 is 100% homologous to the N-terminal sequences of viroid-induced protein P23 and to fungus-induced protein AP24, suggesting that all these proteins might be equivalent.[6,26] Inoculation of *Arabidopsis thaliana* with turnip crinkle virus has been shown to induce genes encoding acidic PR-5 protein.[38]

The induction of TLPs is not, however, restricted to virus infection or to dicots. They are induced in both monocotyledonous and dicotyledonous plants in response to both viral and fungal infection. Bryngelson and Green[39] reported that a 19 kDa TLP was induced when barley leaves were challenged with an incompatible race of *Erysiphe graminis* (powdery mildew). Similarly, when wheat seedlings were treated with spores of *Erysiphe graminis*, TLP mRNA is induced within 14 h.[40] Infection of *Arabidopsis thaliana* with *Pseudomonas syringae* resulted in the induction of TLP mRNA within 6 days[41] and possibly as early as 24 h.[42] Lin et al.[43] carried out an extensive analysis of induction of TLP in oat seedlings infected with stem rust fungus (*Puccinina graminis*). Four distinct mRNAs for TLPs were shown to be induced, some as early as 24 h after infection. Plants infected with an incompatible isolate of the rust fungus accumulated higher levels of TLP mRNAs compared to compatible isolates. Induction of a 16 kDa TLP (and its 0.7 kb mRNA) in response to infiltration of rice leaves with a nonpathogenic bacterium *Pseudomonas syringae* pv. *syringae* was reported by Reimmann and Dudler.[44] Rice plants respond somewhat differently to fungal infection. Velazhahan et al.[45] demonstrated that infection of rice plants with the sheath blight fungus, *Rhizoctonia solani* resulted in the induction of two TLPs of sizes 24 and 25 kDa within 1 to 2 days after infection and continued to accumulate for several days. The corresponding mRNAs (1.1 kb) were also induced and peaked around 4 days. There was no evidence of induction of the 16 kDa TLP or its 0.7 kb mRNA in *R. solani*-infected rice plants reported by Reimmann and Dudler.[44] Apparently, the regulation of the larger rice TLPs is somewhat different from that of the smaller TLP. But it is abundantly clear that TLPs can be induced in response to viral, bacterial, and fungal infection.

5.5.2 Osmotic Stress

Selective members of the TLP family are induced by osmotic stress. Osmotin, the basic tobacco TLP, accumulates as dense inclusion bodies in the vacuoles of cultured tobacco cells during adaptation to salt (NaCl or KCl) stress.[10,46,47] As much as 12% of total cellular protein can be osmotin under these conditions. Osmotin also accumulates when tobacco cells are subjected to drought stress such as that induced by the inclusion of 25 to 30% polyethyleneglycol in the culture medium. Even though suspension cells approaching stationary phase also accumulate osmotin, the levels are much lower than the levels seen in salt-adapted cells.[9] It was also found that tomato cells in suspension accumulated a 26 kDa protein in response to high salt (NaCl and KCl), and PEG.[9] The induction of the 26 kDa osmotin-like protein was also demonstrated when tomato seedlings were grown hydroponically in saline solutions for 10 days. Immunologically related 26 kDa proteins were also detected in alfalfa and green bean but whether or not they are induced by salt stress was not investigated.[9] Subsequent characterization of the cDNA clones for both osmotin and the tomato protein (NP24) predicted the sizes of the mature proteins to be around 24 kDa.

5.5.3 Abscisic Acid (ABA) and Ethylene

When unadapted tobacco suspension cells were treated with ABA, a 26 kDa protein indistinguishable from osmotin (except for a difference in pI) was detected within 24 h,[47] but the synthesis declined rapidly within 4 days. The induction of a TLP by ABA was observed in several plants (millet, soybean, carrot, cotton, potato, and tomato). However, in a detailed study Singh et al.[48] reported that osmotin mRNA levels increased about 8-fold 24 h after ABA treatment, but this level was only half of the level in salt-adapted cells. However, there was no detectable increase in osmotin levels over that found in unadapted cells as determined by Western blot analysis, indicating additional regulation at the translational level.

To elucidate the role of ABA in inducing osmotin mRNA and protein, Grillo et al.[49] carried out experiments using ABA-deficient tomato mutants. Osmotin gene expression and protein accumulation were enhanced in the leaves of wild-type tomato plants in response to exposure to NaCl, abscisic acid (ABA), or after dehydration to 80% of original plant fresh weight.[49] Moreover, accumulation of osmotin mRNA was accompanied by a large increase in endogenous ABA level. No induction of osmotin mRNA was observed in the ABA-deficient mutant, *sitiens,* when subjected to salt or water stress. Treatment of this ABA-deficient mutant with exogenous ABA resulted in accumulation of osmotin mRNA, indicating a role for ABA in osmotin mRNA accumulation. On the other hand, in another ABA-deficient mutant, *flacca*, a twofold induction was seen with ABA application, but simultaneous application of both NaCl and ABA resulted in an 8-fold stimulation of osmotin mRNA levels. In transgenic tomato plants (with normal ABA accumulation upon water or salt stress) that were transformed with osmotin promoter/β-glucuronidase (GUS) fusion gene, induction of GUS activity by NaCl could be partially blocked by the ABA inhibitor, fluoridone, indicating that osmotin gene expression may be triggered in part through ABA. Ethylene may also be involved in osmotin induction because norbornadiene, an ethylene inhibitor, also partially inhibited osmotin mRNA accumulation. When norbornadiene was administered along with fluoridone, the inhibition of salt-induced accumulation of osmotin mRNA was greater than with either inhibitor alone, indicating that ABA and ethylene mediate osmotin induction in wild-type tomato plants.[49] Koiwa et al.[11] also reported induction of osmotin and a neutral TLP upon ethylene treatment.

Campillo and Lewis[50] investigated the accumulation of proteins in bean abscission zones after induction of abscission by ethylene treatment. Among the proteins that accumulated were several PR proteins, including a 22 kDa TLP starting as early as 24 h and continuing during the 72-h ethylene treatment. The effect of indole acetic acid (IAA), an inhibitor of abscission, was also studied. IAA led to a substantial reduction in the levels of ethylene-induced TLP accumulation. It is possible that TLP may be part of a preemptive defense response by the plant against pathogen attack during programmed senescence.

5.5.4 Salicylate, Methyl Jasmonate, and Elicitors

Osmotin and related proteins are induced not only by the hormones, ethylene and ABA, the signal molecules, salicylic acid and jasmonic acid, but also by the abiotic elicitor, 2,6-dichloroisonicotinic acid indicating that a variety of stress signals can induce the TLPs. When *Arabidopsis* plants were sprayed with dichloroisonicotinic acid or salicylic acid, there was a rapid accumulation of mRNA encoding a TLP (size of encoded mature protein is 23 kDa) within 1 day.[41] Similarly, Koiwa et al.[11] reported the induction of an acidic TLP (PR-S), but not of osmotin or neutral TLP in Samsun NN tobacco plants sprayed with salicylic acid. In a detailed study of induction of osmotin expression in response to several hormones and signal molecules, Xu et al.[19] found that salicylic acid, methyl jasmonate, and ethylene induced osmotin mRNA in tobacco seedlings. However, combinations of these signal molecules were much more potent (a synergistic effect) than individual compounds. Most effective was the combination of ethylene and methyl jasmonate, which mimicked the levels seen in osmotic stress. Okadaic acid, an inhibitor of a protein kinase system, also resulted in induction of osmotin mRNA, suggesting that a protein kinase signal transduction cascade may be involved in osmotin mRNA induction. Thus, available evidence suggests the operation of multiple pathways for activation of TLP responding to single, or combinations of environmental signals. The DNA sequence elements that govern these responses will be discussed in more detail in the next section.

5.5.5 Wounding

Neale et al.[51] studied the expression of osmotin upon wounding of tobacco leaves. Within 8 h after wounding, the levels of osmotin mRNA in wounded leaves increased dramatically (14-fold) and persisted at least until 24 h. Interestingly, the levels of osmotin mRNA in an unwounded leaf of the wounded plant were also elevated to a similar level, indicating that a systemic signal was involved in osmotin gene regulation. Similarly, Brederode et al.[52] reported that the basic forms of TLPs (but not the acidic forms) are induced upon wounding in tobacco plants.

5.6 cDNAs AND GENES FOR TLPs

5.6.1 cDNAs

TLP clones have been isolated from cDNA libraries prepared from poly(A)-containing RNA from TMV-infected tobacco plants, salt-adapted tobacco cells, tomato roots, fungus-infected wheat, oat, barley, and rice plants, floral meristems of *Arabidopsis, Atriplex nummularia,* ripe cherries, mercuric chloride-treated maize, flower buds of *Brassica napus,* and premeiotic tomato flowers. See Table 5.1 for a list of some of the clones isolated and references. Included in the table are data for some TLPs for which cDNA clones have not yet been described. As expected, the cDNAs encode TLPs with two size ranges, consistent with the data obtained from analyses of the proteins. It is also clear that multiple TLPs are expressed even within the same plant exposed to a single pathogen, as in the case of tobacco, oat, and *Arabidopsis*. In some cases, specific cDNAs are induced in response to different pathogens as was demonstrated in the case of rice.[44,45]

5.6.2 Genes

Southern blot analyses are consistent with the presence of a small family of TLP genes in several plants. However, the interpretations are somewhat complicated by the finding that, because of substantial divergence, a single TLP cDNA probe often fails to detect all genes of the TLP family, especially when stringent conditions of hybridization are used.[43,45] Genes encoding TLPs (and related osmotin and OLPs) have been described only from tobacco and studied extensively. These studies are summarized below.

Two genes for TLPs, *E2* and *E22*, have been isolated from tobacco.[53] These correspond to the major and minor forms of tobacco TLPs.[1] They do not have introns. Deletion analysis of the promoter region of one PR-5 gene, *E2*, linked to *GUS* reporter gene, indicated that the region between –1364 and –718 is important for induction of expression of the PR-5 gene in transgenic plants in both local and systemic induction by TMV.[54] A comparison of this region of the PR-5 gene promoter with other PR-protein gene promoters failed to indicate any sequence similarities.

A gene for tobacco osmotin was isolated and characterized.[55,56] This gene also does not have an intron. Studies with transgenic plants expressing an osmotin promoter–*GUS* gene fusion indicated that this promoter is responsive to both ABA and ethylene. The promoter was much more active in the root tissues than in the shoot tissue. A more detailed study of the osmotin promoter was carried out by Ragothama et al.,[57] who demonstrated the presence of several sequence elements in the promoter region that bound to nuclear factors presumably involved in transcriptional regulation. One of these factors bound to DNA sequences that resembled the ABA-responsive sequence TAAGA/CGCCGCC found in other genes. Three distinct regions were identified in the osmotin promoter. One region from –108 to –248 was found to be essential for the gene activity. This region was responsible for regulation by ABA, ethylene, salt, desiccation, and wounding. The other two promoter elements had either an enhancer-like activity or a negative regulatory activity on the expression of the osmotin gene.

A gene for the neutral osmotin-like protein (OLP) closely related to the cDNA reported by Takeda et al.[58] was isolated from tobacco.[59] This gene also has no introns and is regulated by ethylene. The promoter region of this gene contained two copies of the sequence element AGCCGCC, which is found in several basic PR-protein genes. When transgenic tobacco plants containing the OLP promoter–*GUS* gene fusion were analyzed for the reporter gene expression, it was found that mutations in these sequences resulted in loss of ethylene-inducible GUS expression. Furthermore, EREBP2, a nuclear protein factor that specifically binds to the AGCCGCC sequence in a tobacco β-1,3-glucanase gene was also shown to bind to this sequence element in OLP gene promoter, leading the authors to suggest that ethylene-induced expression of this gene is regulated by the binding of this factor to the AGCCGCC sequence.

The regulation of osmotin gene promoter by two elicitors, arachidonic acid and a cellulase preparation from *Aspergillus niger,* was studied using osmotin promoter–*GUS* fusions.[18] The promoter was a 1.6 kb fragment from osmotin gene.[55] In transgenic tobacco seedlings containing this fusion gene, *GUS* gene expression was stimulated by either elicitor (cellulase was more potent), but together they had a synergistic effect. Similar synergism was seen with osmotin mRNA accumulation in the same plants, especially in the roots. The involvement of ethylene in the induction of osmotin gene by these elicitors was demonstrated by the finding that norbornadiene, an ethylene action inhibitor, completely blocked this induction. As reported by Ragothama et al.[57] for other environmental and hormonal stimuli, the –248 to –108 region was required for induction by arachidonic acid.

5.7 TRANSGENIC PLANTS OVER-EXPRESSING PR-5 PROTEINS HAVE ENHANCED DISEASE RESISTANCE

The isolation and characterization of cDNA and genomic clones for PR-5 proteins has led to the generation of transgenic plants expressing these proteins at high levels in several laboratories. The earliest studies involving transgenic tobacco plants expressing the tobacco PR-S cDNA under the control of CaMV 35 promoter resulted in constitutive expression of both RNA and protein.[60] However, over-expression of this protein did not result in protection against TMV infection. Further, transgenic tobacco plants with antisense constructs of this gene had reduced levels of PR-5 protein (<20% of the levels found in control plants upon TMV infection), but no significant differences in the symptoms or number of lesions were found.[60] When it was subsequently shown that zeamatin and other related proteins from cereals, as well as osmotin and osmotin-like proteins, had antifungal activity against a variety of fungi, including *P. infestans, Candida albicans, Neurospora crassa,* and *Trichoderma reesei* in *in vitro* assays,[16,26,27] attempts were made to produce transgenic plants that over-express osmotin and osmotin-like protein constitutively to study the role of osmotin in plant defense.[61,62] Transgenic tobacco and potato plants over-expressing the tobacco osmotin gene have been developed.[61] The transgenic tobacco and potato plants constitutively over-expressed osmotin to a level of 2% of total cellular proteins, which is equivalent to that induced by fungal infection in nontransformed plants. Constitutive accumulation of osmotin in transgenic potato plants led to enhanced resistance to *P. infestans,* the potato late blight pathogen, indicating that osmotin can play a defensive role during fungal infection when they are over-expressed in a heterologous system. In contrast, transgenic tobacco plants over-expressing tobacco osmotin did not exhibit any change in the development of disease symptoms when inoculated with *Phytophthora parasitica* var. *nicotianae,* which is a pathogen of tobacco plants. Purified osmotin is more effective against *P. infestans* than against *P. parasitica.* Thus, the failure of osmotin over-expression to offer protection against *P. parasitica* may be due to a lower effectiveness of osmotin against this fungus. Alternatively, the lack of effectiveness at this concentration of osmotin could be the result of pathogen/host coevolution as postulated by Lamb et al.[63] For example, constitutive, high-level expression of an acidic form of tobacco PR-5 protein in transgenic tobacco plants did not affect the colonization by the vesicular–arbuscular mycorrhizal fungus, *Glomus mosseae.*[64]

Zhu et al.[62] developed several transgenic potato plants constitutively expressing sense or antisense RNAs from chimeric gene constructs in which the CaMV 35S promoter controlled the expression of a cDNA, pA13, for an osmotin-like protein (OLP) from potato. In the antisense transformants, fungal-infection-induced accumulation of sense OLP mRNAs and the corresponding proteins (OLP) was diminished, but there was no alteration in disease susceptibility compared to controls. However, sense transgenic potato plants expressing high levels of potato OLP showed delayed development of disease symptoms when inoculated with *P. infestans.* These results point out that even when a potato OLP is over-expressed in potato, there is protection against the potato late blight pathogen. Thus, it is unlikely that the lack of protection seen against a tobacco pathogen by Liu et al.[61] in transgenic tobacco plants over-expressing tobacco osmotin is due to coevolution of host/pathogen, but is rather due to the specificity of interaction between the particular PR-5 protein being expressed and the pathogen used for the challenge.

Zhu et al.[62] tested the effect of over-expression of OLP on freezing tolerance because the accumulation of OLP mRNAs has been reported to be associated with the induction of freezing tolerance in potato cell cultures when cold acclimated at 4°C.[65] Over-expression of sense or antisense OLP RNAs had no effect on the development of freezing tolerance in transgenic plants. It is possible that specific members of the PR-5 group may have specialized functions, such as freezing tolerance, or may possess antifungal activity against selective pathogens.

Transgenic rice plants constitutively expressing a 23 kDa rice TLP at high levels have been obtained. The accumulation level of this protein varied among individual transformants, but several plants accumulated TLPs to a level of 0.2 to 0.5% of total proteins. Several lines of TLP-transformants were challenged with the sheath blight pathogen, *Rhizoctonia solani.*[66] The plants expressing high levels of TLPs had significantly fewer and smaller lesions compared to nontransgenic controls, indicating that over-expression of a rice TLP can result in protection against the spread of the disease even though the onset of disease was not prevented (see Chapter 13).

5.8 EXTRACELLULAR TARGETING OF PR-5 PROTEINS

The results of Liu et al.,[61] Zhu et al.,[62] and Datta et al.[66] discussed above indicated that over-expression of TLPs can enhance the resistance to at least some fungal pathogens. It is conceivable that the protection offered by these proteins, which normally accumulate in the vacuoles, would prevent the spread of the pathogen rather than the onset of the disease. Several groups have attempted to target these TLPs to a different cellular location by manipulation of these genes. The *AP24* gene-encoded osmotin has an N-terminal stretch of 25 amino acids which resembles a signal peptide sequence. Comparison of the amino acid sequence of mature AP24, obtained by sequencing peptide fragments of AP24 with the sequence predicted from cDNA clones, indicated that posttranslational processing of this protein involves removal of the 25 amino acid-long N-terminal signal peptide and an 18 amino acid-long C-terminal propeptide. An examination of the C-terminal amino acid sequences of vacuolar TLPs, namely, AP24 from tobacco, NP24 from tomato, and thaumatin and the extracellular PR-5 protein (PR-R) of tobacco, indicated that the intracellular PR-5 proteins contain extensions of between 8 and 21 amino acids beyond the C-terminus of the extracellular PR-5 protein. To determine the possible role of the C-terminal amino acid extension of AP24 in vacuolar sorting, wild-type and a C-terminus deletion mutant of *ap24* gene were introduced and expressed in transgenic tobacco plants. In transgenic tobacco plants expressing the wild-type construct, the encoded AP24 protein accumulated in the vacuoles. In contrast, the protein produced in plants expressing the mutant construct was secreted to an extracellular location, suggesting that the C-terminal propeptide is necessary for targeting AP24 to the vacuoles. Nevertheless, the truncated protein caused sporangial lysis of *Phytophthora infestans,* indicating that it has retained its biological activity. Similarly a C-terminal deletion of a neutral PR-5 protein (OLP) has been reported to result in secretion of OLP into the extracellular space in transgenic tobacco plants, suggesting that C-terminal sequences of several TLP proteins may be involved in vacuolar targeting,[67] as in the case of several other PR-proteins.

Liu et al.[68] generated transgenic tobacco and potato plants over-expressing an osmotin gene with a deletion of the region coding for a 20 amino acid-stretch at the C-terminus. The osmotin from the truncated osmotin gene was secreted into the extracellular matrix. The truncated osmotin purified from transgenic tobacco plants retained their antifungal activity against *Trichoderma longibrachiatum* comparable to the full-length protein. In the early stages of infection, transgenic plants expressing the osmotin extracellularly appeared to have fewer disease symptoms than the plants expressing osmotin in the vacuoles. However, at later stages, the two sets of transgenic plants were indistinguishable in their disease resistance, but both sets of transgenic plants had fewer symptoms than nontransgenic plants. Therefore, it is likely that in order to achieve a more complete inhibition of fungal pathogens, proper targeting of PR proteins and control of the timing of their expression may be needed.

5.9 DO OTHER DEFENSE PROTEINS HAVE PR-5 PROTEIN DOMAINS?

In addition to the inducible TLPs with sizes of 16 to 26 kDa, other proteins immunologically related to thaumatin with larger sizes have been detected in floral tissues. Using monospecific anti-thaumatin antibodies, Richard et al.[13] detected proteins with molecular masses of 42.6, 31.6, and 26.3 kDa in immature and mature flower organs and other proteins with sizes of 46.7, 41.7, and 27.5 kDa in thin cell layer explants forming flowers. Immunolocalization indicated that these TLPs were synthesized in newly differentiated floral meristems, tracheids, and parenchyma cells. Further evidence for the presence of larger protein(s) with sequence similarity to TLPs has come from the isolation of a gene from *Arabidopsis thaliana.*[69] During a search for a gene for S-locus receptor kinase (SRK) using a kinase domain probe related to SRK, these authors isolated a gene encoding a protein with an extracellular domain related to PR-5 proteins, a central transmembrane domain, and a C-terminal domain resembling serine/threonine kinases. This protein was 665 amino acids long, significantly larger than the typical size of about 200 amino acids for most TLPs. This gene was found to be expressed in rosette leaves and siliques and, at higher levels, in inflorescence stems and roots of *Arabidopsis*. However, this gene was not expressed in leaves of *Arabidopsis* plants inoculated with turnip crinkle virus under conditions when normal-size TLPs were induced by this virus, suggesting a role as a preexisting pathogen surveillance mechanism rather than as a defense response. The sequence similarity between the extracellular domain of this receptor kinase and PR-5 proteins suggests that a common microbial epitope may be recognized by this kinase and TLPs.[69]

5.10 OSMOTIN INTERFERES WITH A SIGNAL TRANSDUCTION PATHWAY

Studies carried out with several laboratory strains of yeast differing in their sensitivity to osmotin have indicated a role for cell wall-associated proteins, called PIR proteins, in conditioning sensitivity to osmotin. Over-expression of the PIR proteins, which were immunolocalized in the cell walls of yeast cells, was shown to result in greater resistance to osmotin.[83] Further, removal of the cell walls rendered the resulting spheroplasts from both resistant and sensitive yeast equally sensitive to osmotin, suggesting that the cell wall acts as a barrier to prevent osmotin from reaching the targets in the plasma membrane. More recent experiments have revealed that osmotin interferes with a yeast signal transduction pathway to overcome the defensive role of cell wall barrier proteins (PIR proteins).[84] The pathway involves several (STE4, STE18, STE20, STE5, STE11, STE7, FUS3, KSS1, and STE12) of the signal transduction components regulating mating pheromone response, which utilizes a heterotrimeric G-protein and a MAP kinase cascade. Deletions of the genes encoding components of the pheromone response signal transduction pathway enhanced resistance to osmotin. However, a mutation in the gene for a negative regulator of the pathway, *SST2*, resulted in supersensitivity to osmotin. Phosphorylation of STE7 was enhanced within 5 minutes of osmotin treatment, indicating that the activation of the pathway precedes cell death. These results point out an additional role of osmotin in altering signal transduction pathways of the pathogen to enhance its sensitivity. These investigators also observed that osmotin treatment resulted in dramatic vacuolization, membrane blebbing, and nuclear and cytoplasmic condensation, which are classic symptoms of apoptosis, suggesting that osmotin may be a signal for programmed cell death.

5.11 CONCLUSION

Infection of plants with pathogens often leads to an increase in the amounts of proteins belonging to several classes of PR proteins with antimicrobial activity. While PR proteins such as chitinases (PR-3 group) and β-1,3-glucanases (PR-2 group) have the potential to hydrolyze the cell wall components, chitin and β-1,3-glucan, respectively, cell walls of most agronomically important fungi in the taxonomic class *Oomycetes* have β-1,3-glucan but contain no chitin. Oomycetes fungi such as *Phytophthora cactorum, Pythium ultimum,* and *Pythium aphanidermatum* are also insensitive to a mixture of chitinase and β-1,3-glucanase.[70] The PR-5 proteins exhibit antifungal action against many fungi, including *P. infestans,* a member of the class *Oomycetes*.[26] The finding that TLPs have a membrane-permeabilizing function with selectivity to fungal pathogens makes this class of proteins very attractive as components of antifungal defenses that can be deployed against pathogens that are not controlled by other PR proteins. Greater success in controlling devastating fungal diseases may be achieved by a combination of defense strategies involving multiple groups of PR proteins which have different cellular targets. Our knowledge of how PR-5 proteins inhibit the fungal growth is still rudimentary. Much work still needs to be done to understand the mechanism of antifungal action of PR-5 proteins so that the full potential of this class of PR proteins as defense proteins can be realized.

REFERENCES

1. Pierpoint, W. S., Tatham, A. S., and Pappin, D. J. C., Identification of the virus-induced protein of tobacco leaves that resembles the sweet-tasting protein thaumatin, *Physiol. Mol. Plant Path.,* 31, 291, 1987.
2. Van Loon, L. C., Gerritsen, Y. A. M., and Ritter, C. E., Identification, purification, and characterization of pathogenesis-related proteins from virus-infected Samsun NN tobacco leaves, *Plant Mol. Biol.,* 9, 593, 1987.
3. Kauffmann, S., Legrand, M., and Fritig, B., Isolation and characterization of six pathogenesis-related (PR) proteins of Samsun NN tobacco, *Plant Mol. Biol.,* 14, 381, 1990.
4. Daniell, W. F., On the *Synsepalum dulcificum,* Decand; or miraculous berry of Western Africa, *Pharm. J.,* 11, 445, 1852.
5. Stintzi, A., Heitz T., Kauffmann, S., Legrand, M., and Fritig, B., Identification of a basic pathogenesis-related, thaumatin-like protein of virus-infected tobacco as osmotin, *Physiol. Mol. Plant Pathol.,* 38: 137, 1991.

6. Rodrigo, J., Vera, P., Frank, K., and Conejero, V., Identification of the viroid-induced tomato pathogenesis-related (PR) protein P23 as the thaumatin-like tomato protein NP24 associated with osmotic stress, *Plant Mol. Biol.,* 16, 931, 1991.
7. Dudler, R., Mauch, F., and Reimmann, C., Thaumatin-like proteins, in *Thaumatin,* Witty, M. and Higginbotham, J. D., Eds., CRC Press, Boca Raton, FL, 1994, 193.
8. Pierpoint, W. S., Jackson, P. J., and Evans, R. M., The presence of a thaumatin-like protein, a chitinase and a glucanase among the pathogenesis-related proteins of tobacco (*Solanum tuberosum*), *Physiol. Molec. Plant Pathol.,* 36, 325, 1990.
9. King, G. J., Hussey, C. E., Jr., and Turner, V. A., A protein induced by NaCl in suspension cultures of *Nicotiana tabacum* accumulates in whole plant roots, *Plant Mol. Biol.,* 7, 441, 1986.
10. LaRosa, P. C., Singh, N. K., Hasegawa, P. M., and Bressan, R. A., Stable NaCl tolerance of tobacco cells is associated with enhanced accumulation of osmotin, *Plant Physiol.,* 91, 855, 1989.
11. Koiwa, H., Sato, F., and Yamada, Y., Characterization of accumulation of tobacco PR-5 proteins by IEF-immunoblot analysis, *Plant Cell Physiol.,* 35, 821, 1994.
12. Cheong, N. E., Choi, Y. O., Kim, W. Y., Chang, S., Cho, M. J., and Lee, S. Y., Purification of an antifungal PR-5 protein from flower buds of *Brassica campestris* and cloning of its gene, *Physiol. Plantarum,* 101, 583, 1997.
13. Richard, L., Arro, M., Hoebeke, J., Meeks-Wagner, R., and Van, K. T. T., Immunological evidence of thaumatin-like proteins during tobacco floral differentiation, *Plant Physiol.,* 98, 337, 1992.
14. Vu, L. and Huynh, Q. K., Isolation and characterization of a 27-kDa antifungal protein from the fruits of *Diospyros texana, Biochem. Biophys. Res. Commun.,* 202 666, 1994.
15. Fils-Lycaon, B. R., Wiersma, P. A., Eastwell, K. C., and Sautiere, P., A cherry protein and its gene, abundantly expressed in ripening fruit, have been identified as thaumatin-like, *Plant Physiol.,* 111, 269, 1996.
16. Roberts, W. K. and Selitrennikoff, C. P., Zeamatin, an antifungal protein from maize with membrane-permeabilizing activity, *J. Gen. Microbiol.,* 136, 1771, 1990.
17. Vigers, A. J., Roberts, W. K., and Selitrennikoff, C. P., A new family of plant antifungal proteins, *Mol. Plant-Microbe Interact.,* 4, 315, 1991.
18. Chang, P. F. L., Cheah, K. T., Narasimhan, M. L., Hasegawa, P. M., and Bressan, R. A., Osmotin gene expression is controlled by elicitor synergism, *Physiol. Plant.,* 95, 620, 1995.
19. Xu, Y., Chang, P. F. L., Liu, D., Narasimhan, M. L., Raghothama, K. G., Hasegawa, P. M., and Bressan, R. A., Plant defense genes are synergistically induced by ethylene and methyl jasmonate, *Plant Cell,* 6, 1077, 1994.
20. De Vos, A., Hatada M., van der Wel, H., Krabbendam, H., Peerdeman, A. F., and Kim, S. H., Three dimensional structure of thaumatin I, an intensely sweet protein, *Proc. Natl. Acad. Sci. U.S.A.,* 82, 1406, 1985.
21. Ogata, C. M., Gordon, P. F., de Vos, A. M., and Kim, S. H., Crystal structure of a sweet tasting protein thaumatin I, at 1.65 Å resolution, *J. Mol. Biol.,* 228, 893, 1992.
22. Van der Wel, H. and Loeve, K., Isolation and characterization of thaumatin I and II, the sweet tasting proteins from *Thaumatococcus daniellii* Benth, *Eur. J. Biochem.,* 31, 221, 1972.
23. Batalia, M. A., Monzingo, A. F., Ernst, S., Roberts, W., and Robertus, J. D., The crystal structure of the antifungal protein zeamatin, a member of the thaumatin-like, PR-5 protein family, *Nature Structural Biol.,* 3, 19, 1996.
24. Hellekant, G., The physiology of thaumatin, in *Thaumatin,* Witty, M. and Higginbotham, J, Eds. CRC Press, Boca Raton, FL, 1994, 99.

25. Kim, S. H., de Vos, A., and Ogata, C., Crystal structures of two intensely sweet proteins, *Trends in Biochemical Sciences,* 13, 13, 1988.
26. Woloshuk, C. P., Muelenhoff, J. S., Sela-Buurlage, M., Elzen, P. J. M., and Cornelissen, B. J. C., Pathogen-induced proteins with inhibitory activity towards *Phytophthora infestans, Plant Cell,* 3, 619, 1991.
27. Vigers, A. J., Wiedemann, S., Roberts, W. K., Legrand, M., Selitrennikoff, C. P., and Fritig, B., Thaumatin-like pathogenesis-related proteins are antifungal, *Plant Sci.,* 83, 155, 1992.
28. Hejgaard, J., Jacobsen, S., and Svendsen, I., Two antifungal thaumatin-like proteins from barley grain, *FEBS Letters,* 291, 127, 1991.
29. Stintzi, A., Heitz, T., Prasad, V., Wiedmann-Merdinoglu, Kauffmann, S., Geoffroy, P., Legrand, M., and Fritig, B., Plant "Pathogenesis-related" proteins and their role in defense against pathogens, *Biochimie.*, 75, 687, 1993.
30. Abad, L. R., D'Urzo, M. P., Liu, D., Narasimhan, M. L., Reuveni, M., Zhu, J. K., Niu, X., Singh, N. K., Hasegawa, P. M., and Bressan, R. A., Antifungal activity of tobacco osmotin has specificity and involves plasma membrane permeabilization, *Plant Sci.,* 118, 11, 1996.
31. Hon, W. C., Griffith, M., Mlynarz, A., Kwok, Y. C., and Yang, D. S. C., Antifreeze proteins in winter rye are similar to pathogenesis-related proteins, *Plant Physiol.,* 109, 879, 1995.
32. Richardson, M., Valdes-Rodriguez, S., and Blanco-Labra, A., A possible function for thaumatin and a TMV-induced protein suggested by homology to a maize inhibitor, *Nature*, 327, 432, 1987.
33. Huynh, Q. K., Borgmeyer, J. R., and Zobel, J. F., Isolation and characterization of a 22 kDa protein with antifungal properties from maize seeds, *Biochem. Biophys. Res. Commun.,* 182, 1, 1992.
34. Pierpoint, W. S., The pathogenesis-related proteins of tobacco leaves, *Phytochemistry,* 25, 1595, 1986.
35. Sehgal, O. P. and Mohammed, F., Pathogenesis-related proteins, in *Plant Viruses,* Vol. II., Mandahar, C. L., Ed., CRC Press, Inc., Boca Raton, FL, 1990, 65.
36. Sehgal, O. P., Rieger, R., and Mohammed, F,. Induction of bean PR-4d-type protein in divergent plant species after infection with tobacco ring spot virus and its relationship with tobacco PR-5, *Phytopathology,* 81, 215, 1991.
37. Ruiz-Medrano, R., Jimenez-Moraila, B., Herrera-Estrella, L., and Rivera-Bustamante, R. F., Nucleotide sequence of an osmotin-like cDNA induced in tomato during viroid infection, *Plant Mol. Biol.,* 20, 1199, 1992.
38. Dempsey, D. A., Wobbe, K. K., and Klessig, D. F., Resistance and susceptible responses of *Arabidopsis thaliana* to turnip crinkle virus, *Phytopathology,* 83, 1021, 1993.
39. Bryngelsson, T. and Green, B., Characterization of a pathogenesis-related, thaumatin-like protein isolated from barley challenged with an incompatible race of mildew, *Physiol. Mol. Plant Pathol.,* 35, 45, 1989.
40. Rebmann, G., Mauch, F., and Dudler, R., Sequence of a wheat cDNA encoding a pathogen-induced thaumatin-like protein, *Plant Mol. Biol.,* 17, 283, 1991.
41. Uknes, S., Mauch-Mani, B., Moyer, M., Potter, S., Williams, S., Dincher, S., Chandler, D., Slusarenko, A., Ward, E., and Ryals, J., Acquired resistance in *Arabidopsis, Plant Cell,* 4, 645, 1992.
42. Hu, X. and Reddy, A. S. N., Cloning and expression of a PR5-like protein from *Arabidopsis*: inhibition of fungal growth by bacterially expressed protein, *Plant Molec. Biol.,* 34, 949, 1997.

43. Lin, K. C., Bushnell, W. R., Szabo, L. J., and Smith, A. G., Isolation and expression of a host response gene family encoding thaumatin-like proteins in incompatible oat-stem rust fungus interactions, *Mol. Plant Microbe Interact.*, 9, 511, 1996.
44. Reimmann, C. and Dudler, R., cDNA cloning and sequence analysis of a pathogen-induced thaumatin-like protein from rice (*Oryza sativa*), *Plant Physiol.*, 101, 1113, 1993.
45. Velazhahan, R., Cole, K. C., Anuratha, C. S., and Muthukrishnan, S., Induction of thaumatin-like proteins (TLPs) in *Rhizoctonia solani*-infected rice and characterization of two new cDNA clones, *Physiol. Plant*, 102, 21, 1998.
46. Singh, N. K., Handa, A. K., Hasegawa, P. M., and Bressan, R. A., Proteins associated with adaptation of cultured tobacco cells to NaCl, *Plant Physiol.*, 79, 126, 1985.
47. Singh, N. K., Bracker, C. A., Hasegawa, P. M., Handa, A. K., Buckel, S., Hermodson, M. A., Pfankock, E., Regnier, F. E., and Bressan, R. A., Characterization of osmotin, *Plant Physiol.*, 85, 529, 1987.
48. Singh, N. K., Nelson, D. E., Kuhn, D., Hasegawa, P. M., and Bressan, R. A., Molecular cloning of osmotin and regulation of its expression by ABA and adaptation to low water potential, *Plant Physiol.*, 90, 1096, 1989.
49. Grillo, S., Leone, A., Xu, Y., Tucci, M., Francione, R., Hasegawa, P. M., Monti, L., and Bressan, R. A., Control of osmotin gene expression by ABA and osmotic stress in vegetative tissues of wild-type and ABA-deficient mutants of tomato, *Physiol. Plant.*, 93, 498, 1995.
50. Campillo, E. and Lewis, L. N., Identification and kinetics of accumulation of proteins induced by ethylene in bean abscission zones, *Plant Physiol.*, 98, 955, 1992.
51. Neale, A. D., Wahleithner, J. A., Lund, M., Bonnett, H. T., Kelly, A., Meeks-Wagner, D. R., Peacock, W. J., and Dennis, E. S., Chitinase, β-1,3-glucanase, osmotin, and extensin are expressed in tobacco explants during flower formation, *Plant Cell*, 2, 673, 1990.
52. Brederode, F. T., Linthorst, H. J. M., and Bol, J. F., Differential induction of acquired resistance and PR gene expression in tobacco by virus infection, ethephon treatment, UV light and wounding, *Plant Mol. Biol.*, 17, 1117, 1991.
53. Van Kan, J. A. L., Van de Rhee, M. D., Zuidema, D., Cornelissen, B. J. C., and Bol, J. F., Structure of tobacco genes encoding thaumatin-like proteins, *Plant Mol. Biol.*, 12, 153, 1989.
54. Albrecht, H., van de Rhee, M. D., and Bol, J. F., Analysis of *cis*-regulatory elements involved in induction of a tobacco PR-5 gene by virus infection, *Plant Mol. Biol.*, 18, 155, 1992.
55. Nelson, D. E., Ragothama, K. G., Singh, N. K., Hasegawa, P. M., and Bressan, R. A., Analysis of structure and transcriptional activation of an osmotin gene, *Plant Mol. Biol.*, 19, 577, 1992.
56. Melchers, L. S., Sela-Buurlage, M. B., Vloemans, S. A., Woloshuk C. P., Van Roekel, J. S. C., Pen, J., van den Elzen, P. J. M., and Cornelissen, B. J. C., Extracellular targeting of the vacuolar tobacco proteins AP24, chitinase and β-1,3-glucanase in transgenic plants, *Plant Mol. Biol.*, 21, 583, 1993.
57. Ragothama, K. G., Liu, D., Nelson D. E., Hasegawa, P. M., and Bressan, R. A., Analysis of an osmotically regulated pathogenesis-related osmotin gene promoter, *Plant Mol. Biol.*, 23, 1117, 1993.
58. Takeda, S., Sato, F., Ida, K., and Yamada, Y., Nucleotide sequence of a cDNA for osmotin-like protein from cultured tobacco cells, *Plant Physiol.*, 97, 844, 1991.

59. Sato, F., Kitajima, S., Koyama, T., and Yamada, Y., Ethylene-induced gene expression of osmotin-like protein, a neutral isoform of tobacco PR-5, is mediated by the AGC-CGCC cis-sequence, *Plant Cell Physiol.,* 37, 249, 1996.
60. Linthorst, H. J. M., Meuwissen, R. L., Kauffmann, S., and Bol, J. F., Constitutive expression of pathogenesis-related PR-1, GRP and PR-S in tobacco has no effect on virus infection, *Plant Cell,* 1, 285, 1989.
61. Liu, D., Ragothama, K. G., Hasegawa, P. M., and Bressan, R. A., Osmotin overexpression in potato delays development of disease symptoms, *Proc. Natl. Acad. Sci. U.S.A.,* 91, 1888, 1994.
62. Zhu, B., Chen, T. H. H., and Li, P. H. Analysis of late-blight disease resistance and freezing tolerance in transgenic potato plants expressing sense and antisense genes for an osmotin-like protein, *Planta,* 198, 70, 1996.
63. Lamb, C. J., Ryals, J. A., Ward, E. R., and Dixon, R. A., Emerging strategies for enhancing crop resistance to microbial pathogens, *Bio/Technology,* 10, 1436, 1992.
64. Vierheilig, H., Alt, M., Lange, J., Gut-Rella, M., Wiemken, A., and Boller, T., Colonization of transgenic tobacco constitutively expressing pathogenesis-related proteins by the vesicular-arbuscular mycorrhizal fungus *Glomus mosseae, Appl. Environ. Microbiol.,* 61, 3031, 1995.
65. Zhu, B., Chen, T. H. H., and Li, P. H., Expression of an ABA-responsive osmotin-like gene during the induction of freezing tolerance in *Solanum commersonii, Plant Mol. Biol.,* 21, 729, 1993.
66. Datta, K., Tu, J., Alam, M. F., Oliva, N., Vasquez, A., Abrigo E., Mew, T., Muthukrishnan, S., and Datta, S. K., Expression of chitinase and thaumatin like protein gene in transgenic rice for sheath blight resistance (Abstract), in General Meeting of the International Program on Rice Biotechnology, Malacca, Sept. 15–17, 1997, 353.
67. Sato, F., Koiwa, H., Sakai, Y., Kato, N., and Yamada, Y., Synthesis and secretion of tobacco neutral PR-5 protein by transgenic tobacco and yeast, *Biochem. Biophys. Res. Commun.,* 211, 909, 1995.
68. Liu, D., Rhodes, D., Paino D'Urzo, M., Xu, Y., Narasimhan, M. L., Hasegawa, P. M., Bressan, R. A., and Abad, L., *In vivo* and *in vitro* activity of truncated osmotin that is secreted into the extracellular matrix, *Plant Sci.,* 121, 123, 1996.
69. Wang, X., Zafian, P., Chowdhary, M., and Lawton, M., The PR5K receptor protein kinase from *Arabidopsis thaliana* is structurally related to a family of plant defense proteins, *Proc. Natl. Acad. Sci. U.S.A.,* 93, 2598, 1996.
70. Mauch, F., Mauch-Mani, B., and Boller, T., Antifungal hydrolases in pea tissue. II. Inhibition of fungal growth by combinations of chitinase and β-1,3-glucanase, *Plant Physiol.,* 88, 936, 1988.
71. Iyengar, R. B., Smits, P., Van der Ouderaa, F., Van der Wel, H., Brouwershaven, J., Ravestein, P., Richters, G., and Van Wassenaar, P. D., The complete amino acid sequence of the sweet protein thaumatin I, *Eur. J. Biochem.,* 96, 193, 1979.
72. Cornelissen, B. J. C., Hooft van Huijsduijnen, R. A. M., and Bol, J. F., A tobacco mosaic virus-induced tobacco protein is homologous to the sweet-tasting protein thaumatin, *Nature,* 321, 531, 1986.
73. Payne, G., Middlesteadt, W., Williams, S., Desai, N., Parks, T. D., Dincher, S., Carnes, M., and Ryals, J., Isolation and nucleotide sequence of a novel cDNA clone encoding the major form of pathogenesis-related protein R, *Plant Mol. Biol.,* 11, 223, 1988.
74. King, G. J., Turner, V. A., Hussey, C. E., Jr., Wurtele, E. S., and Lee, S. M., Isolation and characterization of a tomato cDNA clone which codes for a salt-induced protein, *Plant Mol. Biol.,* 10, 401, 1988.

75. Hu, X. and Reddy, A. S. N., Nucleotide sequence of a cDNA cone encoding a thaumatin-like protein from *Arabidopsis, Plant Physiol.,* 107, 305, 1995.
76. Chen, R., Wang, F., and Smith, A. G., A flower-specific gene encoding an osmotin-like protein from *Lycopersicon esculentum, Gene,* 179, 301, 1996.
77. Malehorn, D. E., Borgmeyer, J. R., Smith, C. E., and Shah, D. M., Characterization and expression of an antifungal Zeamatin-like protein (Zlp) gene from *Zea mays, Plant Physiol.,* 106, 1471, 1994.
78. Casas, A. M., Nelson, D. E., Raghothama, K. G., Paino D' Urzo, M., Singh, N. K., Bressan, R. A., and Hasegawa, P. M., Expression of osmotin-like genes in the halophyte *Atriplex nummularia* L., *Plant Physiol.,* 99, 329, 1991.
79. Hahn, M., Jungling, S., and Knogge, W., Cultivar-specific elicitation of barley defense reactions by the phytotoxic peptide NIP1 from *Rhynchosporium secalis, Molec. Plant Microbe Interactions,* 6, 745, 1993.
80. Pierpoint, W. S., Jackson, P. J., and Evans, R. M., The presence of a thaumatin-like protein, a chitinase and a glucanase among the pathogenesis proteins of potato (*Solanum tuberosum*), *Physiol. Molec. Plant Pathol.,* 36, 325, 1990.
81. Tattersall, D. B., van Heeswijck, R., and Hoj, P. B., Identification and characterization of a fruit-specific, thaumatin-like protein that accumulates at very high levels in conjunction with the onset of sugar accumulation and berry softening in grapes, *Plant Physiol.,* 114, 759, 1997.
82. Graham, J. S., Burkhart, W., Xiong, J., and Gillikin, J. W., Complete amino acid sequence of soybean leaf P21, *Plant Physiol.,* 98, 163, 1992.
83. Yun, D.-J., Zhao, Y., Pardo, J. M., Narasimhan, M. L., Damsz, B., Lee, H., Abad, L. R., D'Urzo, M. P., Hasegawa, P. M., and Bressan, R. A., Stress proteins on the yeast cell surface determine resistance to osmotin, a plant antifungal protein, *Proc. Natl. Acad. Sci. U.S.A.,* 94, 7082, 1997.
84. Yun, D.-J., Ibeas, J. I., Lee, H., Coca, M. A., Narasimhan, M. L., Uesono, Y., Hasegawa, P. M., Pardo, J. M., and Bressan, R. A., Osmotin, a plant antifungal protein, subverts signal transduction to enhance fungal cell susceptibility, *Molec. Cell,* 1, 807, 1998.

6 The PR-6 Family: Proteinase Inhibitors in Plant-Microbe and Plant-Insect Interactions

Thierry Heitz, Pierette Geoffroy, Bernard Fritig, and Michael Legrand

CONTENTS

6.1 Introduction 131
6.2 Occurrence and Structure of Plant Proteinase Inhibitors with Potential Defensive Capabilities 133
 6.2.1 Serine Proteinase Inhibitors 135
 6.2.2 Cysteine Proteinase Inhibitors 135
 6.2.3 Aspartyl- and Metallo-Proteinase Inhibitors 136
6.3 Proteinases and Proteinase Inhibitors in Plant–Microbe Interactions 136
 6.3.1 Induction of Proteinase Inhibitors in Infected Plants 136
 6.3.2 Proteinase Inhibitors Active on Microbial Proteinases 137
6.4 Proteinases and Proteinase Inhibitors in Plant–Insect Interactions 139
 6.4.1 Induction of Proteinase Inhibitors by Herbivore Insects and/or Mechanical Wounding 139
 6.4.2 Proteinase Inhibitors Active on Insect Digestive Proteinases 140
6.5 Regulation of Proteinase Inhibitors as Inducible Plant Defense Proteins 143
6.6 Conclusion and Future Prospects 147
References 148

6.1 INTRODUCTION

Pathogenesis-related (PR) proteins were first defined as "proteins encoded by the host plant but induced only in pathological or related situations."[1] After being originally detected in soluble extracts of tobacco leaves reacting hypersensitively to tobacco mosaic virus (TMV),[2,3] 10 major acidic PR proteins were isolated from this material and grouped into five families on the basis of their relative mobility on native alkaline polyacrylamide gels and on their serological relationships.[4] These

0-8493-0697-3/99/$0.00+$.50

five major groups were then found to generally occur in many plant species, including dicots and monocots. The biological function of PR proteins has remained mysterious for 15 years, but extensive investigation in the past decade has demonstrated the antimicrobial capabilities of this class of defense proteins. For a detailed description of the structures and properties of PR proteins, their role in plant defense, and the regulation of their genes, the reader is referred to the introductory chapter of this volume and to several comprehensive reviews published in this field.[5-7]

As more work was being conducted to characterize novel inducible defense proteins, enzymatic proteins were isolated whose activities and primary structures did not fit into the framework of the five families of PRs defined initially for the tobacco model system. Yet, their biological activities and induction patterns suggested these novel proteins may work in concert with the well-established PR proteins to protect plants from invading pathogens. A definition of additional families of PR proteins was needed to accommodate the newly characterized proteins and complete the description of the antimicrobial repertoire of plant defense proteins.

The recommendations allowing classification of PR proteins were updated, and the definition of a PR protein has been broadened.[8] The main criterion for a protein to qualify as a PR remains the inducibility of its gene in pathological or related situations. Pathological situations "refer to all types of infected states, including parasitic attack by nematodes, phytophagous insects, and herbivores."[8] Related situations include "the application of chemicals that mimic the effect of pathogen infection or induce some aspects of the host response, as well as wound responses that give rise to proteins that are also induced during infections."[8] This latter aspect will be important in this chapter. We feel that the physicochemical properties and quantitative aspects of PR proteins, on which their initial detection and isolation were based, should be taken into account in addition to inducibility for defining a protein as pathogenesis related. These properties include a high stability at low pH and a remarkable resistance to endogenous and exogenous proteinases, two characteristics which make these proteins well adapted to the harsh conditions prevailing in the cellular compartments where they occur — the vacuole, the cell wall, and/or the apoplast.[6]

Proteinase inhibitors (PIs) are highly stable defensive proteins of plant tissues that are both developmentally regulated and induced in response to insect and pathogen attacks.[9] They were first isolated from seeds of *Leguminosae, Gramineae,* and *Solanaceae* (for an early review, see reference 10) before the discovery of PR proteins. More recently, novel PIs with defensive capabilities were described in many other plant families.[9]

In 1994, the family 6 of PR proteins was created along with five new families,[8] and the category was allotted to a well-characterized family of PIs first described by Green and Ryan.[11] By definition, a PR protein family groups proteins with extensive sequence similarities, serological relationships, and/or the same biological activity.[8] Similarly, the PR-6 family was created for PIs related to the "tomato inhibitor I," which was the most extensively characterized inhibitor and for which homologs have been found in several plant species.[12] The definition of the PR-6 family is debatable, since it is limited to a sole subclass of inhibitors of serine proteinases. As it will appear below, an examination of the PIs potentially involved in plant defense shows that members of virtually all known types of plant proteinase inhibitors can interact with proteinases from plant-attacking organisms. This review is thus not restricted to the tomato/potato I subclass of inhibitors but attempts to give an update on our knowledge on PIs of different subclasses implicated in plant–microbe and plant–insect interactions. However, it does not list exhaustively the early work in the field, nor does it consider any particular plant PI. Rather, it focuses on recent examples in which data on gene induction, inhibition of pathogen proteinases, or enhanced resistance of transgenic plants suggest a role in plant defense.

In the context of PR proteins, several aspects of the biology of proteinase inhibitors distinguish them from the "historic" PR protein families. As already pointed out, families 1 to 5 group proteins with marked sequence similarity, whereas PIs form structurally unrelated subclasses of proteins with various regulations, but having in common the property to bind proteinases. Furthermore, PIs control proteinase activity, a general biochemical function likely involved in many physiological processes during the development of a plant. The involvement of PIs in defense may therefore only represent one aspect of their function in the plant cell.

Pathogenic organisms colonizing plant tissues rely on a set of proteinases as part of their virulence factor. These proteinases belong to the four widespread mechanistic classes of proteolytic enzymes, i.e. serine-, cysteine-, aspartic-, and metallo-proteinases. In parallel, plants have evolved genes encoding inhibitors that inactivate some of these proteinases and thus may reduce the ability of the aggressor to digest host proteins and therefore limit the availability of amino acid source for the invader. This is considered part of the continuous battle taking place during the infection process where the two living organisms try to overcome each other's lytic enzymes. Other examples include the polygalacturonase-inhibiting proteins secreted by plants to prevent hydrolysis of pectic components of their cell walls by fungal polygalacturonases.[13] Recently, evidence was presented that fungal pathogens secrete proteins that selectively inhibit some plant endo-β-1,3-glucanases of the PR-2 family.[14] Further work may reveal other enzyme-inhibitor systems whose balance could influence the outcome of attempted infections.

As already pointed out, plants have evolved isoforms of proteinase inhibitors that have been proven to efficiently inactivate proteinases from microbes, parasites, or insects, and these can be considered genuine defense proteins. The following paragraphs will review recent data on the involvement of PIs in natural plant defense.

6.2 OCCURRENCE AND STRUCTURE OF PLANT PROTEINASE INHIBITORS WITH POTENTIAL DEFENSIVE CAPABILITIES

While more than 10 unrelated subclasses of proteinase inhibitors are known in plants,[9] data clearly suggesting a role in defense responses are available only in a limited number of cases. In this chapter, we chose to review examples of PIs which were recently shown to be active on microbial or insect proteinases, or whose genes are induced in infected or wounded plants. Other interesting examples will be

TABLE 6.1
Examples of Plant Proteinase Inhibitors with Potential Roles in Defense Against Pathogens

Proteinase type	Proteinase inhibitor (PI) type	PIs induced in plant–pest interactions	PIs active against microbial proteinases	PIs active against insect proteinases
SERINE	Potato I	Green and Ryan[11]	Geoffroy et al.[29]	Markwick et al.[72]
		Geoffroy et al.[29]		
		Pautot et al.[27]		
		Heitz et al.[30]		
		Cordero et al.[31]		
	Potato II	Pautot et al.[27]		Johnson et al.[81]
		Rickauer et al.[116]		Duan et al.[84]
		Pearce et al.[21]		
		Balandin et al.[18]		
		Gadea et al.[32]		
	Bowman-Birk	Rohrmeier and Lehle[62]	Terras et al.[49]	Ceciliani et al.[73]
		McGurl et al.[64]	Marchetti et al.[45]	Hilder et al.[80]
		Snowden et al.[63]	Lilley et al.[52]	Xu et al.[83]
		Stevens et al.[33]		Santos et al.[91]
	Kunitz	Bradshaw et al.[65]	Marchetti et al.[45]	
		Ishikawa et al.[59]		
		Saarikoski et al.[66]		
		Yeh et al.[67]		
		Suh et al.[126]		
	Uncharacterized	Peng and Black[26]	Roby et al.[28]	Macintosh et al.[90]
		Roby et al.[28]	Lorito et al.[47]	
CYSTEINE		Hildmann et al.[57]	Kondo et al.[36]	Hines et al.[77]
		Botella et al.[68]	Urwin et al.[53]	Orr et al.[78]
			Atkinson et al.[127]	Bolter and Jongsma[88]
			Lilley et al.[52]	Irie et al.[85]
				Kuroda et al.[76]
				Zhao et al.[79]
ASPARTIC		Hansen and Hannapel[58]		
		Hildmann et al.[57]		
METALLO-		Graham and Ryan[69]		Markwick et al.[72]
		Martineau et al.[56]		

considered in which inhibitors isolated from seeds or tubers can efficiently inactivate proteinases from pests, either in artificial diets or when overexpressed in transgenic plants. In addition, such genes encoding seed- or tuber-specific PIs were shown in several instances to be wound- or pathogen-inducible in aerial parts of plants.

Members of the serine and, to a lesser extent, the cysteine proteinase inhibitor families show a wide diversity in structure and regulation, while aspartic acid and metalloprotease inhibitors are the less documented classes. One possible explanation may be that plant pathogens rely mainly on serine and cysteine proteinases to digest plant proteins and that during the course of plant–parasite coevolution, plants have acquired several means of inhibiting such proteinases. Table 6.1 summarizes the classes of proteinase inhibitors considered in this review.

6.2.1 Serine Proteinase Inhibitors

Potato I type inhibitors are small 8 kDa proteins possessing one reactive site, which shows a great variability, thus allowing them to react with many different microbial or animal serine proteinases, such as trypsin, chymotrypsin, or subtilisin, which differ by their cleavage specificity. Laskowski et al.[15] showed that the "reactive sites" of serine proteinase inhibitors are mutating faster than amino acids in the rest of the proteins, implying that their roles in defense against pathogens may exert a strong selective pressure to conserve the reactive sites of proteinase inhibitors. In contrast, one may argue that a high mutating rate in the active site facilitates the adaptation to changes introduced in the target proteinase.

Potato II type inhibitors form an interesting subclass because they present a variable number of homologous active domains. The "historical" members which defined this subclass were isolated from tomato and potato and possessed two active sites, one directed against trypsin and the other against chymotrypsin. Their mature forms consist of 12 kDa proteins with dual specificity. Smaller inhibitory peptides from potato bearing a single catalytic domain are also known.[16] Recently, cDNAs for new members were isolated from tobacco and tomato and were shown to encode proteins with three putative inhibitory domains.[17-19] Finally, a cDNA encoding a protein with homology to tomato PI-II was isolated from *Nicotiana alata* stigmas.[20] The predicted protein contains six repeated domains, each with a potential reactive site. Homologous peptides were isolated from wounded tobacco leaves[21] and may arise from a multimeric precursor after proteolytic processing. The high structural diversity of this gene family has likely evolved from duplication of an ancestral gene. Thus, it appears that plants have acquired the capacity to synthesize large precursors to increase the amounts of inhibitory peptides that can be rapidly released when needed. This may have been necessitated by the fact that one molecule of inhibitor irreversibly interacts with one proteinase molecule: thus, the molecular ratio has to be in favor of the inhibitor to achieve total inhibition of proteinase activity.

Two other subclasses of serine proteinase inhibitors have been studied in detail. Kunitz type and Bowman–Birk type inhibitors, both isolated in 1946 from soybean and homologs have since been characterized from many plant species.[22] Kunitz-type inhibitors are single headed 20 to 22 kDa proteins. Bowman–Birk inhibitors are 8 to 16 kDa proteins and are composed of one or two active sites (single- or double-headed inhibitors) in monocots, and two sites in dicot plants.[23]

6.2.2 Cysteine Proteinase Inhibitors

Inhibitors of cysteine proteinases are called "cystatins" and are proteins in the 12 to 15 kDa molecular mass range which define three distinct subclasses, some of which resemble inhibitors from animals.[24] One exception is a potato 85 kDa tuber protein that contains eight homologous cystatin domains.[25] Cystatins have been found in many plant species and their presence may be ubiquitous.[9]

6.2.3 Aspartyl- and Metallo-Proteinase Inhibitors

These two latter classes have not been analyzed extensively, and members have been described only from tomato and potato.[9] Inhibitors of metallo-proteinases isolated from plants inactivate exo-peptidases like pancreatic carboxypeptidases A and B.

6.3 PROTEINASES AND PROTEINASE INHIBITORS IN PLANT–MICROBE INTERACTIONS

6.3.1 Induction of Proteinase Inhibitors in Infected Plants

Induction of proteinase inhibitor activity was described in a number of plant–pathogen interactions. Peng and Black[26] first described an increase in trypsin and chymotrypsin inhibitory activity in tomato leaves infected with the fungus *Phytophthora infestans,* the induction being stronger in resistant than in susceptible cultivars. The molecular species responsible for the activity were not determined, but in a more recent study, Pautot et al.[27] demonstrated the accumulation of serine proteinase inhibitor I and II mRNAs in leaves of both disease-susceptible and disease-resistant cultivars of tomato infected by the bacterial pathogen *Pseudomonas syringae* pv. *tomato*. Roby et al.[28] described the increase of PI activity in melon plants infected by the fungal pathogen *Colletotrichum lagenarium.* The partially purified fraction was found to be active against a proteinase produced by the fungus and also against trypsin. The microbial proteinase inhibitor of tobacco is one of the best characterized pathogen-induced proteinase inhibitors. A strong inhibitory activity is induced in tobacco during the hypersensitive reaction to tobacco mosaic virus (TMV). The purified inhibitor is an 8 kDa protein belonging to the potato inhibitor I family and displays a strict specificity for serine proteinases of microbial origin, such as subtilisin or proteinase K.[29] Two closely related cDNAs were cloned and shown to encode two active inhibitors.[30] Both mRNAs and proteins accumulate in tobacco in response to TMV with the same kinetics as those described for the "initial" 5 families of PR proteins, and this property contributed, along with their overall stability, to the inclusion of PIs as a novel family of PR proteins.[8] TMV-infected tobacco leaves also express a gene with sequence similarity to potato II type PIs.[18] In maize germinating embryos, a PI mRNA of the potato I family accumulates in response to infection by *Fusarium moniliforme* in parallel to a PR-1 mRNA. However, the two gene families were shown to be activated by different plant signaling molecules.[31] In a search for tomato genes induced by citrus exocortis viroid that are also ethylene responsive, Gadea et al.[32] isolated a cDNA coding for a three-domain inhibitor of the potato II type which is similar in sequence and structure to a TMV-induced mRNA.[18] Finally, a gene that is related to the Bowman–Birk type of serine PIs is strongly induced along with other PR genes during the incompatible interaction of barley seedlings with the fungus *Stagonospora nodorum.*[33]

It appears clearly from this overview that our knowledge of the contribution of proteinase inhibitors in pathogen-induced plant defense is still fragmentary. In many cases in which an inhibitory activity has been detected, the inhibitor protein has not been isolated and characterized, nor has its specificity been precisely addressed. Moreover, the cognate target proteinase is generally unknown. The search for the inhibitor target is usually hampered by the fact that the specific recognition of a pathogen by the plant triggers the activation of a very broad spectrum of antimicrobial responses leading ultimately to the hypersensitive reaction. Some of these responses, including proteinase inhibitors, may not find a matching target in every inducing pathogen and, therefore, their potential role in defense cannot be easily assessed.

6.3.2 Proteinase Inhibitors Active on Microbial Proteinases

To colonize host plant tissues, microbial pathogens rely on a set of hydrolytic enzymes that are coordinately expressed during the course of infection. The specific involvement of proteinases in these processes is difficult to evaluate, since the relative importance of proteolytic activity in pathogenicity may be highly variable and when required, proteinase activity may be involved at specific steps of the infection cycle. Some of the microbial proteinases are only produced *in planta,* and very few have been characterized in detail. In the replication cycle of some plant viruses encoding polyproteins, proteolytic processing is required to release functional proteins and is catalyzed by enzymes encoded by the viral genome. For instance, comoviruses encode a single proteinase,[34] whereas some potyviruses encode three proteinases belonging to several classes.[35] Viral proteinases usually present a strict substrate specificity for motifs present in the cognate precursor to be cleaved. Although they are indirectly required for the replication and spread of these viruses, they are not considered pathogenicity factors degrading structures of the host. Interestingly, oryzacystatin, a cysteine proteinase inhibitor from rice seeds, was found to reduce poliovirus replication in cell cultures and herpes simplex virus replication in infected mice.[36,37] However, these studies did not determine if the antiviral effect was the result of viral cysteine proteinase inhibition. In fact, plant proteinase inhibitors induced in response to a viral infection and blocking the viral cycle by inhibiting polyprotein processing of the inducing virus have not yet been described.

Bacterial and fungal plant pathogens secrete proteinases that are generally not individually required for growth on synthetic media, but may contribute to their aggressiveness when colonizing plant tissues. Extracellular enzymes (pectin-degrading enzymes, cellulases, proteinases) secreted by phytopathogenic bacteria have been studied mainly in the genera *Erwinia* and *Xanthomonas*. Bacterial mutants pleiotropically defective in the export of plant cell wall depolymerizing enzymes are essentially nonpathogenic on the different plant species tested.[38,39] Concerning proteolytic enzymes in particular, *Erwinia carotovora* and *E. chrysanthemi* species secrete mainly metallo-proteinases whose roles in pathogenicity have not been individually tested and for which no plant inhibitor is known.[40] The role of an extracellular proteinase has been studied in *Xanthomonas campestris* pv. *oryzae*, the agent of rice leaf blight, by transposon mutagenesis. The populations of a proteinase-defective mutant in rice plants were 10- to 100-fold smaller than those observed for the parental strain, suggesting an active role for proteinase in this disease.[41] Similarly, a proteinase-deficient mutant of *Xanthomonas campestris* pv. *campestris*, the black rot pathogen, which lacked a serine- and a metallo-proteinase activity, showed considerable loss of virulence in pathogenicity tests when bacteria were introduced into mature turnip leaves through cut vein endings.[38]

Relatively few studies have characterized the physiological roles of proteinases secreted by fungal phytopathogens, for which efficient genetic transformation systems are frequently lacking. The early steps of the colonization of several plant host species by *Botrytis cinerea* imply an aspartic proteinase. When fungal spores were pretreated with pepstatin, a synthetic inhibitor of this class of enzymes, the level of infection was markedly reduced.[42] Remarkable data were also obtained for the interaction between *Pyrenopeziza brassicae* and oilseed rape. An apparent strict correlation exists between pathogenicity levels on detached cotyledons and extracellular cysteine proteinase production *in vitro.*[43] When a proteinase-deficient mutant was transformed with a genomic library from wild-type *P. brassicae,* a transformant was obtained which showed concomitant restoration of pathogenicity and proteolytic activity. However, studies on different pathosystems suggest that some protease-deficient mutants maintain a normal aggressiveness level when compared to the wild-type strain on a natural plant host species.[44]

In addition to the limited knowledge of the role of microbial proteinases themselves in pathogenicity, the effect of plant PIs on the activity of enzymes from microbial pathogens has been the subject of relatively little research. Several early reports describe inhibitors that specifically inhibit proteolytic enzymes from microorganisms and not digestive proteases of animals.[9] These inhibitors are common in legume seeds, and although they do not meet the inducibility criterion to define them as PR proteins, they deserve interest because of their demonstrated defensive capabilities. The well-characterized soybean Kunitz and Bowman–Birk inhibitors were recently found to inhibit proteinases from nonpathogenic fungi and bacteria, in addition to being active on trypsin, chymotrypsin, and elastase.[45] Inhibitors isolated from kidney bean seeds specifically inactivate a serine proteinase from the fungal pathogen *Colletotrichum lindemuthianum,* but they are inactive against trypsin.[46] A preparation of four trypsin and chymotrypsin inhibitors from cabbage foliage whose subclass was not determined showed antifungal activity in an *in vitro* bioassay. This preparation inhibited spore germination and germ tube elongation and also elicited cytoplasmic leakage in the pathogenic fungi *Botrytis cinerea* and *Fusarium solani.*[47] However, the effect was not observed for *Alternaria brassicicola,* a cabbage pathogen; nor was it seen with soybean trypsin inhibitor tested under the same conditions on the three fungal species. These data indicate that a high specificity exists in the interactions of the plant PIs with the fungal enzymes. According to the authors, one explanation for the observation of cytoplasmic leakage would be that the cabbage inhibitors may block the synthesis of chitin in the fungal cell wall. It has been reported that chitin synthase, an important enzyme for fungal growth, is produced as a zymogen and is activated by endogenous trypsin.[48] Therefore, selected proteinase inhibitors may function as repressors of the activation of chitin synthesis, which would be a novel site of action for a fungicide. Trypsin inhibitors from barley showed synergistic antimicrobial activity on filamentous fungi and gram-positive bacteria in combination with wheat and barley thionins.[49] Thionins are known to permeabilize hyphal membranes, but the specific synergistic action of PIs is unclear. As PIs are frequently induced coordinately with other defense responses,[6] possible synergistic effects with other PR proteins should be investigated.

Evidence has recently emerged on detrimental effects of PIs on plant-parasitic nematodes. Plant defense responses to cyst-nematode infestation were suggested to obey a gene-for-gene interaction,[50] and a gene for resistance to *Heterodera schachtii* was recently isolated from sugar beet.[51] Lilley et al.[52] investigated the presence of proteinases in cryostat sections of the soybean cyst-nematode, *Heterodera glycines,* using synthetic substrates for cysteine and serine proteinases. Detected activity was restricted to the intestine and could be inhibited with a combination of cowpea cystatin (CCPI) and Oc-IΔD86, an engineered variant of oryzacystatin with an enhanced affinity, suggesting that both cysteine and serine proteinases were present within the intestine. In a related study, Urwin et al.[53] could profoundly disturb the life cycle of the free-living nematode *Caenorhabditis elegans* using wild-type oryzacystatin (Oc-I), Oc-IΔD86, and cowpea cystatin. After being hatched on a normal medium, no larvae developed to the adult stage when transferred to Oc-I-, Oc-IΔD86-, or CCPI-containing medium for 6, 12, or 24 h after hatching. These larvae could not recover following transfer to fresh, non-supplemented medium and eventually died. Finally, the behavior of the plant parasitic nematode *Globodera pallida* was studied in transgenic tomato roots over-expressing oryzacystatins. It was found that the expression of Oc-I and Oc-IΔD86 in hairy roots affected the growth of *G. pallida.* Individuals feeding on Oc-IΔD86-expressing roots failed to increase in size between 4 and 6 weeks, while adult females growing on control plants increased their cross-sectional area by threefold.[53] The authors suggest that this age-sensitive effect may arise from stage-specific differences in proteinase activity in *G. pallida* or from accumulation of Oc-IΔD86 in the parasite over time.

6.4 PROTEINASES AND PROTEINASE INHIBITORS IN PLANT–INSECT INTERACTIONS

6.4.1 Induction of Proteinase Inhibitors by Herbivore Insects and/or Mechanical Wounding

The first evidence that environmental factors could affect the regulation of PIs in plants came from the observation that Colorado potato beetles feeding on tomato and potato leaves induced the accumulation of large quantities of PIs.[11] It was then found that any severe mechanical injury of the leaves induced the response throughout the plant, both locally and systemically. Subsequent research has shown that the regulation of proteinase inhibitor proteins in response to insect damage is widespread in nature, and wound-inducible PIs have been described in numerous plant families.[12] In most of the studies, the inducibility by genuine insect attacks was not investigated. The most extensive work was carried out in tomato and potato, which were found to have wound-inducible genes encoding inhibitors of the four mechanistic classes of proteinases described above.[12,54-59] Among them, type I and type II serine proteinase inhibitors from tomato and potato were then used as marker genes to dissect biochemically the signaling pathways leading to the expression of defense protein in response to insect attacks.[60,61]

A cDNA encoding a type I PI was isolated from maize embryos, and the accumulation of its mRNA was shown to be wound- and pathogen-inducible.[31] At least three examples of wound-inducible Bowman–Birk type inhibitor genes have been characterized in maize, wheat, and alfalfa. The *wip1* gene is activated in cut etiolated maize seedlings.[62] The *wali3* and *wali5* genes were isolated as aluminum-responsive genes in wheat and were subsequently shown to be induced in wounded leaves and roots.[63] In these two cases, it is not known if these genes encode functional inhibitors. The *ati* gene is mechanically inducible in alfalfa leaves and stems and is expressed in roots in response to microorganisms present in the soil.[64] Interestingly, two genes from poplar (*win3*) and willow (*swin1.1*) encode proteins similar to Kunitz-type inhibitors and are systemically induced in wounded trees.[65,66] Similarly, three cDNAs similar to the Kunitz type of inhibitors were isolated from potato tubers, and one of the corresponding genes, *pkpi-A,* was found to be wound-inducible in leaves.[59] Finally, the storage protein sporamin, which accumulates in sweet potato tubers, was recently found to display sequence similarity with Kunitz-type inhibitors. The recombinant protein expressed in *E. coli* has strong trypsin inhibitory activity, and its gene was demonstrated to be systemically induced by wounding in leaves.[67] Wound-induced genes encoding cysteine proteinase inhibitors are known in potato[57] and in soybean.[68] In both cases, the accumulation of transcripts occurred locally, and also in distal, unwounded leaves. Similar expression profiles were described for aspartyl proteinase inhibitor genes of cultivated potato[57] and of a non-tuber-bearing potato species.[58] A metallocarboxypeptidase inhibitor activity has also been shown to increase in potato leaves after wounding.[69] In tomato leaves, wounding causes a dramatic increase in a metallocarboxypeptidase inhibitor mRNA without a concomitant increase of the corresponding protein, suggesting a control at the post-transcriptional level.[56]

6.4.2 Proteinase Inhibitors Active on Insect Digestive Proteinases

The activation of plant PI genes by mechanical wounding is believed to mimic the chewing action of herbivore insects, and many inhibitors found to be inducible by mechanical damage likely have putative target proteinases in the digestive tracts of specific insect predators. Studies on the effect of PIs on insect physiology began when Lipke et al.[70] found that a protein fraction from soybean inhibited growth, as well as proteolytic activity *in vitro*, of the meal worm, *Tribolium confusum.* Evidence has accumulated during the last decades on the interactions between insect proteinases and plant proteinase inhibitors.[9]

Digestive proteolytic enzymes from insects have been studied mainly in the orders Lepidoptera and Coleoptera, which both include a number of crop pests.[9] Proteolytic activity in the digestive tracts of Lepidopteran insects is mostly accounted for by serine proteinases, whose activity is optimal in the 9 to 11 pH range, a value generally found in the midguts of these insects. In contrast, Coleopteran insects use mainly cysteine proteinases for food digestion, and their midguts have a mildly acidic (pH~5) pH, which is adapted to the optimal requirements of these proteinases. Plant proteinase inhibitors inactivating both types of enzymes are known and were shown to have antinutritional effects on insects, with consequences on growth and development.[9] The physiological effects of the inhibitors are not simply limited to a reduction of the proteolytic activity in the digestive tracts of insects. Feedback mechanisms exist that compensate the loss of proteolytic activity by a hyperproduction of proteinases, which in turn can lead to the depletion of essential amino acids and finally to retarded growth rates.[71] In support of this hypothesis, it was found that the negative effects of soybean trypsin inhibitor and potato inhibitor II in diets of *Heliothis zea* and *Spodoptera exigua* could be reversed by the addition of methionine in the insect food. Potent inhibition of insect proteinases has been demonstrated for several types of constitutive or wound-inducible PIs. When added to artificial diets, potato PI-I was the most effective single PI in reducing the growth rate of larvae of the codling moth, *Cydia pomonella,* followed by soybean trypsin inhibitor.[72] The most significant reductions in growth rates occurred with a combination of potato PI-I and carboxypeptidase inhibitor. A single-headed Bowman–Birk trypsin inhibitor from snail medic seeds (*Medicago scutellata*) was shown to inhibit the trypsin-like activity of the crop pests *Adoxophyes orana, Hyphantria cunea, Lobesia botrana,* and *Ostrinia nubilalis.*[73] Insects from the order Coleoptera are generally insensitive to serine proteinase inhibitors. For example, larvae of the cowpea weevil can consume cowpea tissues containing several percent of the total protein as serine proteinase inhibitors without adverse effects on digestion.[74] In contrast, several plant cystatins were demonstrated to be active on digestive proteinases from various Coleopteran species. Oryzacystatin purified from rice seeds was shown to inhibit almost completely proteolytic activity in midgut extracts of the rice weevil and of the red flour beetle, two stored grain insects[75] and to cause growth retardation or death of two bean insect pests of the Coleoptera and Hemiptera orders.[76] A similar activity was also described for a cystatin in soybean seeds.[77] The 85 kDa protein crystals identified in potato tubers as multicystatin effectively caused growth inhibition of corn rootworm larvae.[78] One constitutive (L1) and two wound- and methyl-jasmonate-inducible (N2 and R1) soybean leaf cystatins were studied for their inhibition capabilities on digestive enzymes. The inhibition constants of the inducible inhibitors against papain were about 1000 times lower than the inhibition constant of the constitutive inhibitor.[79] In addition, the wound-inducible N2 and R1 cystatins had substantially greater inhibitory activities than L1 against gut cysteine proteinases of the third-instar larvae of western corn rootworm and Colorado potato beetle.

These *in vitro* effects of PIs were confirmed *in vivo* on several plant–insect systems by the use of transgenic plants. Hilder et al.[80] demonstrated that tobacco plants expressing a foreign cowpea trypsin inhibitor gene (Bowman–Birk subclass) were more resistant to attacks by larvae of the Lepidopteran *Heliothis virescens* than untransformed control plants. Johnson et al.[81] transformed tobacco with genes coding for serine PIs of the potato I (chymotrypsin inhibitory activity) and potato II subclasses (having trypsin and chymotrypsin inhibitory activity). Plants expressing high levels of inhibitor II protein caused severe growth retardation or even death when fed to larvae of *Manduca sexta,* the tobacco hornworm. In contrast, larvae grew similarly on plants accumulating inhibitor I and on untransformed plants, indicating that the trypsin-inhibitor activity of inhibitor II was solely responsible for the inhibition of growth. The defensive role of the wound-inducible PIs was elegantly demonstrated by Orozco-Cardenas et al.[82] Expression in transgenic tomato of a prosystemin gene (encoding a systemic signal for wound-responsive genes, see below) in antisense orientation resulted in a reduced induction of inhibitor I and II in response to *Manduca sexta* feeding. In consequence, the growth rates of the hornworm larvae feeding on these plants were 3 times higher than growth rates of larvae feeding on untransformed control plants. However, as systemin was shown to be a master signal for defense against attacking herbivores in tomato,[61] the reduced induction of other defense genes besides the two aforementioned PI genes (like polyphenoloxidase) in prosystemin antisense plants may contribute to the observed decrease in insect resistance.

Rice is one of the most important crops for which efficient transformation–regeneration systems have been developed recently. Several foreign PI genes have been introduced in rice with significant effects on insect resistance. The constitutive expression of the cowpea trypsin inhibitor gene,[83] or the wound-inducible expression of the potato PI-II gene[84] in rice both confer resistance to the stem borer, which is a major Lepidopteran rice pest. In another study, a corn cystatin (CC) gene was introduced in the rice genome and the CC protein was shown to accumulate to 2% of total soluble seed protein, which is more than 20 times higher than the level of the endogenous oryzacystatin (OC).[85] Moreover, CC showed a wider inhibition spectrum than OC. Crude extracts of transgenic rice plants efficiently inhibited cysteine proteinase activity in the midguts of the rice pest *Sitophilus zeamais,* whereas similar amounts of protein fraction from untransformed seeds had no significant effect.

However, the limitations of combating insect digestive proteolysis processes with a single PI have been emphasized recently.[86] Jongsma et al.[87] studied growth and gut proteinases of *Spodoptera exigua* larvae feeding on tobacco expressing constitutively a potato PI-II gene. Although considerable levels of transgene expression could be demonstrated, the growth of the larvae was not affected. Analysis of the composition of trypsin gut activity revealed that only 18% of the proteinase activity of insects reared on these transgenic plants was sensitive to inhibition by PI-II, while 78% was sensitive in insects reared on control plants. Larvae had compensated for the loss of trypsin activity due to inhibition by a 2.5-fold induction of new activity that was insensitive to inhibition by PI-II. The authors suggested that induction of new proteinase activities represents the mechanism by which herbivore insects overcome plant proteinase inhibitor defense. A similar adaptation was described for Colorado potato beetles feeding on potato plants treated with methyl jasmonate to induce high levels of cysteine PIs.[88]

An alternative approach consisting of the expression of insect-encoded PIs in transgenic plants was recently shown to present a high potential for crop protection. Tobacco was transformed with a *Manduca sexta* gene engineered to encode trypsin, chymotrypsin, or elastase inhibitors.[89] When these plants were tested against *Bemisia tabaci,* the sweet potato white fly type B, insect reproduction was reduced by as much as 98% compared to controls. These results suggest that perturbation of proteolytic activities involved in insect development can be used to control pest populations.

Another interesting strategy to achieve an increased resistance level to insect attacks would be to take advantage of putative synergistic effects between PIs and the endotoxin of the bacterium *Bacillus thuringiensis* (*cryA* gene). In artificial diets, serine proteinase inhibitors present at very low levels (4 μ*M*) enhanced the insecticidal activity of *B. thuringiensis* by 2- to 20-fold.[90] However, Santos et al.[91] obtained contrasting results when testing transgenes for engineering insect resistance in *Arabidopsis*. When expressed in transformed plants, the *B. thuringiensis* endotoxin gene was more efficient than the cowpea trypsin inhibitor (*cpTI*) gene in controlling four caterpillar species. In plants expressing both transgenes, no synergistic effect was observed. Instead, these plants had a resistance level which was intermediate between plants expressing the *cryIA* gene and plants expressing the *cpTI* gene. The authors suggested that this may arise either from an effect of *cryIA* gene copy number, which varied in the different genotypes, or alternatively from an antagonistic interaction between the products of the two transgenes.

Finally, a possible regulatory effect of grazing-induced plant PIs on mammals was described by Seldal et al.[92] A tight positive association was found between grazing-induced trypsin inhibitor activity in grasses and fluctuations in lemming populations. Lemmings feeding on a sedge and on a grass induce an accumulation of PIs in the plant tissue. These high levels of antidigestive proteins in the rodents' food persist for a long period, and lemmings from a declining population showed pancreatic hypertrophy, which is a well-documented pathology caused by a prolonged dietary intake of trypsin inhibitors. Adverse effects on the biological cycle of the animals result in multiannual cyclic variations in their population.

6.5 REGULATION OF PROTEINASE INHIBITORS AS INDUCIBLE PLANT DEFENSE PROTEINS

PI genes are regulated at various stages of plant development in storage organs such as seeds or tubers, flowers or fruits, and are induced by environmental cues such as insect predation or microbial infections. Consistent with their expression under diverse physiological conditions, proteinase inhibitor genes have been found to be sensitive to a number of factors, including hormones, sugars and oligosaccharides, lipid derivatives, and various elicitor molecules. For example, a cDNA encoding a three-domain type II trypsin inhibitor was isolated from tomato, and its gene was found to be induced by exogenous auxin in roots[17] and downregulated by gibberellins in shoots.[19] However, the levels of auxin were shown to decrease in tobacco leaves after wounding.[93] In fact, the specific implication of these two hormones in the expression of PI genes in a given physiological response in not known. In contrast, induction of PI genes in defense reactions has been the subject of numerous studies and a densifying network of molecular events is being uncovered. The response to wounding or insect attacks is clearly the most characterized induction pathway for

PI genes with the tomato and potato systems serving as prototypes for studying the other plant species where induction by mechanical damage has been demonstrated.

Oligouronides (derived from pectin in plant cell walls) and chitosan (released from fungal cell walls) have been known for a long time to be efficient inducers of PI synthesis in tomato.[94] These oligosaccharidic signals are generated by hydrolytic enzymes released by the two partners upon pathogen attack and are believed to be local signals for defense induction because they are not readily transported in plants.[95] The search for systemic signals has led to the finding that four types of factors could contribute to the response in unwounded parts of wounded plants, including systemin, abscisic acid (ABA), hydraulic signals (variation potentials), or electrical current (activation potentials).[60,96-98] Alternative pathways may exist that allow for redundancy in the systemic transduction of the wound signal.

The systemin-octadecanoid pathway worked out in tomato presents the most convincing cumulative data describing the sequential events leading to the systemic activation of proteinase inhibitor genes after insect attacks. Systemin is an 18 amino acid peptide which was isolated from tomato leaves in a screen for proteinase inhibitor-inducing activity,[99] and it represents the most powerful known inducer of inhibitors I and II. Systemin was shown to be mobile in the phloem with a velocity comparable to sucrose and to the endogenous wound signal.[11,100] The essential role of systemin in the signaling of the defense response was demonstrated by studying plants that have been transformed with an antisense gene encoding prosystemin, the prohormone precursor of systemin. When compared to control plants, these transgenic plants showed a substantial reduction in the systemic induction of PIs by wounding[101] and a drastic reduction in the ability to resist insect attacks.[82] Conversely, plants over-expressing a sense prosystemin transgene displayed a constitutive expression of PIs, polyphenoloxidase (PPO), and other systemic wound-response proteins.[61,102,103] PPO activity is believed to contribute to the decrease of the nutritional quality of attacked plants by converting soluble phenolic compounds to quinones that eventually alkylate side chains of amino acids and prevent the digestion of proteins. The data are in favor of systemin acting as a master signal controlling the induction of a set of defense genes to deter attacking herbivores. A second series of arguments points to the activation by systemin of a lipid-derived intracellular signaling cascade, analogous to the prostaglandin signaling system found in animals.[61,102,103] As systemin was being characterized, it was found that exogenous jasmonic acid (JA) and its volatile derivative, methyl jasmonate, were powerful inducers of the PI genes. It was proposed that after systemin perception by an uncharacterized reception system, the fatty acid linolenic acid was released from membranes and converted to the oxylipins phytodienoic acid and jasmonic acid through the octadecanoid pathway.[104] Consistent with this, mechanical wounding or supplying systemin to young tomato plants resulted in a rapid and transient accumulation of linolenic and jasmonic acids.[94,105] In addition, chemical inhibitors of the octadecanoid pathway substantially reduce the induction of PIs by systemin, polygalacturonic acid, and linolenic acid, but not by JA.[61,106] This wound-induced pathway apparently involves several intercellular and intracellular compartments[96,107] which may provide specific sites to regulate the transduction of the wound signal.[108] Recently, evidence was provided that the chewing action of insects may not be limited to the sole mechanical damage, because the magnitude of the defense response can be much greater in plants challenged with insects than in mechanically wounded plants. A molecule called volicitin consisting of linolenic acid derived with L-glutamine was isolated from oral secretions of beet armyworm caterpillars.[109] When applied to damaged leaves of corn seedlings, volicitin induced the seedlings to emit volatile compounds that attract parasitic wasps, natural enemies of the caterpillars. It is not known if volicitin is also an inducer of PI genes nor if its biological activity is triggered through the octadecanoid pathway.

Genetic approaches are confirming biochemical data or defining new effectors of the signaling cascade. A tomato mutant called *def1* was described with a defect in the transduction pathway.[110] Supplementation of these plants with intermediates of the octadecanoid pathway indicated that *def1* plants are affected in a biosynthetic step between hydroperoxylinolenic acid and 12-oxo-phytodienoic acid. In consequence, *def1* plants fail to accumulate PIs and other defense proteins after wounding and are compromised in resistance against *Manduca sexta* larvae attacks. In agreement with these observations, a mutant was obtained in *Arabidopsis* which is deficient in linolenic acid and which displays high sensitivity to attacks by larvae of a common saprophage insect. Application of exogenous methyl jasmonate could reduce mortality of mutant plants to essentially wild-type levels.[111] The signaling steps downstream of JA are presently uncharacterized, but several JA-insensitive mutant classes of *Arabidopsis* have been identified recently[112] and should help in the near future to understand how this hormone activates PI gene expression.

Work in tomato and potato has revealed that jasmonate does not seem to be the only hormone required for wound induction of PI genes. It has been suggested that abscisic acid (ABA) participates in the response. Wounding and systemin caused an increase in endogenous ABA and JA levels preceding PI gene expression.[60] In addition, exogenous ABA was shown to induce an increase in endogenous JA levels, and the authors concluded that the site of action of JA was located downstream of the site of action of ABA. However, the activation of PI genes was not confirmed in another study when ABA was supplied to excised tomato plants through the cut stems.[96] Furthermore, induction of PI genes by wounding or systemin does not occur in ABA-deficient plants, but treatment with linolenic acid could induce the PI genes in these mutants, indicating that ABA may function at a step between systemin perception and the octadecanoid pathway.[113]

Ethylene was also shown to be required for a functional wound response, although this hormone is not an inducer of PI genes when administered alone. By using chemical inhibitors of ethylene action and mutant plants blocked in ethylene signaling, O'Donnell et al.[114] concluded that the presence of this hormone is required to trigger jasmonate-dependent responses after wounding.

The signaling pathways leading to PI gene expression upon microbial attacks have been given much less attention than herbivore-induced responses. For instance, there are no data available on the possible involvement of the systemin pathway in defense reactions of infected plants. It appears that some of the PIs that are wound- and jasmonate-inducible are also activated by microbial signals. Wound-inducible PI-I and PI-II genes were also found to be expressed in tomato plants infected with *Pseudomonas syringae* pv. *tomato*, but the specific signals triggering this induction are not known.[27] The appropriate activation of such a gene was confirmed when tobacco plants were transformed with a construct containing a wound-inducible potato PI-II promoter fused to the coding region of cecropin, an antibacterial peptide of insects. These plants showed an enhanced resistance to infection by *Pseudomonas syringae* pv. *tabaci*.[115]

The sensitivity of many PI genes to methyl jasmonate contrasts with what is observed for the genes encoding the families 1, 2, 3, and 5 of PR proteins, which are unaffected in tobacco when this compound is administered alone (T. Heitz, unpublished results). This differential regulation was also demonstrated in germinating embryos of maize in which a fungal infection induced both a proteinase inhibitor (MPI) gene and a PR-1 gene, while wounding and methyl jasmonate stimulated accumulation of MPI mRNA but not PR-1 mRNA.[31] Similarly, tobacco cells responded to fungal elicitor or methyl jasmonate treatment with an increase in trypsin inhibitor activity.[116] However, in tobacco plants, the TIMP genes encoding inhibitors of microbial proteinases are regulated coordinately with the other families of PR proteins during the hypersensitive reaction (HR) to TMV or in response to a purified glycoprotein elicitor.[117] The TIMP proteins, like other tobacco PR proteins, accumulated in plants treated with salicylic acid, but not in MJ-treated or wounded plants.[30] These PI genes thus appear to be specifically regulated by signals generated during the HR to pathogens. In fact, the recognition of microbes by plants involves multiple molecular communication events which are transduced in a complex array of signaling pathways activating several sets of defensive genes. The reaction to pathogenic infections is much more complex than the response to insect chewing, since most of the known defense signals, including salicylic acid,[118] jasmonic acid,[119] and ethylene,[120] are produced in hypersensitively reacting plants. An inhibitory effect of exogenous salicylic acid on the wound and jasmonate induction of defense proteins was described in tomato.[121] As salicylic and jasmonic acids were shown to be produced in infected *Arabidopsis*,[119] this negative interaction may not be relevant in plants reacting to microbial pathogens (T. Heitz, unpublished results). The elucidation of positive and negative cross-talks between signaling pathways controlling different downstream defense responses is needed to fully understand how plants can induce a broad-spectrum resistance state.

Another level of PI activity regulation may be the turnover of inhibitor proteins. Proteinases cleaving PIs in plant–pathogen systems have not been described, but a cysteine proteinase was isolated from soybean cotyledons which is capable of degrading Kunitz and Bowman–Birk trypsin inhibitors during early seedling growth.[122] Proteinase activity itself is induced in plants as a consequence of pathogen attack. An alkaline proteinase (PR-P69) was isolated from tomato leaves infected with the citrus exocortis virus and shown to be one of the major secreted PR proteins induced in this system.[123] This finding was at the origin of the definition of the family 7 of PR proteins,[8] but similar proteinases have not yet been characterized in other plant–pathogen systems. The recent cloning of the cDNA encoding PR-P69 showed that this calcium-activated proteinase belongs to the subtilisin-related proprotein convertases (SPC) family.[124] In animals and yeast, SPC members appear to be related to the proteolytic maturation of precursor proteins releasing biologically active peptides and proteins. The role of PR-P69 in the development of the plant defense response remains to be elucidated. Tobacco leaves were found to express constitutively an extracellular aspartyl proteinase of 36–40 kDa that cleaves PR proteins under acidic conditions.[125] A similar activity was also detected in tomato, and these proteinases may be involved in the regulation of the turnover of PR proteins which are resistant to degradation by most of the proteolytic activities. It thus appears that upon pathogen attack, both proteinases and proteinase inhibitors may control, by a subtle interplay, some important processes in the regulation of defense responses.

6.6 CONCLUSION AND FUTURE PROSPECTS

The wide distribution of pathogen-inducible PI genes in the plant kingdom strongly suggests that the encoded proteins contribute significantly to the protection of plants against parasite offense. The abundant accumulation of PIs in pathological situations, as well as their remarkable stability in extreme conditions, are typical of pathogenesis-related proteins and justify the definition of proteinase inhibitors as a new family of PRs.[8] The spectrum of plant PIs that are induced in concert with other defense responses has been extended these last years and found to have targets in the four classes of proteinases. The structural and functional diversity in PIs reflects the "arms race" between parasites which degrade plant proteins and colonize plant tissues and their hosts which have evolved proteinase inhibitory counterparts. The defensive capabilities of PIs against insect predation have been clearly demonstrated, both in artificial diets and in transgenic plants. Whereas most of the earlier studies concerned serine PIs, more recent data indicate that inhibitors directed against each class of proteinases can profoundly disturb the life cycle of several taxonomic groups of insects. Nematodes were recently shown to be particularly sensitive to cystatins, perhaps because they rely on a limited number of digestive proteinases and have no efficient compensatory mechanism for producing inhibitor-insensitive proteinases. The search for plant PIs which inactivate proteinases that are essential for microbial development has so far given more diverse results and awaits more intense genetic engineering of the two partners, either by knocking out microbial genes encoding secreted proteinases or by overexpressing new isoforms of PIs in plants. Conversely, the evaluation of the resistance level against a wide array of phytopathogens in mutant plants affected in the production of active PIs should provide clues about the role of these PIs in defense. Future work should also consider the hypothesis of alternative modes of action of PIs in plant defense, since unknown proteolytic processes may be critical for a parasite to colonize plant tissues.

The study of the signaling pathways governing the induction of PI genes upon pathogen attack has highlighted, in some instances, regulatory processes that are distinct from those controlling the activation of other PR gene families. These PIs may represent, along with other PRs, valuable markers to investigate the interplay of complex signaling networks operating during the defense reactions of plants to pathogens.

REFERENCES

1. Antoniw, J. F., Ritter, C. E., Pierpoint, W. S., and Van Loon, L. C., Comparison of three pathogenesis-related proteins from plants of two cultivars of tobacco infected with TMV, *J. Gen. Virol.,* 47, 79, 1980.
2. Gianinazzi, S., Martin, C., and Vallée, J., Hypersensibilité aux virus, température et protéines solubles chez le *Nicotiana* Xanthi-nc. Apparition de nouvelles macromolécules lors de la répression virale, *C. R. Acad. Sc. Paris,* 270, 2383, 1970.
3. Van Loon, L. C. and Van Kammen, A. V., Polyacrylamide disc electrophoresis of the soluble leaf proteins from *Nicotiana tabacum* var. "Samsun" and "Samsun NN." Changes in protein constitution after infection with tobacco mosaic virus, *Virology,* 40, 199, 1970.
4. Van Loon, L. C., Gerritsen, Y. A. M., and Ritter, C. E., Identification, purification, and characterization of pathogenesis-related proteins from virus-infected Samsun NN tobacco leaves, *Plant Mol. Biol.,* 9, 593, 1987.
5. Linthorst, H., Pathogenesis-related proteins of plants, *Crit. Rev. Plant Sci.,* 10, 123, 1991.
6. Stintzi, A., Heitz, T., Prasad, V., and Wiedemann-Merdinoglu, S., Plant pathogenesis-related proteins and their role in defense against pathogens, *Biochimie,* 75, 687, 1993.
7. Kombrink, E. and Somssich, I., Pathogenesis-related proteins and plant defense, in *The Mycota V Part A Plant Relationships,* Carroll, Tudzinski, Eds.: Springer Verlag, Berlin 1997, 107.
8. Van Loon, L. C., Pierpoint, W. S., Boller, T., and Conejero, V., Recommendations for naming plant pathogenesis-related proteins, *Plant Mol. Biol. Rep.,* 12, 245, 1994.
9. Ryan, C. A., Protease inhibitors in plants: genes for improving defenses against insects and pathogens, *Annu. Rev. Phytopathol.,* 28, 425, 1990.
10. Ryan, C. A., Proteolytic enzymes and their inhibitors in plants, *Annu. Rev. Plant Physiol.,* 24, 173, 1973.
11. Green, T. R. and Ryan, C. A., Wound-induced proteinase inhibitor in plant leaves: a possible defense mechanism against insects, *Science,* 175, 776, 1972.
12. Ryan, C. A., The search for the proteinase inhibitor inducing factor, PIIF, *Plant Mol. Biol.,* 19, 123, 1992.
13. Cervone, F., DeLorenzo, G., Degri, L., Salvi, G., and Bergami, M., Purification and characterization of a polygalacturonase-inhibiting protein from *Phaseolus vulgaris* L., *Plant Physiol.,* 85, 631, 1987.
14. Ham, K. S., Wu, S. C., Darvill, A., and Albersheim, P., Fungal pathogens secrete an inhibitor protein that distinguishes isoforms of plant pathogenesis-related endo-β-1,3-glucanases, *Plant J.,* 11, 169, 1997.
15. Laskowski, M. J., Kato, I., Kohr, W., Park, S. J., Tashiro, M., and Whatley, H. E., Positive darwinian selection in evolution of protein inhibitors of serine proteinases, *Cold Spring Harbor Symp. Quant. Biol.,* 52, 545, 1988.
16. Hass, G. M., Hermodson, M. A., and Ryan, C. A., Primary structures of two low molecular weight proteinase inhibitors from potatoes, *Biochemistry,* 21, 2, 1982.
17. Taylor, B. H., Young, R. Y., and Scheuring, C. F., Induction of a proteinase inhibitor II-class gene by auxin in tomato roots, *Plant Mol. Biol.,* 23, 1005, 1993.
18. Balandin, T., van der Does, C., Bellés, A. J. M., Bol, J. F., and Linthorst, H. J. M., Structure and induction pattern of a novel proteinase inhibitor class II gene of tobacco, *Plant Mol. Biol.,* 27, 1197, 1995.

19. Jacobsen, S. E. and Olszewski, N. E., Gibberellins regulate the abundance of RNAs with sequence similarity to proteinase inhibitors, dioxygenases and dehydrogenases, *Planta*, 198, 78, 1996.
20. Atkinson, A. H., Heath, R. H., Simpson, R. J., Clarke, A. E., and Anderson, M. A., Proteinase inhibitors in *Nicotiana alata* stigmas are derived from a precursor protein which is processed into five homologous inhibitors, *Plant Cell,* 5, 203, 1993.
21. Pearce, M. J., Johnson, S., and Ryan, C. A., Purification and characterization from tobacco (*Nicotiana tabacum*) leaves of six small, wound-inducible, proteinase isoinhibitors of the potato inhibitor II family, *Plant Physiol.,* 102, 639, 1993.
22. Ikenara, T. and Nokiora, S., Bowman-Birk family serine proteinase inhibitors, in *Proteinase Inhibitors,* Barrett, A. J. and Salvesen, G., Eds., Elsevier Science Publishers, Amsterdam, 1986, 361-374.
23. Prakash, B., Selvaraj, S., Murthy, M. R. N., Sreerama, Y. N., Rao, D., and Gowda, L. R., Analysis of the amino acid sequences of plant Bowman-Birk inhibitors, *J. Mol. Evol.,* 42, 560, 1996.
24. Barrett, A. J., Rawlings, N. D., Davies, M. E., Machleidt, W., Salvesen, G., and Turk, V., Cysteine proteinase inhibitors of the cystatin superfamily, in *Proteinase Inhibitors,* Barrett, A. J. and Salvesen, G., Eds., Elsevier Science Publishers, Amsterdam, 1986, 515.
25. Walsh, T. A. and Strickland, J. A., Proteolysis of the 85-kilodalton crystalline cysteine proteinase inhibitor from potato releases functional cystatin domains, *Plant Physiol.,* 103, 1227, 1993.
26. Peng, J. H. and Black, L. L., Increased proteinase inhibitor activity in response to infection of resistant tomato plants by *Phytophthora infestans, Phytopathology,* 66, 958, 1976.
27. Pautot, V., Holzer, F. M., and Walling, L. L., Differential expression of tomato proteinase inhibitor I and II genes during bacterial pathogen invasion and wounding, *Mol. Plant-Microbe Interact.,* 4, 284, 1991.
28. Roby, D., Toppan, A., and Esquerré-Tugayé, M. T., Cell surfaces in plant microorganism interactions. VIII. Increased proteinase inhibitor activity in melon plants in response to infection by *Colletotrichum lagenarium* or to treatment with an elicitor fraction from this fungus, *Physiol. Mol. Plant Pathol.,* 30, 453, 1987.
29. Geoffroy, P., Legrand, M., and Fritig, B., Isolation and characterization of a proteinaceous inhibitor of microbial proteinases induced during the hypersensitive reaction of tobacco to tobacco mosaic virus, *Mol. Plant-Microbe Interact.,* 3, 327, 1990.
30. Heitz, T., Geoffroy, P., Stintzi, A, Fritig, B., and Legrand, M., cDNA cloning and gene expression analysis of the microbial proteinase inhibitor of tobacco, *J. Biol. Chem.,* 268, 16987, 1993.
31. Cordero, M. J., Raventós, D., and San Segundo, B., Expression of a maize proteinase inhibitor gene is induced in response to wounding and fungal infection: systemic wound-response of a monocot gene, *Plant J.,* 6, 141, 1994.
32. Gadea, J., Mayda, M. E., Conejero, V., and Vera, P., Characterization of defense-related genes ectopically expressed in viroid-infected tomato plants, *Mol. Plant-Microbe Interact.,* 6, 790, 1996.
33. Stevens, C., Titarenko, E., Hargreaves, J. A., and Gurr, S. J., Defense-related gene activation during an incompatible interaction between *Stagonospora (Septoria) nodorum* and barley (*Hordeum vulgare* L.) coleoptile cells, *Plant Mol. Biol.,* 31, 741, 1996.

34. Vos, P., Verner, J., Jaegle, M., Wellink, J., van Kammen, A., and Goldbach, R., Two viral proteins involved in the proteolytic processing of the cowpea mosaic virus polyproteins, *Nucl. Acids Res.,* 16, 1967, 1988.
35. Riechmann, J. L., Lain, S., and Garcia, J. A., Highlights and prospects of potyvirus molecular biology, *J. Gen. Virol.,* 73, 1, 1992.
36. Kondo, H., Ijiri, S., Abe, K., Maeda, H., and Arai, S., Inhibitory effect of oryzacystatins and a truncation mutant on replication of poliovirus in infected vero cells, *FEBS Lett.,* 299, 48, 1992.
37. Aoki, H., Akaike, T., Abe, K., Kuroda, M., Arai, S., Okamura, R. I., Negi, A., and Maeda, H., Antiviral effect of oryzacystatin, a proteinase inhibitor in rice, against herpes simplex virus type 1 *in vitro* and *in vivo, Antimic. Agents Chemother.,* 39, 846, 1995.
38. Dow, J. M., Clarke, B. R., Milligan, D. E., Tang, J. L., and Daniels, M., Extracellular proteases from *Xanthomonas campestris* pv. *campestris, Applied Environ. Microbiol.,* 56, 2994, 1990.
39. Frederick, R. D., Chin, J., Bennetzen, J. L., and Handa, A. K., Identification of a pathogenicity locus, *rpfA*, in *Erwinia carotovora* subsp. *carotovora* that encodes a two-component sensor regulator protein, *Mol. Plant-Microbe Interact.,* 10, 407, 1997.
40. Kyosto, S. R. M., Cramer, C. L., and Lacy, G. H., *Erwinia carotovora* ssp. *carotovora* extracellular protease: characterization and nucleotide sequence of the gene, *J. Bacteriol.,* 173, 6537, 1991.
41. Xu, G. W. and Gonzales, C. F., Evaluation of TN4431-induced protease mutants of *Xanthomonas campestris* pv. *oryzae* for growth *in planta* and pathogeneticity, *Phytopathology,* 79, 1210, 1989.
42. Movahedi, S. and Heale, J. B., The roles of aspartic proteinase and endo-pectin lyase enzymes in the primary stages of infection and pathogenesis of various host tissues by different isolates of *Brotrytis cinerea* Pers ex. Pers., *Physiol. Mol Plant Pathol.,* 36, 303, 1990.
43. Ball, A. M., Ashby, A. M., Daniels, M. J., Ingram, D. S., and Johnstone, K., Evidence for the requirement of extracellular protease in the pathogenic interaction of *Pyrenopeziza brassicae* with oilseed rape, *Physiol. Mol. Plant Pathol.,* 38, 147, 1991.
44. Gholson, R. K. and Essenberg, M., An extracellular protease-deficient mutant of *Xanthomonas campestris* pv. *malvacearum* maintains its pathogenicity to cotton, *Phytopathology,* 76, 1140, 1986.
45. Marchetti, S., Pitotti, A., Giordano, A., Chiaba, C., and Fogher, C., Partial inactivation of microbial proteinases with soybean Kunitz and Bowman-Birk inhibitors, *J. Sci. Food Agric.,* 69, 423, 1995.
46. Mosolov, V. V., Loginova, M. D., Malova, E. L., and Benken, I. I., A specific inhibitor of *Colletotrichum lindemuthianum* protease from kidney bean (*Phaseolus vulgaris*) seeds, *Planta,* 144, 265, 1979.
47. Lorito, M., Broadway, R. M., Hayes, C. K., Woo, S. L., Noviello, C., Williams, D. L., and Harman, G. E., Proteinase inhibitors from plants as a novel class of fungicides, *Mol. Plant-Microbe Interact.,* 7, 525, 1994.
48. Machida, S. and Saito, M., Purification and characterization of membrane-bound chitin synthase, *J. Biol. Chem.,* 268, 1702, 1993.
49. Terras, F. R. G., Schoofs, H. M. E., Thevissen, K., Osborn, R. W., Vanderleyden, J., Cammue, B. P. A., and Broekaert, W., Synergistic enhancement of the antifungal activity of wheat and barley thionins by radish and oilseed rape 2 S albumins and by barley trypsin inhibitors, *Plant Physiol.,* 103, 1311, 1993.

50. Roberts, P. A., Conceptual and practical aspects of variability in root-knot nematodes related to host plant resistance, *Annu. Rev. Phytopathol.*, 33, 199, 1995.
51. Cai, D., Kleine, M., Kifle, S., Harloff, H. J., Sandal, N. N., Marcker, K. A., Klein-Lankhorst, R. M., Salentijin, E. M. J., Lange, W., Stiekema, W. J., Wyss, U., Grundler, F. M. V., and Jung, C., Positional cloning of a gene for nematode resistance in sugar beet, *Science*, 275, 832, 1997.
52. Lilley, C. J., Urwin, P. E., McPherson, M. J., and Atkinson, H. J., Characterization of intestinally active proteinases of cyst-nematodes, *Parasitology*, 113, 415, 1996.
53. Urwin, P. E., Atkinson, H. J., Waller, D. A., and McPherson, M. J., Engineered oryzacystatin-I expressed in transgenic hairy roots confers resistance to *Globodera pallida, Plant J.*, 8, 121, 1995.
54. Graham, J. S., Pearce, G., Merryweather, J., Titani, K., Ericsson, L. H., and Ryan, C. A., Wound-induced proteinase inhibitors from tomato leaves. I. The cDNA-deduced primary structure of pre-inhibitor I and its post-translational processing, *J. Biol. Chem.*, 260, 6555, 1985.
55. Graham, J. S., Pearce, G., Merryweather, J., Titani, K., Ericsson, L. H., and Ryan, C. A., Wound-induced proteinase inhibitors from tomato leaves. II. The cDNA-deduced primary structure of pre-inhibitor II, *J. Biol. Chem.*, 260, 6561, 1985.
56. Martineau, B., McBride, K. E., and Hauck, C. A., Regulation of metallocarboxypeptidase inhibitor gene expression in tomato, *Mol. Gen. Genet.*, 228, 281, 1991.
57. Hildmann, T., Ebneth, M., Peña-Cortés, H., Sanchez-Serrano, J. J., Willmitzer, L., and Prat, S., General roles of abscisic and jasmonic acids in gene activation as a result of mechanical wounding, *Plant Cell*, 4, 1157, 1992.
58. Hansen, J. D. and Hannapel, D. J., A wound-inducible potato proteinase inhibitor gene expressed in non-tuber-bearing species is not sucrose inducible, *Plant Physiol.*, 100, 164, 1992.
59. Ishikawa, A., Ohta, S., Matsuoka, K., Hattori, T., and Nakamura, K., A family of potato genes that encode Kunitz-type proteinase inhibitors: structural comparisons and differential expression, *Plant Cell Physiol.*, 35, 303, 1994.
60. Peña-Cortés, H., Fisahn, J., and Willmitzer, L., Signals involved in wound-induced proteinase inhibitor II gene expression in tomato and potato plants, *Proc. Natl. Acad. Sci. U.S.A.*, 92, 4106, 1995.
61. Bergey, D. R., Howe, G. A., and Ryan, C. A., Polypeptide signaling for plant defensive genes exhibits analogies to defense signaling in animals, *Proc. Natl. Acad. Sci. U.S.A.*, 93, 12053, 1996.
62. Rohrmeier, T. and Lehle, L., *WIP1*, a wound-inducible gene from maize with homology to Bowman-Birk proteinase inhibitors, *Plant Mol. Biol.*, 22, 783, 1993.
63. Snowden, K. C., Richards, K. D., and Garner, R. C., Aluminium-induced genes. Induction by toxic metals, low calcium, and wounding and pattern of expression in root tips, *Plant Physiol.*, 107, 341, 1995.
64. McGurl, B., Mukherjee, S., Kahn, M., and Ryan, C. A., Characterization of two proteinase inhibitor (ATI) cDNAs from alfalfa leaves (*Medicago sativa* var. Vernema): the expression of ATI genes in response to wounding and soil microorganisms, *Plant Mol. Biol.*, 27, 995, 1995.
65. Bradshaw, H. D., Hollick, J. B., Parsons, T. J., Clarke, H. R. G., and Gordon, M. P., Systemically wound-responsive genes in poplar trees encode proteins similar to sweet potato sporamins and legume Kunitz trypsin inhibitors, *Plant Mol. Biol.*, 14, 51, 1989.
66. Saarikoski, P., Clapham, D., and von Arnold, S., A wound-inducible gene from *Salix viminalis* coding for a trypsin inhibitor, *Plant Mol. Biol.*, 31, 465, 1996.

67. Yeh, K. W., Chen, J. C., Lin, M. I., Chen, Y. M., and Lin, C. Y., Functional activity of sporamin from sweet potato (*Ipomea batatas* Lam): a tuber storage protein with trypsin inhibitory activity, *Plant Mol. Biol.*, 33, 565, 1997.
68. Botella, M. A., Xu, Y., Prabha, T. N., Zhao, Y., Narasimhan, M. L., Wilson, K. A., Nielsen, S. S., Bressan, R. A., and Hasegawa, P. M., Differential expression of soybean cysteine proteinase inhibitor genes during development and in response to wounding and methyl jasmonate, *Plant Physiol.*, 112, 1201, 1996.
69. Graham, J. S. and Ryan, C. A., Accumulation of a metallo-carboxypeptidase inhibitor in leaves of wounded potato plants, *Biochem. Biophys. Res. Commun.*, 101, 1164, 1981.
70. Lipke, H., Fraenkel, G. S., and Liener, I., Effect of soybean inhibitors on growth of *Tribolium confusum, Agric. Food Chem.*, 2, 410, 1954.
71. Broadway, R. A. and Duffey, S. S., Plant proteinase inhibitors: mechanism of action and effect on the growth and digestive physiology of larval *Heliothis zea* and *Spodoptera exigua, J. Insect Physiol.*, 32, 827, 1986.
72. Markwick, N. P., Reid, S. J., Laing, W. A., and Christeller, J. T., Effects of dietary protein and protease inhibitors on codling moth (*Lepidoptera: Tortricidae*), *J. Econ. Entomol.*, 88, 33, 1995.
73. Ceciliani, F., Tava, A., Iori, R., Mortarino, M., Odoardi, M., and Ronchi, S., A trypsin inhibitor from snail medic seeds active against pest proteases, *Phytochemistry*, 44, 393, 1997.
74. Gatehouse, A. M. R. and Boulter, D., Assessment of the antimetabolic effects of trypsin inhibitors from cowpea (*Vigna unguiculata*) and other legumes on development of the brucid beetle *Callosobruchus maculatus, J. Sci. Food Agric.*, 34, 345, 1983.
75. Liang, C., Brookhart, G., Feng, G. H., Reeck, G. R., and Kramer, K. J., Inhibition of digestive proteinases of stored grain coleoptera by oryzacystatin, a cysteine proteinase inhibitor from rice seed, *FEBS Lett.*, 278, 139, 1991.
76. Kuroda, M., Ishimoto, M., Suzuki, K., Kondo, H., Abe, K., and Arai, S., Oryzacystatins exhibit growth-inhibitory and lethal effects on different species of bean insect pests, *Callosobruchus chinensis* (Coleoptera) and *Riportus clavatus* (Hemiptera), *Biosci. Biotechnol. Biochem.*, 60, 209, 1996.
77. Hines, M. E., Osuala, C. I., and Nielsen, S. S., Isolation and partial characterization of a soybean cystatin cysteine proteinase inhibitor of coleopteran digestive proteolytic activity, *J. Agric. Food Chem.*, 39, 1515, 1991.
78. Orr, G. L., Strickland, J. A., and Walsh, T. A., Inhibitory of *Diabrotica* larval growth by a multicystatin from potato tubers, *J. Insect Physiol.*, 40, 893, 1994.
79. Zhao, Y., Botella, M. A., Subramanian, L., Niu, X., Nielsen, S., Bressan, R. A., and Hasegawa, P. M., Two wound-inducible soybean, cysteine proteinase inhibitors have greater insect digestive proteinase inhibitory activities than a constitutive homolog, *Plant Physiol.*, 111, 1299, 1996.
80. Hilder, V. A., Gatehouse, A. M. R., Sheerman, S. E., Barker, R. F., and Boulter, D., A novel mechanism of insect resistance engineered into tobacco, *Nature*, 330, 160, 1987.
81. Johnson, R., Narvaez, J., An, G., and Ryan, C., Expression of proteinase inhibitors I and II in transgenic tobacco plants: effects on natural defense against *Manduca sexta* larvae, *Proc. Natl. Acad. Sci. U.S.A.*, 86, 9871, 1989.
82. Orozco-Cardenas, M., McGurl, B., and Ryan, C. A., Expression of an antisense prosystemin gene in tomato plants reduces resistance toward *Manduca sexta* larvae, *Proc. Natl. Acad. Sci. U.S.A.*, 90, 8273, 1993.

83. Xu, D., Xue, Q., McElroy, D., Mawal, Y., Hilder, V. A., and Wu, R., Constitutive expression of a cowpea trypsin inhibitor gene, *CpTi*, in transgenic rice plants confers resistance to two major rice insect pests, *Mol. Breeding,* 2, 167, 1996.
84. Duan, X., Li, X., Xue, Q., Abo-El-Saad, M., Xu, D., and Wu, R., Transgenic rice plants harboring an introduced potato proteinase inhibitor II gene are insect resistant, *Nature Biotechnol.,* 14, 494, 1996.
85. Irie, K., Hosoyama, H., Takeuchi, T., Iwabushi, K., Watanabe, H., Abe, M., Abe, K., and Arai, S., Transgenic rice established to express corn cystatin exhibits strong inhibitory activity against gut proteinases, *Plant Mol. Biol.,* 30, 149, 1996.
86. Jongsma, M. A., Stiekema, W. J., and Bosch, D., Combatting inhibitor-insensitive protease of insect pests, *Trends Biotechnol.,* 14, 331, 1996.
87. Jongsma, M. A., Bakker, P. L., Peters, J., Bosch, D., and Stiekema, W. J., Adaptation of *Spodoptera exigua* larvae to plant proteinase inhibitors by induction of gut proteinase activity insensitive to inhibition, *Proc. Natl. Acad. Sci. U.S.A.,* 93, 8041, 1995.
88. Bolter, C. J. and Jongsma, M. A., Colorado beetles (*Leptinotarsa decemlineata*) adapt to proteinase inhibitors induced in potato leaves by methyl jasmonate, *J. Insect Physiol.,* 41, 1071, 1995.
89. Thomas, J. C., Adams, D. G., Keppenne, V. D., Wasmann, C. C., Brown, J. K., Kanost, M. R., and Bohnert, H. J., *Manduca sexta* encoded protease inhibitors expressed in *Nicotiana tabacum* provide protection against insects, *Plant Physiol. Biochem.,* 33, 611, 1995.
90. Macintosh, S. C., Kishore, G. M., Perlak, F. J., Marrone, P. G., Stone, T. B., Sims, S. R., and Fuchs, R. L., Potentiation of *Bacillus thuringiensis* insecticidal activity by serine proteinase inhibitors, *J. Agric. Food Chem.,* 38, 1145, 1990.
91. Santos, M. O., Adang, M. J., All, J. N., Boerma, H. R., and Parrott, W. A., Testing transgenes for insect resistance using *Arabidopsis, Mol. Breeding,* 3, 183, 1997.
92. Seldal, T., Andersen, K. J., and Högstedt, G., Grazing induced proteinase inhibitors: a possible cause for lemming population cycle, *Oikos*, 70, 3, 1994.
93. Thorburg, R. W. and Li, X., Wounding *Nicotiana tabacum* leaves causes a decline in endogenous indole-3-acetic acid, *Plant Physiol.,* 96, 802, 1991.
94. Doares, S. H., Syrovets, T., Weiler, E. W., and Ryan, C. A., Oligogalacturonides and chitosan activate plant defensive genes through the octadecanoid pathway, *Proc. Natl. Acad. Sci. U.S.A.,* 92, 4095, 1995.
95. Baydoun, E. A. H. and Fry, S. C., The immobility of pectic substances in injured tomato leaves and its bearing on the identity of the wound hormone, *Planta,* 165, 269, 1985.
96. Schaller, A. and Ryan, C. A., Systemin — a polypeptide defense signal in plants, *BioEssays*, 18, 27, 1995.
97. Malone, M. and Alarcon, J. J., Only xylem-borne factors can account for systemic wound signaling in the tomato plant, *Planta*, 196, 740, 1995.
98. Wildon, D. C., Thain, J. F., Minchin, P. E. H., Gubb, I. R., Reilly, A. J., Skipper, Y. D., Doherty, H. M., O'Donnell, P. J., and Bowles, D. J., Electrical signalling and systemic proteinase inhibitor induction in the wounded plant, *Nature*, 360, 62, 1992.
99. Pearce, G., Strydom, D., Johnson, S., and Ryan, C. A., A polypeptide from tomato leaves activates the expression of proteinase inhibitor genes, *Science*, 253, 895, 1991.
100. Narváez-Vásquez, J., Pearce, G., Orozco-Cardenas, M. L., Franceschi, V. R., and Ryan, C. A., Autoradiographic and biochemical evidence for the systemic translocation of systemin in tomato plants, *Planta*, 195, 593, 1995.
101. McGurl, B., Pearce, G., Orozco-Cardenas, M., and Ryan, C. A., Structure, expression, and antisense inhibition of the systemin precursor gene, *Science*, 255, 1570, 1992.

102. McGurl, B., Orozco-Cardenas, M., Pearce, G., and Ryan, C. A., Overexpression of the prosystemin gene in transgenic tomato plants generates a systemic signal that constitutively induces proteinase inhibitor synthesis, *Proc. Natl. Acad. Sci. U.S.A.*, 91, 9799, 1994.
103. Constabel, C. P., Bergey, D. R., and Ryan, C. A., Systemin activates synthesis of wound-inducible tomato leaf polyphenoloxidase via the octadecanoid defense signaling pathway, *Proc. Natl. Acad. Sci. U.S.A.*, 92, 407, 1995.
104. Farmer, E. E. and Ryan, C. A., Octadecanoid precursors of jasmonic acid activate the synthesis of wound-inducible proteinase inhibitors, *Plant Cell*, 4, 129, 1992.
105. Conconi, A., Miquel, M., Browse, J. A., and Ryan, C. A., Intracellular levels of free linolenic and linoleic acids increase in tomato leaves in response to wounding, *Plant Physiol.*, 111, 797, 1996.
106. Koiwa, H., Bressan, R. A., and Hasegawa, P. M., Regulation of proteinase inhibitors and plant defense, *Trends Plant Sci.*, 2, 379, 1997.
107. Heitz, T., Bergey, D. R., and Ryan, C. A., A gene encoding a chloroplast-targeted lipoxygenase in tomato leaves is transiently induced by wounding, systemin, and methyl jasmonate, *Plant Physiol.*, 114, 1085, 1997.
108. Harms, K., Atzorn, R., Brash, A., Kühn, H., Wasternack, C., Willmitzer, L. and Peña-Cortés, H., Expression of a flax allene oxide synthase cDNA leads to increased endogenous jasmonic acid (JA) levels in transgenic potato plants but not to a corresponding activation of JA-responding genes, *Plant Cell*, 7, 1645, 1995.
109. Alborn, H. T., Turlings, T. C. J., Jones, T. H., Stenhagen, G., Loughrin, J. H., and Tumlinson, J. H., An elicitor of plant volatiles from beet armyworm oral secretion, *Science*, 276, 945, 1997.
110. Howe, G. A., Lightner, J., Browse, J., and Ryan, C. A., An octadecanoid pathway mutant (JL5) of tomato is compromised in signalling for defense against insect attack, *Plant Cell*, 8, 2067, 1996.
111. McConn, M., Creelman, R. A., Bell, E., Mullet, J. E., and Browse, J., Jasmonate is essential for insect defense in *Arabidopsis*, *Proc. Natl. Acad. Sci. U.S.A.*, 94, 5473, 1997.
112. Creelman, R. A. and Mullet. J. E., Biosynthesis and action of jasmonates in plants, *Annu. Rev. Plant Physiol. Plant Mol. Biol.*, 48, 355, 1997.
113. Peña-Cortés, H., Prat, S, Atzorn, R., Wasternack, C. and Willmitzer, L., Abscisic acid-deficient plants do not accumulate proteinase inhibitor following systemin treatment, *Planta*, 198, 447, 1996.
114. O'Donnell, P. J., Calvert, C., Atzorn, R., Wasternack, C., Leyser, H. M. O., and Bowles, D. J., Ethylene as a signal mediating the wound response of tomato plants, *Science*, 274, 1914, 1996.
115. Huang, Y., Nordeen, R. O., Di, M., Owens, L. D., and McBeath, J. H., Expression of an engineered cecropin gene cassette in transgenic tobacco plants confers disease resistance to *Pseudomonas syringae* pv. *tabaci*, *Phytopathology*, 87, 494, 1997.
116. Rickauer, M., Bottin, A., and Esquerré-Tugayé, M. T., Regulation of proteinase inhibitor production in tobacco cells by fungal elicitors, hormonal factors and methyl jasmonate, *Plant Physiol. Biochem.*, 30, 579, 1992.
117. Dorey, S., Baillieul, F., Pierrel, M. A., Saindrenan, P., Fritig, B., and Kauffmann, S., Spatial and temporal induction of cell death, defense genes, and accumulation of salicylic acid in tobacco leaves reacting hypersensitively to a fungal glycoprotein elicitor, *Mol. Plant-Microbe Interact.*, 5, 646, 1997.

118. Mauch-Mani, B. and Slusarenko, A. J., Production of salicylic acid precursors is a major function of phenylalanine ammonia-lyase in the resistance of Arabidopsis to *Peronospora parasitica, Plant Cell,* 8, 203, 1996.
119. Penninckx, I. A. M. A., Eggermont, K., Terras, F. R. G., Thomma, B P. H. J., De Samblanx, G. W., Buchala, A., Métraux, J. P., Manners, J. M., and Broekaert, W. F., Pathogen-induced systemic activation of a plant defense gene in Arabidopsis follows a salicylic acid-independent pathway, *Plant Cell,* 8, 2309, 1996.
120. Hammond-Kosack, K. E., Silverman, P., Raskin, I., and Jones, J. D. G., Race-specific elicitors of *Cladosporium fulvum* induce changes in cell morphology and the synthesis of ethylene and salicylic acid in tomato plants carrying the corresponding *Cf* disease resistance gene, *Plant Physiol.,* 110, 1381, 1996.
121. Doares, S. H., Narvaez-Vasquez, J., Conconi, A., and Ryan, C. A., Salicylic acid inhibits synthesis of proteinase inhibitors in tomato leaves induced by systemin and jasmonic acid, *Plant Physiol.,* 108, 1741, 1995.
122. Papastoitsis, G. and Wilson, K. A., Initiation of the degradation of the soybean Kunitz and Bowman-Birk trypsin inhibitors by a cysteine protease, *Plant Physiol.,* 96, 1086, 1991.
123. Vera, P. and Conejero, V., Pathogenesis-related proteins of tomato. P-69 as an alkaline endoproteinase, *Plant Physiol.,* 87, 58, 1988.
124. Tornero, P., Conejero, V., and Vera, P., Primary structure and expression of a pathogen-induced protease (PR-P69) in tomato plants: similarity of functional domains to subtilisin-like endoproteases, *Proc. Natl. Acad. Sci. U.S.A.,* 93, 6332, 1996.
125. Rodrigo, I., Vera, P., van Loon, L. C., and Conejero, V., Degradation of tobacco pathogenesis-related proteins. Evidence for conserved mechanisms of degradation of pathogenesis-related proteins in plants, *Plant Physiol.,* 95, 616, 1991.
126. Suh, S. G., Stiekema, W. J., and Hannapel, D. J., Proteinase inhibitor activity and wound-inducible gene expression of the 22 kDa potato-tuber proteins, *Planta,* 184, 423, 1991.
127. Atkinson, H. J., Urwin, P. E., Clarke, M. C., and McPherson, M J., Image analysis of the growth of *Globodera pallida* and *Meloidogyne incognita* on transgenic tomato roots expressing cystatins, *J. Nematol.,* 28, 209, 1996.

7 The Response of Plant Cell Wall Hydroxyproline-Rich Glycoproteins to Microbial Pathogens and Their Elicitors

Marie-Thérèse Esquerré-Tugayé, C. Campargue, and D. Mazau

CONTENTS

7.1 Introduction 157
7.2 HRGP-Extensin 158
  7.2.1 Structural Features 158
  7.2.2 Accumulation in Diseased Plants 159
  7.2.3 Gene Expression and Regulation 161
  7.2.4 Role in Defense 163
7.3 Other HYP-Containing Glycoproteins 164
7.4 Conclusion 165
References 166

7.1 INTRODUCTION

The concept of active defense arose at the beginning of the 20th century from the very first experiments of induced resistance against plant pathogens by previous inoculation of their hosts with related or unrelated strains.[1] Later on, it was shown that cross protection can be applied to several plant species, notably against viral, bacterial, and fungal diseases.[2] The term "acquired resistance" is now used to define the ability of unchallenged plant tissues to gain a certain level of resistance once other plant parts have been induced by a given stimulus.[3]

Inherent in the concept of protection against diseases is the likeliness that *defense molecules* account for the limitation of pathogen spread. The discovery that low-

0-8493-0697-3/99/$0.00+$.50

molecular-weight (MW) secondary metabolites, thereafter designated *phytoalexins*, display strong antimicrobial activities was the first illustration that defense molecules can be induced in plants.[4] Evidence that other classes of compounds, notably proteins, participate in the defense of plants was obtained much later. The finding that small proteins endowed with proteinase inhibitory activity are synthesized upon mechanical and insect wounding suggested that they might protect plants against pests and against microbial digestion by proteolytic enzymes.[5] Proteinases and other hydrolases play important roles in microbial pathogenicity. *Colletotrichum lagenarium*, a fungal pathogen of melon and other cucurbits, for instance, synthesizes serine proteinases.[6] Looking at melon *cell wall proteins* as possible substrates for these enzymes led us to show that, instead of being diminished, the cell walls were highly enriched in proteins, notably in the amino acid *hydroxyproline* (Hyp), which typifies plant cell wall proteins.[7] After the work of Lamport and Northcote,[8] and Dougall and Shimbayashi,[9] the unequivocal presence of a Hyp-containing protein in plant cell walls was admitted. Lamport called it "extensin,"[10] assuming its involvement in cell wall extension. It was subsequently found that extensin is a glycoprotein consisting of a large number of serine–hydroxyproline$_4$ (Ser–Hyp$_4$) repeats, bearing galactose and oligoarabinosides O-glycosidically linked to Ser and Hyp, respectively.[11] Likewise, the peptides isolated from the cell walls of infected melons were shown to contain the Ser–Hyp$_4$ sequences and the glycosylation pattern of Hyp matched the glycosylation of extensin[12] — hence the demonstration that melon cell walls are highly enriched in this glycoprotein upon infection. The biological significance of this response was addressed by using melon seedlings where the level of Hyp was artificially increased by ethylene treatment or lowered by growing them in the presence of L-trans Hyp, a specific inhibitor of prolyl-hydroxylase. An inverse relationship was found between the amount of cell wall Hyp and the susceptibility to *Colletotrichum lagenarium*.[13] The most obvious interpretation of these experiments was that enrichment in Hyp-rich glycoprotein is a defense response. To avoid any interference with the proposed yet undemonstrated role in extension, this glycoprotein was abbreviated as HRGP.[12]

Since then, cloning of the genes encoding HRGPs from carrot[14] and other plant species (for a review, see reference 15) has confirmed and completed our knowledge of the molecule. The notion of Hyp-containing proteins has been extended to three additional classes of cell surface compounds, namely arabinogalactan proteins (AGPs), Solanaceous lectins, and Hyp/Proline-rich proteins (H/PRPs). Of these four classes, only the response of *HRGP-extensin* is well documented upon infection by pathogens. This review will mainly focus on this glycoprotein.

7.2 HRGP-EXTENSIN

7.2.1 Structural Features

HRGP-extensin, hereafter referred to as HRGP, is probably the most abundant and best characterized plant cell wall protein (for a review, see references 15 and 16). At the structural level, it is typified by several specific markers whose nature was summarized by Kieliszewsky and Lamport.[17] Thus, identification of wall-bound Hyp constitutes a diagnosis for the presence of HRGP, because this glycoprotein is the main cell wall-immobilized, Hyp-containing component. In most dicotyledonous species, Hyp is found throughout the peptide backbone under the form of –Ser–Hyp_4– repeats interspersed by other peptide motifs rich in lysine and tyrosine. Variants of the Ser–Hyp repeats are found in some species, notably in monocots. Submitting cell walls to partial hydrolysis by barium hydroxide and trypsin digestion allows recovery of various glycan and peptide portions of HRGP. Their analysis showed that most Hyp are glycosylated by 1 to 4 L-arabinofuranose residues, and serine by 1 galactose residue. In some of the peptides, a novel tyrosine derivative, isodityrosine (IsodiTyr) was found. This dimer accounts for intra- and possibly intermolecular linkages, and constitutes an additional structural biochemical marker of wall-bound HRGP.

Due to its rapid immobilization once in the cell wall, it has been difficult to solubilize HRGP in the native form. Actually, it was only in 1980 that it was first isolated by Stuart and Varner,[18] and then in other laboratories.[19-21] It soon appeared that, at least in dicots, several distinct HRGPs are present in the cell walls. Although they differ in their MW and peptide pattern, they all contain numerous Ser–Hyp_4 motifs, and their glycan portion represents one third to one half of the molecule. According to the molecular markers which typify HRGPs, a rod-like structure was predicted for these glycoproteins. Indeed, this was confirmed by electron microscopy observation of the solubilized molecule.[22]

Cloning of HRGP genes, first from carrots[14] and then from other plant species,[5] has extended our knowledge of these glycoproteins and confirmed that these are encoded by a small multigene family. Some of the genes are constitutively expressed in relation to plant development, albeit at a generally low level.

7.2.2 Accumulation in Diseased Plants

In response to pathogens and their signals, plant cell walls become highly enriched in HRGPs. Since the first report that they accumulate in *Colletotrichum lagenarium*-infected melon seedlings,[7,23] their *increase* has been shown in a large number of plant–pathogen interactions (Table 7.1). To date, this observation has been reported in more than 30 different systems.[23-41] Many of these correspond to important agricultural diseases on dicots (potato, tomato, tobacco, sunflower, grapevine, bean, etc.) and monocots (wheat, barley) which affect the aerial part as well as the root system of the plant. These are caused by a wide variety of fungi (Oomycetes, Ascomycetes, Basidiomycetes, Deuteromycetes), bacteria (*Pseudomonas, Xanthomonas*), and other pathogens (viruses, nematodes). These are also increased during symbiosis upon interaction of plant roots with mycorrhizal fungi and in *Rhizobium*-induced legume nodules. In race-cultivar specific systems, these accumulate earlier during incompatible interactions (host resistant, pathogen avirulent) than during compatible ones (host susceptible, pathogen virulent).

In order to study HRGP responses at the *cellular level,* antibodies against the glycoprotein or against its peptide portion (HRP) were prepared. The use of ultrastructural cytochemistry and gold labeling techniques showed that HRGPs accumulate particularly in defense structures such as papillae and paramural deposits.[25] Such

TABLE 7.1
Plant-Pathogen Interactions which Affect HRGP-Extensin Levels

Plants	Fungi	Bacteria	Viruses/ Nematodes	References
CUCURBITACEAE				
Melon	*Colletotrichum lagenarium*	*Pseudomonas syr. phaseolicola*		13, 23, 24, 25
	Fusarium oxysporum melonis	*Pseudomonas fluorescens*		
Cucumber	*Pythium sp.*			24, 26
	Cladosporium cucumerinum			
FABACEAE				
Alfalfa	*Colletotrichum trifolii*			24
French bean	*Colletotrichum lindemuthianum*	*Pseudomonas syr. phaseolicola*		24, 25
	Uromyces appendiculatus			
Vigna	*Uromyces vignae*	*Rhizobium NGR 234*		27, 28
Pea	*Glomus versiforme*			29
SOLANACEAE				
Tobacco	*Peronospora tabacina*	*Pseudomonas solanacearum*	YSMV, TMV	24, 30, 31
			Meloidogyne inc. *Globodera tabacum*	32
Potato		*Pseudomonas solanacearum*	Leaf roll virus	24, 33
		Erwinia carotovora		34
Tomato	*Fusarium oxysporum*	*Clavibacter michiganense*	*Meloidogyne inc.*	35, 36, 37
VITACEAE				
Grapevine	*Plasmopara viticola*			24
ASTERACEAE				
Sunflower	*Sclerotinia sclerotiorum*			38
POACEAE				
Wheat	*Erysiphe graminis*			39
	Puccinia graminis			40
	Puccinia recondita			40
Barley	*Erysiphe graminis*			40

TABLE 7.1 *(continued)*
Plant-Pathogen Interactions which Affect HRGP-Extensin Levels

Plants	Fungi	Bacteria	Viruses/ Nematodes	References
LILIACEAE				
Leek	*Glomus versiforme*			29
BRASSICACEAE				
Turnip		*Xanthomonas campestris*		41

structures are formed in close contact to fungi and bacteria, by invagination of the host plasma membrane, and are characterized by massive deposition of defense compounds in the newly expanded plasmalemma–cell wall interface. Among them, electron dense deposits are often observed. In the incompatible interaction between *Phaseolus vulgaris* cv Kievitsboon Koekoek and race β of *Colletotrichum lindemuthianum,* the fungal hyphae appear encased in the papillae and are highly altered.[25] HRGPs are clearly localized inside and at the periphery of the papillae. When plants are infected by obligate biotroph parasites[40] or by symbiots,[29,42] HRGPs are secreted at the host–microbe interface. Together with other newly synthesized compounds, notably AGPs and wall polysaccharides, these contribute to the formation of an extrahaustorial matrix whose protective role against colonization has been proposed.[40]

7.2.3 Gene Expression and Regulation

HRGP cDNA and genomic clones have been obtained from several species and used as homologous or heterologous probes to study HRGP gene expression. HRGPs are encoded by a multigene family whose constitutive level is generally low in healthy plants except in some organs, notably roots, stems, and seed coats, and in tissue culture. In response to pathogen ingress, a strong activation of HRGP genes occur at the transcriptional level, which leads to the induction of several transcripts ranging from about 1.5 to 6 kb.[43,44] In a given plant species, these are differentially expressed, both within a given cultivar and between cultivars. Thus, two mRNA species of 3.6 and 4.1 kb strongly accumulate during the incompatible interaction of *Phaseolus vulgaris* cv. Kievitsboon Koekoek with *Colletotrichum lindemuthianum* race β, whereas a 2.13 kb mRNA accumulates at late stages of the compatible interaction between the same cultivar and race γ.[45] Similar kinetics have been obtained in other systems, notably in sunflower seedlings tolerant vs. susceptible to *Sclerotinia sclerotiorum.*[38] In the compatible interaction between tobacco and the nematode *Meloidogyne javanica,* three transcripts of 1.4, 3.5, and 6 kb are induced in a coordinate manner; none of them is induced preferentially, but their expression appears to be tissue-specific, i.e., in close contact and at a distance from the pathogen as shown by *in situ* hybridization, histochemical GUS assay, and protein immunolocalization studies.[32] This observation confirms the local and systemic induction of HRGP, already reported in melon and bean anthracnose, for example.

The pattern of *HRGP induction* suggested that various *signals* are involved in the regulation of its expression. The concept of elicitors of plant defense and of secondary messengers emerged in the late 1970s and led to showing that fungal *elicitors* (from *Colletotrichum lagenarium*) have the ability to induce HRGP biosynthesis.[46] Since then, the induction of HRGPs by various elicitors has been demonstrated, notably originating from *Colletotrichum lindemuthianum* and *Phytophthora parasitica nicotianae.*[43,47] Elicitors were originally defined as those molecules present at the cell surface or secreted into the culture medium of plant pathogens, which have the same ability to induce the defense system of plants as the pathogens these are derived from.[48] The notion of elicitor has been extended to plant endogenous compounds, notably oligogalacturonides (OGA) derived from the cell wall.[48] Thus, plants contain their own elicitors of defense. It appeared recently that, besides elicitors, plant cell walls contain complex OGAs, which suppress HRGP gene expression in bean seedlings, most particularly the 2.5 kb transcript.[49] These fragments are released from pectic polysaccharides by pectinases, notably the endopolygalacturonase of *Colletotrichum lindemuthianum.* HRGP suppression might result in cell wall architecture remodeling, thus facilitating local degradation by the pathogen and penetration into the host. A similar effect was recently shown to affect three HRGP transcripts in root hairs of *Vigna unguiculata* in response to *Rhizobium* infection.[28]

In addition to the effect of microbial elicitors, the release of endogenous elicitors under stress conditions, such as wounding by pathogens, might account for the HRGP response. Indeed, cell wall enrichment in this glycoprotein upon *wounding* is a general phenomenon which has been reported in dicotyledonous as well as monocotyledonous species.[14,50-52] Again, a differential response is observed in relation to organs, tissues, and genes. Thus, in *Brassica*, ext A, which encodes an HRGP normally expressed in roots, is mainly stimulated in the aerial part of the plant upon wounding.[52] In tomato, it is essentially the 4.0 kb transcript which is induced,[50] and in bean seedlings, the 2.13, 3.6 and 4.1 kb transcripts show differential kinetics.[45] Immunolocalization, *in situ* hybridization, and analysis of transgenic plants containing a construct consisting of the HRGP promotor–reporter *GUS* gene fusion, have shown that enrichment of the cell walls in HRGP occurs at the level of the vascular system in *Zea mays* and *Brassica napus.*[51,53]

Induction of HRGP at a distance from the point of inoculation by pathogens or from the wounded site has led to the search for systemic signals. *Ethylene* was the first signal to be identified.[54,55] This gaseous plant hormone is abundantly synthesized in response to infection, elicitors, and wounding. This might explain why environmental and developmental factors influence HRGP biosynthesis, as well as other defense genes. Treating melon plants with ethylene resulted in enrichment of the cell wall in HRGP and, correlatively, to the induction of resistance against *C. lagenenarium.*[13] Cassab et al.[56] have shown by tissue printing of pea epicotyls that the glycoprotein is increased at the level of the vascular system and also in cortical and epidermal cells. A functional analysis of the pDC5A1 HRGP gene from carrot allowed identification of nucleotide sequences in the promoter, and nuclear proteins involved in negative or positive regulation of the gene in response to wounding and ethylene.[57-59] This is consistent with the induction and suppression of HRGP gene expression, notably with the negative regulation effects of complex OGAs and *Rhizobium* reported above. By using promoter–reporter gene fusion constructs consisting of different 5′ parts of the bean 4.1 kb transcript-encoding gene and glucuronidase as reporter gene, Wycoff et al.[60] have shown that tissue specificity and wound induction lie in a region between –94 and –251 relative to the transcription start site and that activation by infection lies outside this region.

In the presence of the AVG and cobalt chloride inhibitors of ethylene biosynthesis, elicitor-induced HRGP accumulation in the cell wall of melon,[46] as well as wound-induced HRGP gene expression in *Brassica napus,*[52] are lowered. However, studying the role of ethylene has shown that this hormone is not the only secondary messenger that regulates HRGP biosynthesis. Elicitor-treated cell suspension cultures provide an easy experimental system to look for other putative signals. Thus, changes in the extracellular pH, ion fluxes (K^+ efflux, Ca^{2+} influx), production of active oxygen species, and synthesis of secondary metabolites have been measured. Among the latter, *jasmonic acid* (JA) derived from polyunsaturated fatty acids via the lipoxygenase (LOX) pathway has been shown to induce HRGP in tobacco.[47] This is a case of convergence with animal systems, because JA is a compound homologous to arachidonic acid derived-prostaglandins. These eicosaenoid compounds play important roles in the inflammatory reaction during which collagen, an HRGP of the extracellular matrix of animal cells, is increased. The mechanism(s) involved in JA recognition and regulatory effects on HRGP gene expression remain to be elucidated.

7.2.4 Role in Defense

The fact that HRGPs generally accumulate under stress conditions suggests that this response is of prime importance for plant survival. Their most likely roles lie in their structure and their agglutinating properties. Thus, it is assumed that HRGPs play a role in *cell wall structure* and self assembly. The peptide backbone is subjected to posttranslational modifications which include hydroxylation of proline, glycosylation, and cross-linking. Glycosylation by oligoarabinosides confers molecular inflexibility and stabilizes the glycoprotein in a rod like structure.

When HRGP monomers are secreted into the cell wall, they readily associate ionically to other polymers and, at this stage, they can be eluted by salt solutions.[18,20] However, after a few hours, these become tightly linked by covalent bonds to the wall network, which renders them highly insoluble. This phenomenon is even more striking upon infection, elicitation, and wounding, through intra-, and probably intermolecular oxidative *cross-linking of HRGP* monomers, in the presence of H_2O_2 and peroxidase, whose production and activity are increased in parallel.[61-63] Schnabelrauch et al.[64] have identified a pI 4.6 anionic peroxidase, whose most simple substrate contains the –Val–Tyr–Lys– motif of HRGP. Oxidative cross-links are supposed to occur between two tyrosine residues located on the same polypeptide backbone or on two adjacent backbones, allowing the formation of isodityrosine (IDT) bridges. IDT and a recently identified tetramer tyrosine derivative (Di–IDT) have been respectively identified in HRGP[65] and cell wall[66] hydrolysates. Such cross-links might create a network of HRGPs interpenetrated by other wall polymers, thus contributing to cell wall reinforcement and repair upon pathogen attack. The fact that it is more difficult to solubilize the cell wall and obtain protoplasts from elicitor-treated than from control cells[61] supports this view.

Protein cross-linking occurs early during incompatibility between soybean and *Pseudomonas syringae glycinea* and later in the compatible interaction.[61] This phenomenon precedes the transcriptional activation of HRGP and other defense genes and thus represents one of the very first events of the hypersensitive response.

Besides having a structural role, HRGPs are also characterized by a high proportion of the basic amino acids lysine and histidine. As such, they behave as *polycations* and display agglutinating activities against negatively charged surfaces. Thus, it was found that two HRGPs, from tobacco and potato,[30,33] agglutinate avirulent strains of *Pseudomonas solanacearum* on these plants, but only weakly virulent ones. In addition, the tobacco glycoprotein whose protein portion contains 12% lysine and 5% histidine agglutinates *Phytophthora parasitica nicotianae* zoospores in a manner which is inhibited by the lipopolysaccharide of *Pseudomonas* or by increasing the ionic strength of the medium.[30] It is believed that the agglutinating properties of HRGPs participate in the immobilization of bacteria observed during incompatibility. Indeed, O'Connell et al.[25] have shown by ultrastructural immunocytochemistry that bacteria appear encased in an HRGP-containing material of host origin during hypersensitive responses.

At the present time, more direct proof of the role(s) played by HRGPs is not available. In this context, HRGP mutants as well as transgenic plants under- or over-expressing HRGPs would be of great value. While such constructs have been introduced in tobacco, the authors did not apparently check the disease phenotype of such plants.[31]

7.3 OTHER HYP-CONTAINING GLYCOPROTEINS

Arabinogalactan proteins (AGPs), Solanaceous lectins, Hyp/Proline-rich proteins (H/PRPs), and other hybrid proteins contain Hyp.

The AGPs are a group of Hyp-rich proteoglycans widely distributed in plants, whose protein portion represents less than 10% of the total mass of the molecule.[15] Hyp, alanine, often encountered as Ala–Hyp repeats, together with serine, proline, and threonine, are the major amino acids. The carbohydrate component consists of arabinogalactan chains O-glycosidically linked to Hyp and Ser. These are predominantly found in the intercellular spaces and in the growth medium of cell suspension cultures.[67] In contrast to the cell wall HRGP-extensin, these are highly soluble even at low ionic strength. These are supposed to be involved in some stages of plant development and as adhesion molecules in cell–cell interactions. In this respect, it should be interesting to look for a possible role in plant–microorganism interactions. The recent cloning of genes encoding the protein portion of several AGP genes[68-70] will be of great help in understanding the function of these compounds.

The *Solanaceous lectins* are found in the cell walls of Solanaceous plants.[15] A well-known example is the potato tuber lectin (PTL).[71,72] PTL is a glycoprotein of molecular mass 50 kDa which consists of 50% protein and 50% carbohydrate and is composed of two domains: a serine and Hyp-rich domain closely resembling the wall HRGP and a glycine and cysteine-rich domain responsible for the chitin-binding activity of the molecule. This activity might be involved in interactions with chitin and N-acetyl glucosamine-containing pathogens. The fact that PTL accumulates in potato tubers upon wounding and viral infection[15,73] and that another Hyp-rich lectin isolated from bean cells is increased in response to a fungal elicitor[74] supports the view that such lectins might interact with plant pathogens, notably those containing chitin. Further experiments are required to check this point.

The *Hyp/Proline rich-proteins (H/PRP)* are a class of cell wall proteins whose molecular characterization has shown the presence of Pro–Pro repeats and Val–Tyr–Lys sequences.[15,75] These are associated with plant development, and the ENOD subclass is present in the legume root nodules formed in response to *Rhizobium* infection.[76] Some of these PRPs have been isolated and found to contain Hyp in the second position of the Pro–Pro repeats.[77] H/PRP gene expression is altered by wounding, ethylene, and elicitor treatment,[77,78] which is consistent with a role in plant–microorganism interactions. Their high lysine content, as well as their insolubilization once in the cell wall, resemble some of the HRGP characteristic features. However, their precise functions are not yet known.

Class I chitinases represent another class of Hyp-containing hybrid molecules. The two isoforms of tobacco are very similar and are characterized by a chitin-binding domain and a catalytic domain linked together by a spacer containing a few residues of threonine and hydroxyproline.[79,80] In contrast to the lectins, they are not glycosylated, and they are only present in the vacuole. They exhibit a high antifungal activity *in vitro*.

7.4 CONCLUSION

Cell surfaces play a key role in plant–microorganism interactions because they are the source of signal molecules which elicit defense responses in the host and the site at which elicitor-induced defense molecules accumulate. One of the most general and best characterized responses is the accumulation of hydroxyproline-rich glycoproteins. It fulfills the triple criteria of defense in that it is early and locally induced at sites where the pathogen is restricted during hypersensitivity; it is delayed and more extended in compatible interactions and might contribute to lesion limitation; it participates in systemic acquired resistance in that it is elicited by systemic signals and is involved in plant protection against diseases.

Considering the structural features and polycationic properties of HRGPs, one may assume that varying the level of HRGP results in wall remodeling and modifications of the net charge of the cell surface. The likely interaction of polycationic HRGPs with polyanionic pectic polysaccharides or other molecules might be highly altered in these conditions. Engineering plants which over- or under-express the HRGP genes of interest will certainly improve our understanding of the role of HRGPs in plant–pathogen interactions in the future.

REFERENCES

1. Chester, K. S., The problem of acquired physiological immunity in plants, *Quar. Rev. Biol.,* 8, 129, 1933.
2. Kuc, J., Plant immunization-mechanisms and practical implications, in *Active Defense Mechanisms in Plants,* Wood, R. K. S., Ed., Plenum Press, NATO-ASI, New York, 1982, 157.
3. Lawton, K., Uknes, S., Friedrich, L., Gaffney, T., Alexander, D., Goodman, R., Métraux, J. P., Kessmann, H., Ahl-Goy, P., Gut Rella, M., Ward, E., and Ryals, J., The molecular biology of systemic acquired resistance, in *Mechanisms of Plant Defense Responses,* Fritig, B. and Legrand, M., Eds., Kluwer Academic Publishers, Netherlands, 1993, 422.
4. Cruickshank, I. A. M., Phytoalexins, *Annu. Rev. Phytopathol.,* 1, 351, 1963.
5. Ryan, C. A., Proteolytic enzymes and their inhibitors in plants, *Annu. Rev. Plant Physiol.,* 24, 173, 1973.
6. Pladys, D., Esquerré-Tugayé, M. T., and Touzé, A., Purification and partial characterization of proteolytic enzymes in melon seedlings infected by *Colletotrichum lagenarium, Phytopathol. Z.,* 102, 266, 1981.
7. Esquerré-Tugayé, M. T., Influence d'une maladie parasitaire sur la teneur en hydroxyproline des parois cellulaires d'épicotyles et de pétioles de plantes de melon, *C.R. Acad. Sci.,* 276, 525, 1973.
8. Lamport, D. T. A. and Northcote, D. H., Hydroxyproline in primary cell walls of higher plants, *Nature,* 188, 665, 1960.
9. Dougall, D. K. and Shimbayashi, K., Factors affecting growth of tobacco callus tissue and its incorporation of tyrosine, *Plant Physiol.,* 35, 396, 1960.
10. Lamport, D. T. A., The protein component of primary cell walls, *Adv. Bot. Res.,* 2, 151, 1965.
11. Lamport, D. T. A., The glycopeptide linkage of extensin: O-D-galactosyl serine and O-L-arabinosyl hydroxyproline, in *Biogenesis of Plant Cell Wall Polysaccharides,* Loewus, L., Ed., Academic Press, New York, London, 1973, 149.
12. Esquerré-Tugayé, M. T. and Lamport, D. T. A., Cell surfaces in plant-microorganism interactions. I. A structural investigation of cell wall hydroxyproline-rich glycoproteins which accumulate in fungus-infected plants, *Plant Physiol.,* 64, 314, 1979.
13. Esquerré-Tugayé, M. T., Lafitte, C., Mazau, D., Toppan, A., and Touzé, A., Cell surfaces in plant-microorganism interactions. II. Evidence for the accumulation of hydroxyproline-rich glycoproteins in the cell wall of diseased plants as a defense mechanism, *Plant Physiol.,* 64, 320, 1979.
14. Chen, J. and Varner, J. E., An extracellular matrix protein in plants: characterization of a genomic clone for carrot extensin, *EMBO J.,* 4, 2145, 1985.
15. Showalter, A. M., Structure and function of plant cell wall proteins, *Plant Cell,* 5, 9, 1993.
16. Cassab, G. I. and Varner, J. E., Cell wall proteins, *Annu. Rev. Plant Physiol. Plant Mol. Biol.,* 39, 321, 1988.
17. Kieliszewski, M. J. and Lamport, D. T. A., Extensin: repetitive motifs, functional sites, post-translational codes, and phylogeny, *Plant J.,* 5, 157, 1994.
18. Stuart, D. A. and Varner, J. E., Purification and characterization of a salt-extractable hydroxyproline-rich glycoprotein from aerated carrot discs, *Plant Physiol.,* 66, 787, 1980.

19. Leach, J. E., Cantrell, M. A., and Sequeira, L., Hydroxyproline-rich bacterial agglutinin from potato, *Plant Physiol.,* 70, 1353, 1982.
20. Smith, J. J., Muldoon, E. P., and Lamport, D. T. A., Isolation of extensin precursors by direct elution of intact tomato cell suspension cultures, *Phytochemistry,* 23, 1233, 1984.
21. Mazau, D., Rumeau, D., and Esquerré-Tugayé, M. T., Two different families of hydroxyproline-rich glycoproteins in melon callus, *Plant Physiol.,* 86, 540, 1988.
22. Heckman, J. W., Jr., Terhune, B. T., and Lamport, D. T. A., Characterization of native and modified extensin monomers and oligomers by electron microscopy and gel filtration, *Plant Physiol.,* 86, 848, 1988.
23. Esquerré-Tugayé, M. T. and Mazau, D., Effect of a fungal disease on extensin, the plant cell wall glycoprotein, *J. Exp. Bot.,* 25, 509, 1974.
24. Mazau, D. and Esquerré-Tugayé, M. T., Hydroxyproline-rich glycoprotein accumulation in the cell walls of plants infected by various pathogens, *Physiol. Mol. Plant Pathol.,* 29, 147, 1986.
25. O'Connell, R. J., Brown, I. R., Mansfield, J. W., Bailey, J. A., Mazau, D., Rumeau, D., and Esquerré-Tugayé, M. T., Immunocytochemical localization of hydroxyproline-rich glycoprotein accumulating in melon and bean at sites of resistance to bacteria and fungi, *Mol. Plant-Microbe Interact.,* 3, 33, 1990.
26. Hammerschmidt, R., Lamport, D. T. A., and Muldoon, E. P., Cell wall hydroxyproline enhancement and lignin deposition as an early event in the resistance of cucumber to *Cladosporium cucumerinum, Physiol. Plant Pathol.,* 24, 43, 1984.
27. Stark-Urnau, M. and Mendgen, K., Sequential deposition of plant glycoproteins and polysaccharides at the host-parasite interface of *Uromyces vignae* and *Vigna sinensis.* Evidence of endocytosis and secretion, *Protoplasma,* 186, 1, 1995.
28. Arsenijevic-Maksimovic, I., Broughton, W. J., and Krause, A., Rhizobia modulate root-hair-specific expression of extensin genes, *Mol. Plant-Microbe Interact.,* 10, 95, 1997.
29. Bonfante-Fasolo, P., Tamagnone, L., Peretto, R., Esquerré-Tugayé, M. T., Mazau, D., Mosiniak, M., and Vian, B., Immunocytochemical location of hydroxyproline rich glycoproteins at the interface between a mycorrhizal fungus and its host plant, *Protoplasma,* 165, 127, 1991.
30. Mellon, J. E. and Helgeson, J. P., Interaction of a hydroxyproline-rich glycoprotein from tobacco callus with potential pathogens, *Plant Physiol.,* 70, 401, 1982.
31. Memelink, J., Swords, K. M. M., de Kam, R. J., Schilperoort, R. Z., Hoge, J. H. C., and Stachelin, L. A., Structure and regulation of tobacco extensin, *Plant J.,* 4, 1011, 1993.
32. Niebel, A., de Almeida Engler, J., Tiré, C., Engler, G., Van Montagu, M., and Choyeen, G., Induction patterns of an extensin gene in tobacco upon nematode infection, *Plant Cell,* 5, 1698, 1993.
33. Leach, J. E., Cantrell, M. A., and Sequeira, L., A hydroxyproline-rich bacterial agglutinin from potato: its localization by immunofluorescence, *Physiol. Plant Pathol.,* 21, 319, 1982.
34. Rumeau, D., Maher, E. A., Kelman, A., and Showalter, A. M., Extensin and phenylalanine ammonia-lyase gene expression altered in potato tubers in response to wounding, hypoxia and *Erwinia carotovora* infection, *Plant Physiol.,* 93, 1134, 1990.
35. Benhamou, N., Mazau, D., Grenier, J., and Esquerré-Tugayé, M. T., Time-course study of the accumulation of hydroxyproline-rich glycoproteins in root cells of susceptible and resistant tomato plants infected by *Fusarium oxysporum* f. sp. *radicis-lycopersici, Planta,* 184, 196, 1991.

36. Benhamou, N., Cell surface interactions between tomato and *Clavibacter michiganense* subsp. *michiganense*: localization of some polysaccharides and hydroxyproline-rich glycoproteins in infected host leaf tissues, Physiol. *Mol. Plant Pathol.,* 38, 15, 1991.
37. Van der Eycken, W., de Almeida Engler, J., Inzé, D., Van Montagu, M., and Gheysen, G., A molecular study of root-knot nematode-induced feeding sites, *Plant J.,* 9, 45, 1996.
38. Mouly, A., Rumeau, D., and Esquerré-Tugayé, M. T., Differential accumulation of hydroxyproline-rich glycoproteins transcripts in sunflower plants infected with *Sclerotinia sclerotiorum* or treated with oxalic acid, *Plant Sci.,* 85, 51, 1992.
39. Clarke, J. A., Lisker, N., Lamport, D. T. A., and Ellingboe, A. H., Hydroxyproline enhancement as a primary event in the successful development of *Erysiphe graminis* in wheat, *Plant Physiol.,* 67, 188, 1981.
40. Hippe-Sanwald, S., Marticke, R. H., Kieliszewski, M. J., and Sommerville, S. C., Immunogold localization of THRGP-like epitopes in the haustorial interface of obligate, biotrophic fungi on monocots, *Protoplasma,* 178, 138, 1994.
41. Davies, H. A., Daniels, M. J., and Maxwell, D. J., Induction of extracellular matrix glycoproteins in *Brassica* petioles by wounding and in response to *Xanthomonas campestris, Mol. Plant-Microbe Interact.,* 10, 812, 1997.
42. Benhamou, N., Lafontaine, P. J., Mazau, D., and Esquerré-Tugayé, M. T., Differential accumulation of hydroxyproline-rich glycoproteins in bean root nodule cells infected with a wild-type strain or a C_4-dicarboxylic acid mutant of *Rhizobium leguminosarum* bv. *phaseoli, Planta,* 184, 457, 1991.
43. Showalter, A. M., Bell, J. N., Cramer, C. L., Bailey, J. A., Varner, J. E., and Lamb, C. J., Accumulation of hydroxyproline-rich glycoprotein mRNAs in response to fungal elicitor and infection, *Proc. Natl. Acad. Sci. U.S.A.,* 82, 6551, 1985.
44. Rumeau, D., Mazau, D., Panabières, F., Delseny, M., and Esquerré-Tugayé, M. T., Accumulation of hydroxyproline-rich glycoprotein mRNAs in infected or ethylene treated melon plants, *Physiol. Mol. Plant Pathol.,* 33, 419, 1988.
45. Corbin, D. R., Sauer, N., and Lamb, C. J., Differential regulation of hydroxyproline-rich glycoprotein gene family in wounded and infected plants, *Mol. Cell Biol.,* 7, 4337, 1987.
46. Roby, D., Toppan, A., and Esquerré-Tugayé, M. T., Cell surfaces in plant-microorganism interactions. V. Elicitors of fungal and of plant origin trigger the synthesis of ethylene and of cell wall hydroxyproline-rich glycoprotein in plants, *Plant Physiol.,* 77, 700, 1985.
47. Rickauer, M., Brodschelm, W., Bottin, A., Véronési, C., Grimal, H., and Esquerré-Tugayé, M. T., The jasmonate pathway is involved differentially in the regulation of different defense responses in tobacco cells, *Planta,* 202, 155, 1997.
48. Darvill, A. G. and Albersheim, P., Phytoalexins and their elicitors. A defense against microbial infection in plants, *Annu. Rev. Plant Physiol.,* 35, 243, 1984.
49. Boudart, G., Dechamp-Guillaume, G., and Lafitte, C., Elicitors and suppressors of hydroxyproline-rich glycoprotein accumulation are solubilized from plant cell walls by endopolygalacturonase, *Eur. J. Biochem.,* 232, 449, 1995.
50. Showalter, A. M., Zhou, J., Rumeau, D., Worst, S. G., and Varner, J. E., Tomato extensin and extensin-like cDNAs: structure and expression in response to wounding, *Plant Mol. Biol.,* 16, 547, 1991.

51. Tagu, D., Walker, N., Ruiz-Avila, L., Burgess, S., Martinez-Izquierdo, J. A., Leguay, J. J., Netter, P., and Puigdomènech, P., Regulation of the maize HRGP gene expression by ethylene and wounding. mRNA accumulation and qualitative expression analysis of the promoter by microprojectile bombardment, *Plant Mol. Biol.,* 20, 529, 1992.
52. Shirsat, A. H., Wieczorek, D., and Kozbial, P., A gene for *Brassica napus* extensin is differentially expressed on wounding, *Plant Mol. Biol.,* 30, 1291, 1996.
53. Shirsat, A. H., Wilford, N., Evans, I. M., Gatehouse, L. N., and Croy, R. R. D., Expression of a *Brassica napus* extensin gene in the vascular system of transgenic tobacco and rape plants, *Plant Mol. Biol.,* 17, 701, 1991.
54. Esquerré-Tugayé, M. T., Mazau, D., Toppan, A., and Roby, D., Elicitation, via l'éthylène, de la synthèse de glycoproteines pariétales associées à la défense des plantes, *Ann. Phytopathology,* 12, 403, 1980.
55. Toppan, A., Roby, D., and Esquerré-Tugayé, M. T., Cell surfaces in plant-microorganism interactions. III. In vivo effect of ethylene on hydroxyproline-rich glycoprotein accumulation in the cell wall of diseased plants, *Plant Physiol.,* 70, 82, 1982.
56. Cassab, G. I., Lin, J. J., Lin, L. S., and Varner, J. E., Ethylene effect on extensin and peroxidase distribution in the sub-apical region of pea epicotyls, *Plant Physiol.,* 88, 522, 1988.
57. Holdsworth, M. J. and Laties, G. G., Site binding of a nuclear factor to the carrot extensin gene is influenced by both ethylene and wounding, *Planta,* 179, 17, 1989.
58. Holdsworth, M. J. and Laties, G. G., Identification of a wound-induced inhibitor of a nuclear factor that binds the carrot extensin gene, *Planta,* 180, 74, 1989.
59. Granell, A., Peretó, J. G., Schindler, U., and Cashmore, A. R., Nuclear factors binding to the extensin promoter exhibit differential activity in carrot protoplasts and cells, *Plant Mol. Biol.,* 18, 739, 1992.
60. Wycoff, K. L., Powell, P. A., Gonzales, R. A., Corbin, D. R., Lamb, C., and Dixon, R. A., Stress activation of a bean hydroxyproline-rich glycoprotein promoter is superimposed on a pattern of tissue-specific developmental expression, *Plant Physiol.,* 109, 41, 1995.
61. Brisson, F. L., Tenhaken, R., and Lamb, C., Function of oxidative cross-linking of cell wall structural proteins in plant disease resistance, *Plant Cell,* 6, 1703, 1994.
62. Bradley, D. J., Kjellbom, P., and Lamb, C. J., Elicitor- and wound-induced oxidative cross-linking of a proline-rich plant cell wall protein: a novel, rapid defense response, *Cell,* 70, 21, 1992.
63. Wojtaszek, P., Trethowan, J., and Bolwell, G. P., Specificity in the immobilization of cell wall proteins in response to different elicitor molecules in suspension-cultured cells of French bean (*Phaseolus vulgaris* L.), *Plant Mol. Biol.,* 28, 1075, 1995.
64. Schnabelrauch, L. S., Kieliszcwski, M., Upham, B. L., Alizedeh, H., and Lamport, D. T. A., Isolation of pl 4.6 extensin peroxidase from tomato cell suspension cultures and identification of Val-Tyr-Lys as putative intermolecular cross-link site, *Plant J.,* 9, 477, 1996.
65. Epstein, L. and Lamport, D. T. A., An intramolecular linkage involving isodityrosine in extensin, *Phytochemistry,* 23, 1241, 1984.
66. Brady, J. D., Sadler, I. H., and Fry, S. C., Di-isodityrosine, a novel tetrameric derivative of tyrosine in plant cell wall proteins: a new potential cross-link, *Biochem J.,* 315, 323, 1996.
67. Gleeson, P. A., McNamara, M., Wettenhall, E. H., Stone, B. A., and Fincher, B. G., Characterization of the hydroxyproline-rich protein core of an arabinogalactan-protein secreted from suspension-cultured *Lolium multiflorum* (Italian ryegrass) endosperm cells, *Biochem J.,* 264, 857, 1989.

68. Du, H., Simpson, R. J., Moritz R. L., Clarke, A. E., and Bacic, A., Isolation of the protein backbone of an arabinogalactan-protein from the styles of *Nicotiana alata* and characterization of a corresponding cDNA, *Plant Cell,* 6, 1643, 1994.
69. Chen, C. G., Pu, Z. Y., and Moritz, R. L., Molecular cloning of a gene encoding an arabinogalactan-protein from pear (*Pyrus communis*) cell suspension culture, *Proc. Natl. Acad. Sci. U.S.A.,* 91, 10305, 1994.
70. Mau, S. l., Chen, C. G., Pu, Z. Y., Moritz, R. L., Simpson, R. J., Bacic, A., and Clarke, A. E., Molecular cloning of cDNA encoding the protein backbones of arabinogalactan-proteins from the filtrate of suspension-cultured cells of *Pyrus communis* and *Nicotiana alata, Plant J.,* 8, 269, 1995.
71. Allen, A. K., Desai, N. N., Neuberger, A., and Creeth, M. J., Properties of potato lectin and the nature of its glycopeptide linkages, *Biochem J.,* 171, 665, 1978.
72. Kieliszewski, M. J., Showalter, A. M., and Leykam, J. F., Potato lectin: a modular protein sharing sequence similarities with the extensin family, the hevein lectin family and snake venom desintegrins (platelet aggregation inhibitors), *Plant J.,* 5, 849, 1994.
73. Scheggia, C., Prisco, A. E., Dey, P. M., Daleo, G. R., and Pont Lezica, R., Alteration of lectin pattern in potato tuber by virus X, *Plant Science,* 58, 9, 1988.
74. Bolwell, G. P., Elicitor induction of the synthesis of a novel lectin-like arabinosylated hydroxyproline-rich glycoprotein in suspension cultures of *Phaseolus vulgaris* L., *Planta,* 172, 184, 1987.
75. Marcus, A., Greenberg, J., and Averyhart-Fullard, V., Repetitive proline-rich proteins in the extracellular matrix of the plant cell, *Physiol. Plant.,* 81, 273, 1991.
76. Franssen, H. J., Hap, J. P., Gloudemans, T., Stiekema, W., Van Dam, H., Govers, F., Louwerse, J., Van Kammen, A., and Bisseling, T., Characterization of cDNA for nodulin-75 of soybean. A gene product involved in early stages of root development, *Proc. Natl. Acad. Sci. U.S.A.,* 84, 4495, 1987.
77. Kleis-San Francisco, S. M. and Tierney, M. L., Isolation and characterization of a proline-rich cell wall protein from soybean seedlings, *Plant Physiol.,* 94, 1897, 1990.
78. Sheng, J., D'Ovidio, R., and Medhy, M. C., Negative and positive regulation of a novel proline-rich protein mRNA by fungal elicitor and wounding, *Plant J.,* 1, 345, 1991.
79. Sticher, L., Hofsteenge, J., Neuhaus, J. M., Boller, T., and Meins, F., Jr., Posttranslational processing of a new class of hydroxyproline-containing proteins. Prolyl hydroxylation and C-terminal cleavage of tobacco (*Nicotiana tabacum*) vacuolar chitinase, *Plant Physiol.,* 101, 1239, 1993.
80. Sticher, L., Hofsteenge, J., Milani, A., Neuhaus, J. M., and Meins, F. Jr., Vacuolar chitinases of tobacco: a new class of hydroxyproline-containing proteins, *Science,* 257, 655, 1992.

8 Induction of Peroxidase During Defense Against Pathogens

Jaishree M. Chittoor, Jan E. Leach, and Frank F. White

CONTENTS

8.1 Introduction 171
8.2 Peroxidase Enzyme Activity Correlates with Resistance 172
8.3 Characterization of Plant Peroxidases 173
 8.3.1 Structural and Functional Properties of the Plant Peroxidase Super-Family 173
 8.3.2 The Region Between F and G Helices May Contribute to Differences in Substrate Specificities 175
 8.3.3 Phylogenetic Clustering of Pathogen-Inducible Peroxidases 175
8.4 Organization of Plant Peroxidase Genes 181
8.5 Expression of Peroxidase Genes 183
 8.5.1 Expression of Peroxidase Genes During Pathogen Stress 183
 8.5.2 Regulation of Plant Peroxidase Genes During Pathogen Stress 184
8.6 Determination of the Potential Role of Peroxidase in Defense Against Pathogens by Suppression or Over-Expression of Peroxidase Genes in Transgenic Rice Plants 185
8.7 Conclusion 186
Acknowledgments 186
References 187

8.1 INTRODUCTION

Plant peroxidases have been studied extensively since the beginning of the 18th century, and the basic biochemical functions of these enzymes have been well characterized. The enzymes are heme-containing glycoproteins that catalyze the oxidation by H_2O_2 of a wide range of organic and inorganic substrates, such as cytochrome c, nitrite, leuco-dyes, ascorbic acid, indole amines, and iodide ion.[1] Peroxidases occur in numerous isoforms in both plants and animals. Plant peroxi-

0-8493-0697-3/99/$0.00+$.50

dases have been implicated in a wide range of physiological processes, such as auxin metabolism, ethylene biosynthesis, lignin formation, respiration, light-mediated processes, growth, and senescence. In addition, peroxidase activity has been correlated with plant defense against pathogens (for a review, see reference 1). However, the specific roles of peroxidase in many of these processes remain inconclusive and are still under investigation. In this review, we summarize the studies on induction of enzyme activity, protein structural features, and gene induction dynamics that have implicated peroxidase in the plant defense response. We also have used phylogenetic clustering of peroxidases to study similarities in pathogen-inducible peroxidases from diverse plant species. Finally, we discuss the methods for determining the role of specific peroxidases in defense.

8.2 PEROXIDASE ENZYME ACTIVITY CORRELATES WITH RESISTANCE

The plant cell wall constitutes one of the first lines of defense against pathogen invasion, and peroxidases are key enzymes in the wall-building processes. These processes include peroxidase-mediated oxidation of hydroxycinnamyl alcohols into free radical intermediates,[2] phenol oxidation,[3] polysaccharide cross-linking,[4] cross-linking of extensin monomers,[5] lignification,[6,7] and suberization.[8,9] Although direct involvement of any one type of peroxidase in defense has not been demonstrated conclusively, extracellular or wall-bound peroxidases have been proposed to enhance resistance by the construction of a cell wall barrier that may impede pathogen ingress and spread.[10-14] The accumulation of cell wall-strengthening materials following infection might be expected to correlate with enhanced resistance.[2,6,13,14]

The accumulation of lignin and phenolic compounds is correlated with resistance in a number of plant–pathogen interactions. The resistance response in wheat cultivar Prelude-Sr5 against an avirulent race of the stem rust fungus *Puccinia graminis* f. sp. *tritici* was correlated with rapid lignification in penetrated host cells.[15-17] In tomato, resistance to the fungal pathogen *Verticillium albo-atrum* was correlated more with a more rapid deposition of suberin and lignin in a resistant isoline than in a susceptible isoline.[18,19] Similarly, infiltration of rice leaves with suspensions of *Xanthomonas oryzae* pv. *oryzae*, the bacterial blight pathogen, caused the deposition of lignin-like polymers at the site of inoculation during resistant interactions.[20] In rice, the spatial and temporal patterns of phenolic polymer deposition were correlated with resistance; that is, the decrease in bacterial multiplication rates and onset of bacteriostasis.[20] No lignin deposition was observed at the site of infiltration following infection with bacteria during a susceptible response.[20]

Extensin, a cell wall-associated hydroxyproline-rich glycoprotein (HRGPs), also accumulates in the walls of a number of plant species during interactions with microbes (for a review, see reference 21). In parsley, HRGPs and peroxidase mRNA have been localized by *in situ* hybridization to the site of fungal infection.[22] HRGPs are thought to act as a matrix for deposition of phenolic materials, and oxidative cross-linking of these proteins into the wall by peroxidases has been proposed to strengthen cell walls. Alternatively, during the process of cross-linking, microbes may be agglutinated and thus immobilized.[21]

Because peroxidases are implicated in the deposition of cell wall-strengthening materials such as lignin, suberin, and extensin, changes in activities of peroxidase during resistant and susceptible interactions have also been monitored in a number of plant-pathogenic race-specific interactions. Increases in peroxidase activity that correlated with resistance between pathogens and their hosts were observed in rice,[23,24] wheat[25] (and the references therein), barley,[26,27] cotton,[28] and sugar cane.[29] In contrast, peroxidase enzyme activity in plants undergoing susceptible interactions was delayed or not induced within the time period studied.

Induction of peroxidase enzyme activity was also correlated with systemically induced resistance in cucumber. Inoculation of the first leaf of cucumber with a hypersensitive reaction-inducing bacterium *Pseudomonas syringae* pv. *syringae*, resulted in systemic resistance and accumulation of peroxidase within 24 h.[30] Activities of at least three acidic extracellular peroxidases increased with the appearance of systemic-induced resistance.[30-33] The same complement of peroxidases also was induced in response to a disease resistance-inducing heat shock treatment.[34] In potato, peroxidase enzyme activity increased during wound-healing responses, but an anionic peroxidase was induced only in cells undergoing suberization.[9]

Insight into the relationship between peroxidase and resistance also has come from studies of the symbiotic relationships of *Rhizobium* and host plants. Inoculation of incompatible strains of *Rhizobium* into white clover plants (heterologous interaction) resulted in a reaction that was similar to resistance. Peroxidase activity was elevated rapidly during these heterologous interactions, and the bacteria failed to induce nitrogen-fixing root nodules.[35] In contrast, during homologous or compatible interactions with *Rhizobium leguminosarum*, peroxidase activity in the root hair cells of white clover was lowered immediately and transiently to about 50% of the activity in uninoculated controls.

Overall, the studies correlating deposition of cell wall-strengthening materials and peroxidase activities are consistent with a role for this enzyme in defense through wall strengthening. However, in addition, the highly reactive oxygen species formed during the deposition of these compounds into the walls by peroxidase activity are likely toxic to pathogens. For example, H_2O_2 generated by wall-bound peroxidase or oxygenase during the oxidative burst, which occurs within minutes after pathogenic interaction, and free radicals, which are produced by extracellular peroxidase oxidative activity during cell wall fortification, are toxic to pathogens[36] (for a review, see reference 12). Alternatively, these active species may act as messengers to activate the plant defense responses that contribute to resistance.[37]

8.3 CHARACTERIZATION OF PLANT PEROXIDASES

8.3.1 Structural and Functional Properties of the Plant Peroxidase Super-Family

Peroxidases have been classified into two large groups, the plant peroxidase super-family and the animal peroxidase super-family. Peroxidases are also found in fungi and bacteria.[38-40] Because the amino acid sequences of peroxidases from fungi and bacteria are related to those of plant peroxidases both in their primary and tertiary structures, those enzymes are classified under the plant peroxidase super-family.[41]

The plant peroxidase super-family can be divided into three broad classes. Class I peroxidases are intracellular and prokaryotic in origin (cytochrome c peroxidase, bacterial peroxidases, chloroplast and cytosol ascorbate peroxidases). Class II includes extracellular fungal peroxidases (*Phanerochaete chrysosporium* lignin peroxidase and *Coprinus cinereus* peroxidase). Class III includes mostly extracellular plant peroxidases (horseradish peroxidase, Hrp C), which are secreted via the endoplasmic reticulum, contain structural Ca^{2+}, and disulfate bridges, and usually have glycosylated carbohydrate side chains. Some class III peroxidases are not extracellular and are targeted to the vacuole by a C-terminal propeptide.[41]

A general structure for peroxidases in the plant super-family was based initially on amino acid sequence similarities with yeast cytochrome c peroxidase, for which high-resolution crystal structure was available (Figure 8.1).[42-44] Subsequent crystallographic studies on Hrp E5, a highly basic isoenzyme of horseradish peroxidase,[45] a peanut peroxidase,[46] and fungal peroxidases[47-49] corroborated the structure of cytochrome c peroxidase as a valid prototype for the peroxidases in this super-family.

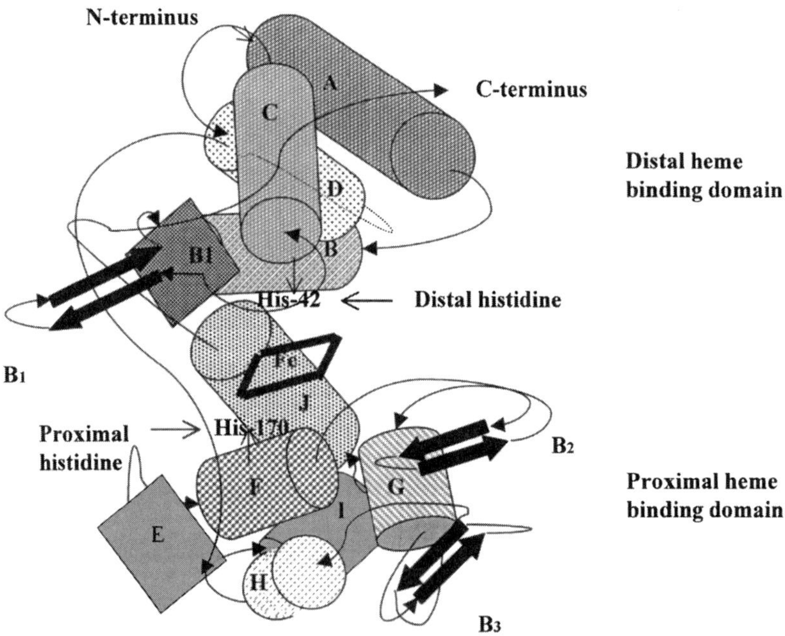

FIGURE 8.1 A diagrammatic representation of the structure of plant peroxidase. Ten helices (A–J) and β-sheets (β1, β2, and β3) are represented by cylinders and parallel arrows, respectively. Reverse turns and coils that connect the helices and the β-sheets are shown as connecting lines. The proximal and distal heme-binding domains and the proximal and distal histidine residues (Hrp C numbering) are indicated. The rectangular box present between the proximal and distal heme-binding regions indicates the position of the central heme prosthetic group. (Adapted from Welinder, K. G., *Eur. J. Biochem.*, 1985, 151, 497–504. With permission.)

In general, they are characterized by two structural domains that sandwich a central heme prosthetic group. The two domains containing the proximal and distal heme-binding region are created from 10 α-helices (A to J) and three β-sheets (1–3). The proximal and distal heme-binding regions may have originated by an early gene duplication event followed by gene fusion.[50] Invariant histidine residues in the proximal and distal heme-binding pockets (his-170 and his-42, using the Hrp C numbering) along with an arginine (arg-38), an asparagine (asn-70), and an aspartate (asp-247) are thought to be involved in peroxidase-specific catalysis. However, the catalytic sites of plant peroxidases, which catalyze a wide range of substrates, are almost invariant. Therefore, the differences in reactivities exhibited by peroxidases with peroxides and heme ligands probably result from other features of the enzyme.[51]

8.3.2 The Region Between F and G Helices May Contribute to Differences in Substrate Specificities

Nuclear magnetic resonance (NMR) and enzyme inactivation studies have indicated that substrate binds in the vicinity of the exposed edge of the heme group at the heme methyl $C_{18}H_3$.[52,53] Amino acid side chains located in the distal heme pocket or peripherally in the substrate access channel provide additional structural components for the substrate binding site, which extends from the heme edge to the protein exterior. In particular, amino acid side chains of an isoleucine residue, two phenylalanine residues (designated as PheA and PheB), and histidine (his-42, using the Hrp C numbering) are involved in substrate binding.[54-56] The exact identity of the isoleucine and phenylalanine residues is unclear. Studies by site-directed mutagenesis combined with NMR analyses implicate Phe179, Phe187, or Phe221 as the most likely candidate for Phe B, and Phe179 or Phe221 as the most likely candidate for Phe A.[57] Both Phe179 and Phe187 occur in the region between the F and G helices.

In that region, the numbers and identities of amino acid residues vary considerably among class III peroxidases and are distinct from those of residues of the class I and II peroxidases.[51] The variable region between the F and G helices is suggested to provide structural detail to the substrate access channel. This structural detail is proposed to account for some of the differences in substrate binding and preferences exhibited both within class III peroxidases and between the three different classes of peroxidase within the plant peroxidase super-family.[57]

8.3.3 Phylogenetic Clustering of Pathogen-Inducible Peroxidases

To determine the phylogenetic relationships between 62 different plant peroxidases (Table 8.1), the amino acid sequences between positions 38 and 286 (Hrp C numbering) were aligned. These sequences span the proximal and distal heme-binding regions (234 amino acids from the total of ~315 amino acids forming protein). The distances and phylogenetic relationships between the protein sequences were derived using the Kimura protein distance correction and neighbor joining method, respectively.[58-60] The N-terminal and C-terminal amino acid residues preceding the proximal and distal heme-binding domains were diverged considerably and, hence, were

TABLE 8.1
List of Peroxidase Genes

Gene locus	Plant	Accession #	Gene and Expression	Reference
MSRNAPE02	Alfalfa	971564	Expression not determined.	97
ALFPXDD	Alfalfa	537319	Unpublished.	98
			CLUSTER 1	
LETAP1*	Tomato	295815	Wound- and abscisic acid-inducible; localized to wound-healing suberizing cells. Induced in R tomato lines and not in near-isogeneic S line by elicitors of *Verticillium albo-atrum.* LETAP1 demonstrated by primer extension studies to be expressed in the periderm of wound-healing tomato.	65
POTPXD	Tomato	169555		82
LETAP2*	Tomato	295816		
			CLUSTER 2	
A				
TAPOX2*	Wheat	732972	Expressed in roots and induced in leaves during S interaction with EG.	70
TAPEROXIG*	Wheat	21831	Constitutively expressed in roots. Not inducible during resistant interaction with EGH or wounding in leaves.	89
TAPOX3*	Wheat	732974	Expressed in leaves. Not induced during S interactions with EG in leaves.	70
TAPERO	Wheat	X56011	Induced in leaves during resistant interactions with EGH.	80
HVPEROXI	Barley	22587	Induced in leaves during resistant interactions with EGH.	79
POXgX9*	Rice	AF014470	Expressed in the roots. Not induced in leaves during resistant interaction with XOO.	77
POX22.3*	Rice	AFO14467	Expressed in roots and leaves. Induced differentially in leaves during R and S interaction with XOO. Not wound-inducible in leaves.	77
POX8.1*	Rice	AFO14468	Induced differentially in leaves during R and S interaction with XOO. Wound-inducible in leaves.	77
TAPOX4*	Wheat	732976	Expressed in roots and leaves. Not induced during S interactions with EG.	70
TAPOX1*	Wheat	732970	Expressed in roots. Not induced in leaves during S interaction with EG.	70
POX5.1	Rice	AFO14469	Expressed in leaves. Wound-inducible. Not induced in leaves during R or S interaction with XOO.	77

B				
ARCPNC1	Peanut	M37636	Expression not determined.	99
SSNCAPE*	*Stylosanthes humilis*	577503	Induced in leaves following S interaction with *CG*, wounding, and treatment with defense regulator methyl jasmonate.	100
SSNPEROXIA	*S. humilis*	L36110	Expressed in leaves and roots. Induced in leaves during S interaction with CG. Wound-inducible in leaves.	67
MTU16727*	Barrel Medic	571484	Induced in roots by non-nodulating strain of *Rhizobium meliloti.*	61
PERX_BRARA	Turnip	464365	Expressed only in winter. Differs significantly in physio-chemical and enzymatic properties from other turnip peroxidases and Hrp C. Regions between F and G helices are most diverged between turnip and Hrp C peroxidases.	101
B56555 (PIR)	Wood tobacco	1076611	Expressed in protoplast at the beginning of the culture. Expression decreases at the onset of cell division. Expressed in roots. Not expressed in response to chemical treatment, heat shock, pathogen attacks, or tumor induction.	102
			CLUSTER 3	
CUSPREPERB	Cucumber	167533	Induced during systemic acquired resistance and by salicylic acid.	33
CUSPREPERA	Cucumber	167531		
CUSPREPER	Cucumber	167529		
CUSCUPER		167517	Induced by ethylene in the cotyledons.	86
			CLUSTER 4	
TOBPXDLF	Tobacco	170316	Abundant in stem. Expressed in lower amounts in roots and leaves. Lignification-associated.	64
TOBPXD	Tobacco	218308	Unpublished.	103
LECEVI1A	Tomato	29691[illegible]	Induced by viroid infection and ethylene.	87
POPPRXA1*	Populus	21799[illegible]	Expression not determined.	104
			CLUSTER 5	
ATHPRECA*	*Arabidopsis*	166807	Expressed strongly in roots.	105, 106
HRAPRXC3	Horseradish	217934	Expressed strongly in roots and detected in stem.	106, 107
ATHPRXCA*	*Arabidopsis*	166827	Detected in leaves, stem, and roots.	105, 106
ATPRXCB	*Arabidopsis*	405611	Expression not determined.	108

TABLE 8.1 *(continued)*
List of Peroxidase Genes

Gene locus	Plant	Accession #	Gene and Expression	Reference
HRAPRXC2	Horseradish	217932	Expressed strongly in roots. Detected in stem. Wound-inducible.	106, 107, 109
			CLUSTER 6	
POPP01	Poplar	1199776	Unpublished.	110
POPPA	Poplar	537604	Unpublished.	111
POPP2*	Poplar	1255663	Differentially expressed in the stem.	68
POPP02*	Poplar	1199778	Unpublished.	110
POPPB	Poplar	485393	Unpublished.	111
			CLUSTER 7	
MSRNAPE1C	Alfalfa	971562	Expression not determined. C-terminal processed.	97
ALFPXDA	Alfalfa	537315	Unpublished. Has C-terminal extension.	98
MSRNAPE1B	Alfalfa	971560	Expression not determined. C-terminal processed.	97
MSRNAPE1A	Alfalfa	971558	Expression not determined. C-terminal processed.	97
ALFPXDC	Alfalfa	537317	Unpublished. Has C-terminal extension.	98
SSNPEROXIB	*S. humilis*	L36111	Abundant in stem; low in leaves. Not induced during S interaction with CG or wounding.	67
GMU41657	Soybean	U41657	Present abundantly in seed coat.	
			CLUSTER 8	
BLYPRX5A	Barley	167081	Barley seed-specific peroxidase. Not expressed in leaves. C-terminal processed.	75
BLYPRX6A*	Barley	167083	Expression not determined. Has C-terminal extension.	112

VIRPRX	Azuki bean	218328	Expressed in leaves only after wounding, salicylate and ethylene treatment. Has C-terminal extension.	85
			CLUSTER 9	
ARCPNC2	Peanut	166475	Isolated from suspension culture cells.	99
LECEVI16G	Tomato	1161566	Constitutively expressed in roots. Induced by ethylene and viroid infection.	76
SSNPEROXIC	*S. humilis*	L36112	Abundantly expressed in leaves, low in roots. Not expressed in stem or old leaves. Induced in leaves during S interaction with CG. Wound-inducible in leaves.	67
			CLUSTER 10	
TOBCPI40KB	Tobacco	575605	Transcripts related to these two peroxidase genes were isolated from suspension culture. Expression increased during cell proliferation but not during cell expansion.	113
TOBCPI38KA	Tobacco	575603	Same as above.	113
TOMTPRX1A	Tomato	678547	Transcripts related to the two tomato peroxidases were expressed constitutively in roots. Expression	114
TOMTPRX2A	Tomato	295355	in roots was induced by NaCl treatment and localized to root epidermal cells.	
RICPEROX	Rice	287401	Unpublished.	
SPPEROXDS	Duck weed	438245	Abscisic acid inducible.	88
RICPRXRPNB	Rice	1097876	Expression not determined.	115
RICPRXRPAA	Rice	1097875	Expression not determined.	115
COTPROXDS	Cotton	167367	Expressed in roots. Induced in cotyledon during embryo development and germination. Also induced in leaves during salt stress.	116

* Peroxidase genes whose genomic sequences were used in determining consensus 5′ and 3′ splice junctions. R, resistant; S, susceptible; Hrp C, Horseradish peroxidase; EG, *Erysiphe graminis*; EGH, *Erysiphe graminis* f. sp *hordei*; XOO, *Xanthomonas oryzae* pv. *oryzae*; CG, *Colletotrichum gloeosporioide.*

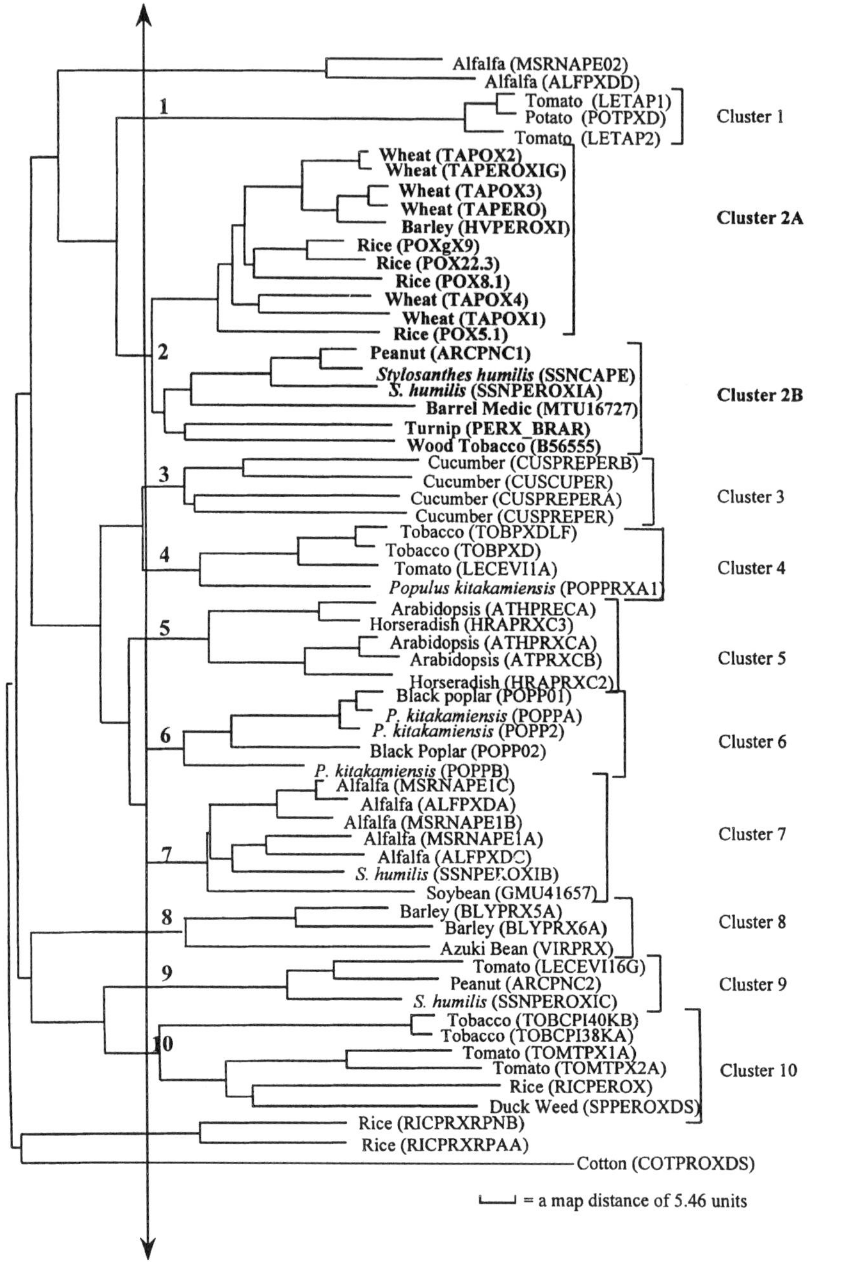

FIGURE 8.2 Phylogenetic analysis of deduced amino acid sequences from 62 plant peroxidases. The phylogenetic map was constructed using PileUp, Distances, and GrowTree programs (GCG, version 8.0). Peroxidase sequences are indicated by the name of the plant of origin and gene locus. Accession numbers are listed in Table 8.1. The vertical arrow was placed to reflect biological relationships among peroxidases. Peroxidases that clustered to a single branch to the right of the vertical line were numbered 1 to 10. Peroxidases in cluster 2, containing monocot (cluster 2A) and dicot (cluster 2B) peroxidases are shown in bold.

not used for the comparison. Phylogenetic analysis showed that peroxidases grouped in several distinct branches (Figure 8.2). Those branches that were located to the right of a vertical line drawn to reflect biological similarities were recognized as clusters. Thus, 10 peroxidase gene clusters were recognized. Although spatial and temporal expressions of cloned peroxidase genes in relation to developmental and environmental cues have not been studied in all cases, most (eight of 13) pathogen-inducible peroxidases are grouped into cluster 2 (Figure 8.2). This cluster included two distinct subgroups, one with peroxidases from monocotyledonous plants and the second with peroxidases from dicotyledonous plants. The remaining five pathogen-inducible plant peroxidases grouped in clusters 1, 3, 4, and 9 (Table 8.1).

Based on the alignment of the 62 peroxidase sequences, the lengths of the peptides between the F and the G helices varied among different peroxidases and were distinctly shorter in cluster 2 peroxidases (Figure 8.3). The amino acid sequences between the F and G helices also were diverged considerably in different peroxidases, but the sequence of these helices in cluster 2 peroxidases was highly similar and distinct from the sequences in peroxidases from other groups (Figure 8.3). Because the region between the F and G helices is suggested to provide structural constraints to the substrate access channel and binding site, cluster 2 peroxidases might catalyze similar substrates. However, whether or not peroxidases within cluster 2 catalyze similar substrates requires experimental characterization of the enzymatic properties of the various cluster 2 peroxidases.

The comparison of exon–intron junctions of 17 of the 62 plant peroxidase genes (indicated in Table 8.1) revealed that eight peroxidase genes contained a unique 5′ splice site at the intron-1-exon junction [TAPOX2 (pox2), TAPOX3 (pox3), POXgX9, POX22.3, POX8.1, TAPOX4 (pox4), TAPOX1 (pox1), SSNCAPE (sphx6b)], and all were from peroxidases that grouped in cluster 2. Only rip1[61] (MTU16727) from cluster 2 did not contain the unusual 5′ splice site in the first intron. Many proteins encoded by genes with these unusual splice sites are associated with reactions involving oxygen and a heme moiety; examples include human cytochrome P-450, human and mouse superoxide dismutase, rat heme oxygenase, chicken and duck α-globulin, earthworm hemoglobin-c, and soybean nodulin-24.[62,63] These findings suggest that cluster 2 peroxidases may have evolved from a common progenitor and probably have similar enzymatic functions. Characterization of the substrate specificities for this group of peroxidase may reveal general strategies for defense against pathogens.

8.4 ORGANIZATION OF PLANT PEROXIDASE GENES

Molecular characterization of class III plant peroxidase genes (which henceforth will be referred to as plant peroxidase genes) began with the cloning of an anionic peroxidase from tobacco.[64] Since then, over 60 peroxidase genes from diverse plant species have been isolated and characterized, including multiple genes from individual species. At least 30 peroxidase genes have been cloned from rice, wheat, tomato, tobacco, and *Stylosanthes humilis* combined (Table 8.1). Within a given species, some of these genes are highly similar, whereas others are relatively more divergent. Highly similar plant peroxidase genes present in multiple copies in the

	F helix			G helix	
	** * *	* * *			* **
Alfalfa (MSRNAPE02)	GAHTLGFSHC	DRF-SNRI--	-QT------P	VDPTLNKQ-Y	AAQL-QQMCP
Alfalfa (ALFPXDD)	GAHTVGFSHC	NKF-TNRVYN	FKT----TSR	VDPTLDLH-Y	AAKL-KSMCP
Barley (BLYPRX5A)	GGHTIGLAHC	SSF-EDRLF-	--------PR	PDPTISP-TF	LSRLKRT-CP
Barley (BLYPRX6A)	GGHTIGLGHC	NSF-EKRLF-	--------PL	PDTTMSP-SF	VARLKRT-CP
Azuki Bean (VIRPRX)	GGHTIGISHC	SSF-NNRLY-	--------PT	QDPVMDK-TF	GKNLRLT-CP
Peanut (ARCPNC2)	GGHTIGTSEC	QFF-SNRLFN	FNG----TAA	ADPAIDP-SF	VSNLQA-LCP
Tomato (LECEVI16G)	GGHTIGTSAC	QFF-SYRLYN	FNS----TGG	PDPSIDA-TF	LSQLQA-LCP
S.humilis (SSNPEROXIC)	GGHTIGTTSC	QLL-SSRLYN	FNG----TNG	PDPTIDP-SF	LPQLKA-LCP
Tobacco (TOBCPI40KB)	GAHTIGVAHC	GAF-SRRLFN	FTG----KGD	VDPSLSS-TY	AESLK-QLCP
Tobacco (TOBCPI38KA)	GAHTIGVAHC	GAF-SRRLFN	FTG----KGD	MDPSLNP-TY	VESLK-QLCP
Tomato (TOMTPX1A)	GAHTIGVSRC	SSF-SERLYN	FTGVVGTQ--	-DPSLDS-EY	ADNLKSRKCR
Tomato (TOMTPX2A)	GAHTIGVSHC	PSF-SSRLYN	FTGVWG-K--	-KSSLDS-EY	AANLKMKKCK
Rice (RICPEROX)	AGHTIGTSHC	FSF-TDRLYN	FTG-LDNAHD	IDPTLEL-QY	MARLRS-KCT
Duck Weed (SPPEROXDS)	GGHTIGNAHC	FTF-TTRLYN	FSG-RGDNSD	TDPSLER-NY	LAKLRA-KCA
Tomato (LETAP1)	GAHTVGFARC	ST-----VC-	----------	TSGNVNP---	AAQL-QCNCS
Potato (POTPXD)	GAHTVGFARC	ST-----VC-	----------	TSGNVNP---	AAQL-QCNCS
Tomato (LETAP2)	GAHTVGFARC	ST-----VC-	----------	TSGNVNP---	AAQL-QCNCS
Rice (RICPRXRPNB)	GAH-IGRASC	TLF-SNRLAN	FTA----SNS	V-PTLDASSL	ASS--QSQVA
Rice (RICPRXRPAA)	GGHTIGRARC	TLF-SNRLS-	-TT----SSS	ADPTLDA-TM	AANL-QSLCA
Wheat (TAPOX2)	**GAHTIGKAQC**	**SNFRT-RIYG**	**----------**	**GDTNINT-AF**	**ATSL-KANCP**
Wheat (TAPEROXIG)	**GAHTIGKAQC**	**SNFRT-RIYG**	**----------**	**GDTNINT-AF**	**ATSL-KANCP**
Wheat (TAPOX3)	**GAHTIGQAQC**	**GTFKD-RIY-**	**----------**	**NETNIDT-TF**	**ATSL-RANCP**
Wheat (TAPERO)	**GAHTIGQAQC**	**GTFKD-RIY-**	**----------**	**NETNIDT-AF**	**ATSL-RANCP**
Barley (HVPEROXI)	**GAHTIGQAQC**	**STFRA-RIYG**	**----------**	**GDTNINT-AY**	**AASL-RANCP**
Rice (POXgX9)	**GAHTIGQAQC**	**QNFRD-RLY-**	**----------**	**NETNIDS-SF**	**ATAL-KANCP**
Rice (POX22-3)	**GAHTIGQAQC**	**QNFRD-RIY-**	**----------**	**NETNIDS-AF**	**ATQR-QANCP**
Wheat (TAPOX4)	**GAHTIGQAQC**	**QNFRD-RLY-**	**----------**	**NETNIDT-AF**	**ATSL-RANCP**
Rice (POX8-1)	**GAHTIGQAQC**	**TNFRG-RIY-**	**----------**	**NETNIDA-GY**	**AASL-RANCP**
Wheat (TAPOX1)	**GAHTIGQSQC**	**RFFRD-RIY-**	**----------**	**NETNINT-TF**	**ATSL-RANCP**
Rice (POX5-1)	**GAHTIGMARC**	**RGFRT-RLY-**	**----------**	**NETNIDA-AF**	**AAAL-KANCP**
Peanut (ARCPNC1)	**GAHTIGQAQC**	**TAFRT-RIY-**	**----------**	**NESNIDP-TY**	**AKSL-QANCP**
***S.humilis* (SSNCAPE)**	**GAHTIGQARC**	**TTFRT-RIY-**	**----------**	**NESNIDP-SY**	**AKSL-QGNCP**
***S.humilis* (SSNPEROXIA)**	**GGHTIGQARC**	**TSFRT-RIY-**	**----------**	**TESNIDP-NF**	**AKSL-QGNCP**
Turnip (PERX_BRAR)	**GAHTIGQSRC**	**VNFRA-RVY-**	**----------**	**NETNINA-AF**	**A-TLRQRSCP**
Wood Tobacco (B65555)	**GAHTIGQAQC**	**FLFRD-RIYS**	**----------**	**NGTDIDA-GF**	**A-STRRRQCP**
Barrel Medic (MTU16727)	**GGHTIGFARC**	**TTFRN-RIY-**	**----------**	**NETNIDP-IF**	**AASLRKT-CP**
Cucumber (CUSPREPERB)	GAHTFGRSRC	-MFFSGRL-N	-NN----PNA	DDSPIDS-TY	ASQL-NQTCQ
Cucumber (CUSCUPER)	GAHTFGKSRC	-QFFDRRL-N	VSN----P--	-DSTLNP-RY	AQQL-RQACS
Cucumber (CUSPREPERA	GAHTFGRSRC	-VFFSGRLSN	FSG----SGQ	PDPTLDP-TY	RQEL-LSACT
Cucumber (CUSPREPER)	GAHTFGKSRC	-MFFSDRLIN	FNG----TGR	PDTTLDP-IY	REQL-RRLCT
Arabidopsis (ATHPRECA)	GGHTFGRAQC	Q-FVTPRLYN	FNG----TNS	PDPSLYP-TY	LVEL-RRLCP
Horseradish (HRAPRXC3)	GGHTFGRAQC	Q-FVTPRLYN	FNG----TNR	PDPTLDP-TY	LVQL-RALCP
Arabidopsis (ATHPRXCA)	GAHTFGKNQC	R-FIMDRLYN	FSN----TGL	PDPTLNT-TY	LQTL-RGQCP
Arabidopsis (ATPRXCB)	GGHTFGKNQC	Q-FILDRFYN	FSN----TGL	PDPTLNT-TY	LQTL-RGLCP
Horseradish (HRAPRXC2)	GGHTFGKNQC	Q-FIMDRLYN	FSN----SGK	PDPTLDK-SY	LSTL-RKQCP
Tobacco (TOBPXDLF)	GAHTFGRARC	GTF-EQRLFN	FNG----SGN	PDLTVDA-TF	LQTL-QGICP
Tobacco (TOBPXD)	GAHTFGRARC	GTF-EQRLFN	FSG----SGN	PDPTVDA-TF	LQTL-QGICP
Tomato (LECEVI1A)	GAHTFGRARC	GTF-QQRLFN	FSG----SGS	PDPTINS-TY	LPTL-QATCP
Black popular (POPPRXA1)	GAHTFGRSQC	QFF-SQRL-N	-------DTN	PDPTLNP-TY	LQTL-RQACP
Poplar (POPP01)	GAHTFGRAKC	STF-DFRLFD	FNS----TGA	PDQSLNT-TL	LADL-QELCP
Poplar (POPPA)	GAHTFGRAKC	STF-DFRLYD	FNS----TGA	PDQSLDP-TL	LAAL-QELCP
Poplar (POPP2)	GAHTFGRAQC	STF-DFRLFD	FNS----TGA	PDPSLDP-TL	LAAL-QELCP
Black Popular (POPP02)	GAHTFGRAQC	RNFID-RLYN	FNN----TGL	PDPTLDT-TY	LATL-QRLCP
Poplar (POPPB)	GAHTFGRAQC	SSF-NLRLYN	FSG----SGN	PDPTLNT-TY	LAEL-QQLCP
Alfalfa (MSRNAPE1C)	GAHTFGRAHC	AQFVS-RLYN	FSS----TGS	PDPTLNT-TY	LQQL-RTICP
Alfalfa (ALFPXDA)	GAHTFGRAHC	AQFVS-RLYN	FSS----TGS	PDPTLNT-TY	LQQL-RTICP
Alfalfa (MSRNAPE1B)	GAHTFGRAHC	SLFVS-RLYN	FSG----TGS	PDPTLNT-TY	LQQL-RTICP
Alfalfa (MSRNAPE1A)	GGHTIGRGQC	RFFVD-RLYN	FSN----TGN	PDSTLNT-TY	LQTL-QAICP
Alfalfa (ALFPXDC)	GAHTIGRGQC	RFFVD-RLYN	FSN----TGN	PDPTLNT-TY	LQTL-RTICP
S.humilis (SSNPEROXIB)	GAHTIGRAQC	RFFSS-RLYN	FSS----SGN	PDPSLNT-TY	LQTL-RSIFP
Soybean (GMU41657)	GGHTSGRARC	STFIN-RLYN	FSN----TGL	IH--LDT-TY	LEVL-RARCP
Cotton (COTPROXDS)	GAHSVGRTHC	VKLV-HRLY-	--------PE	VDPALSPDHV	PHMLHK--CP

FIGURE 8.3 Alignment of amino acid sequences present between the F and G helices. Amino acid sequences of 62 plant peroxidases were aligned using PileUp. Dashes indicate gaps that were introduced to maximize alignment. Peroxidase sequences are indicated by the name of the plant of origin and gene locus. Accession numbers are listed in Table 8.1. Peroxidases that grouped to cluster 2 are in bold. Downward arrow indicates the position of the third intron. Invariable histidine (proximal histidine) and other highly conserved amino acids are indicated by asterisks (*).

genome of a species constitute a gene family. Four copies of lignin-forming peroxidase genes from tobacco,[64] four suberization-associated peroxidase genes from tomato,[65] and at least four peroxidase genes from cucumber[33] are members of gene families in their respective species. The wheat peroxidase gene WIR3[66] and the *S. humilis* peroxidase genes *sphx6* and *sphx5*[67] hybridized to multiple bands in Southern analysis of genomic DNA, and are likely members of multigene families.

Studies on the organization of the peroxidase genes indicate that multigene families of plant peroxidases often are clustered within the genome. In rice, four peroxidase genes, *POXgP2.3, gPOX8.1, gPOX22.3,* and *POXgX9,* are located in tandem over a region of approximately 22 kb near the telomere of chromosome 7. A fifth rice peroxidase gene, *POX5.1,* also mapped to the same locus in the chromosome. Thus, at least five peroxidase genes map to the same locus on chromosome 7 of rice. In rice, as many as 10 to 12 copies of peroxidase genes are estimated to occur in the genome. In the genome of *Populus kitakamiensis,*[68] tomato,[65] horseradish,[69] and wheat,[70] two similar genes for peroxidase are arranged in tandem. Tandem arrangement of peroxidase genes is not limited to plants. Three genes for peroxidase are arranged in tandem in the genome of the white-rot fungus, *Trametes versicolor.*[71] Also, clustering of genes from multigene families is not unique to peroxidases; other examples exist for a number of multigene families in a variety of plant species.[72-74]

8.5 EXPRESSION OF PEROXIDASE GENES

The expressions of peroxidase genes are summarized in Table 8.1. These studies *in toto* demonstrated that the expression of the genes is regulated strictly. Peroxidase genes showed different organ-specific regulation[67,70,75-77] and were induced by factors such as pathogen stress,[33,61,66,67,70,76-84] wounding,[64,67,77,82,84,85] ethylene,[76,85-87] and abscisic acid treatments.[65,88]

8.5.1 Expression of Peroxidase Genes During Pathogen Stress

Although plants contain a number of highly similar peroxidase genes, only some of these are expressed during pathogen stress.[70,77,80,89] In rice, only two (*POX22.3* and *POX8.1*) of four studied peroxidase genes were found to be induced during pathogen stress.[77] Similarly, of the six highly similar wheat peroxidase genes (Table 8.1), only two, *pox2*[70] and *pPOX3*[81,80] were induced during pathogen stress. The presence of multiple copies of highly similar, yet differentially expressed, peroxidase genes may reflect an evolutionary adaptation in plants to fine-tune their responses to varied stimuli and avoid inappropriate gene expression.

Studies on the dynamics of induction of peroxidase genes during pathogen challenge have indicated that the genes were induced more rapidly and to a higher level in resistant interactions than in susceptible interactions.[77-79] In rice, *POX22.3* was induced differentially during infection of leaves with virulent (susceptible interaction) and avirulent (resistant interaction) strains of *Xanthomonas oryzae* pv. *oryzae.*[77] It was induced rapidly in leaves in 12 h during the resistant interaction. In the susceptible interaction, induction of *POX22.3* was delayed and occurred only after 36 h. The level of transcription at 36 h was lower than the level in the resistant interaction. Two barley peroxidase genes, *pBT6-3* and *pCD1311,* also showed differential induction patterns during resistant and susceptible interactions with the powdery mildew pathogen, *Erysiphe graminis* f.sp. *hordei.* Both of these transcripts were also elevated to higher levels during resistant interactions than during susceptible interactions.[79] Transcripts corresponding to anionic potato peroxidase genes were induced by elicitors of *Verticillium albo-atrum* in tomato cell cultures made from a plant cultivar that was resistant to the fungus. In contrast, cells from a near-isogenic susceptible tomato line showed very little induction of the anionic peroxidase gene. In these examples, the dynamics of induction of peroxidase genes were consistent with the induction of peroxidase enzyme activity during resistant and susceptible interactions.

Opposite to the above examples, plant peroxidase genes were induced rapidly during compatible interactions.[61,67,84] The peroxidase gene *rip1* was induced rapidly in the roots of *Medicago truncatula* following inoculation with the symbiont *Rhizobium meliloti.*[61] Similarly, three *S. humilis* peroxidase genes (*sphx6a, sphx6b,* and *sphx2*) were induced rapidly following inoculation with virulent strains of *Colletotrichum gloeosporioides.* The three genes also were induced by wounding.[67,84] Recently, *sphx6a* and *sphx6b* were shown to be induced by methyl jasmonate, which is a signal molecule produced in plants in response to wounding.[84] In both *M. truncatula* and *S. humilis,* induction of peroxidase genes preceded penetration by the pathogen. Attempts to repair the cell wall at the site of infection[35] or formation of specialized cell wall structures[61] have been suggested to explain the increase in peroxidase induction during susceptible interaction.

8.5.2 Regulation of Plant Peroxidase Genes During Pathogen Stress

As can be deduced from the expression studies, peroxidase gene regulation is undoubtedly complex. Some genes with similar promoter sequences are regulated differentially. The nucleotide sequence ~220 bp upstream of the ATG start codon of the putative promoter region of the rice peroxidase gene *POX22.3* is 67 to 70% similar to promoter regions of three wheat peroxidase genes, *pox2, pox3,*[70] and *TAPEROXIG.*[89] In spite of the high degree of similarity, the four peroxidase genes appear to be regulated differently. POX22.3 was expressed in the roots and leaves and was induced differentially during resistant and susceptible interactions with *X. oryzae* pv. *oryzae.*[77] Among the three wheat peroxidase genes (whose promoters are 76% identical between *pox2* and *pox3* and 95% identical between *TAPEROXIG* and *pox2* in the same region), *TAPEROXIG* and *pox2* were expressed in the roots, and *pox3* was expressed only in leaves.[70,89] Only *pox2* was induced in leaves during susceptible interactions with the powdery mildew fungus *E. graminis.*[70]

The putative promoter regions in all four genes contained sequences identical to the G-box element GCACGTG and an upstream highly conserved region. G-box elements are highly conserved *cis*-regulatory regions that are located relatively close to the TATA-box (between position –50 and –250) in a number of environmentally induced plant promoters (for a review, see reference 90). The differences in distances between the G-box element and other potential *cis*-regulatory regions may account for the differences in gene regulation.[90] The presence or absence of certain critical *trans*-acting factors in different tissues or during different developmental stages may also influence gene regulation.

In other instances, peroxidase genes with different putative promoter sequences were regulated similarly. Although the promoter regions of *gPOX22.3* and *gPOX8.1* are distinct (Chittoor et al. unpublished), the genes are expressed rapidly during resistant interactions with *X. oryzae* pv. *oryzae*. One explanation may be that these two genes are induced as a result of two different stimuli occurring during resistant interactions. For example, *POX22.3* may be activated by the specific resistance induction mechanism, whereas wounding or tissue damage that occurs during resistance could be a trigger for activation of *POX8.1*. Molecular characterization of the *cis*-acting elements and the *trans*-acting factors for the promoter regions of these peroxidase genes would provide insight into the regulation of these genes during resistant and susceptible interactions and further our understanding of specific signal transduction pathways leading to gene activation.

8.6 DETERMINATION OF THE POTENTIAL ROLE OF PEROXIDASE IN DEFENSE AGAINST PATHOGENS BY SUPPRESSION OR OVER-EXPRESSION OF PEROXIDASE GENES IN TRANSGENIC RICE PLANTS

Despite the strong correlative evidence for the involvement of peroxidase activity with resistance, a specific role for any peroxidase in resistance has not been defined. One difficulty has been that peroxidases are encoded by closely related multigene families, as discussed above, and one or more genes may contribute to a particular physiological function. Abolition of activities of a highly anionic pathogen- and wound-induced peroxidase in transgenic tomato plants did not result in the inhibition of suberization in wounded periderm.[91] Similarly, no difference in lignin content was observed in transgenic tobacco plants carrying an antisense peroxidase construct, although the peroxidase level was reduced significantly.[92] A possible explanation is that other peroxidases may have substituted for the abolished activity of the peroxidase.[91] Alternatively, suppressed peroxidase may not be involved in suberization of wounded tissue and may have another role. The effects of the suppression of peroxidase on pathogen stress were not reported for either of these cases.

An analogous situation in loblolly pine reveals the difficulties in determining the functional role of an enzyme.[93] Cinnamyl alcohol dehydrogenase (CAD) converts coniferaldehyde to coniferyl alcohol, the primary lignin percursor in pines. However, plants severely depleted in CAD grew normally. Examination of lignin monomers in mutant plants showed that dihydroconiferyl alcohol, which is not the usual precursor in the lignin biosynthetic pathway, is the major component of lignin in the mutant plants.[93] Thus, extensive variation in the lignin composition in plants occurs without affecting the functional properties of lignin. This work also draws our attention to the complexity and potential flexibility of important biosynthetic pathways. Alterations in the levels of other enzymes of the lignin biosynthetic pathway, which includes peroxidases, may not significantly affect the physiological function of the end product.

Insight into the biological function of an enzyme also can be gained by over-expression of the gene(s) encoding the enzyme in transgenic plants.[94] Over-expression of anionic tobacco peroxidase in transgenic tomato plants resulted in a 20-fold increase in lignin content in fruit after wounding compared to a twofold increase in wounded fruit from nontransformed tomato plants. Because the transgenic tomato seedlings were still susceptible to four tomato pathogens,[95] the results were not conclusive in determining a role for peroxidases. The failure of transgenic tomato plants expressing the tobacco anionic peroxidase to resist infection was attributed partly to the age of the plants tested (seedlings) and the excessive amount of inoculum used. Similarly, over-expression of defense-related, intercellular barley peroxidase in transgenic tobacco did not result in enhanced resistance to the tobacco powdery mildew pathogen *Erisyphe cichoracearum.*[96]

8.7 CONCLUSION

Physiological and molecular characterizations of plant peroxidases have provided evidence that the enzymes are involved in plant defense. Induction of peroxidase genes with concomitant increases in peroxidase enzyme activity correlated temporally and spatially with resistance in a number of plant–pathogen interactions. Phylogenetic analyses and comparisons of amino acid and nucleotide sequences of plant peroxidases have suggested that pathogen-inducible plant peroxidases may show preferences for similar substrates and may function similarly. However, determination of the role of peroxidase in defense responses through transgenic studies, such as over-expression or abolition of peroxidase enzyme activities, has remained inconclusive. Studies regarding the role of peroxidase may be confounded by the presence of multiple peroxidase genes even within a single plant species. Until the contributions of specific peroxidase gene family members are evaluated, the role of peroxidase in resistance will remain unknown.

ACKNOWLEDGMENTS

We thank Diana Pavlisko for preparing the manuscript and Dr. Emmanuel Hiliare for reviewing it. J.M.C acknowledges support from the SPIC Science Foundation (Chennai, India) and the Rockefeller Foundation through a predoctoral fellowship. This work was supported in part by USDA grant 94-37303-0548 and a grant from the Rockefeller Foundation to F.F.W and J.E.L. Contribution no. 97-196-B from the Kansas Agricultural Experiment Station, Manhattan, KS, U. S. A.

REFERENCES

1. Gaspar, T., Penel, C., Thorpe, T., and Greppin, H., *Peroxidases: A Survey of Their Biochemical and Physiological Roles in Higher Plants,* University of Geneva Press, Geneva, 1982.
2. Gross, G. G., The biochemistry of lignification, *Adv. Bot. Res.,* 8, 25, 1980.
3. Schmid, P. S. and Feucht, W., Tissue-specific oxidative browning of polyphenols by peroxidase in cherry shoots, *Gartenbauwissenschaft,* 45, 68, 1980.
4. Fry, S. C., Cross-linking of matrix polymers in the growing cell walls of angiosperms, *Annu. Rev. Plant Physiol.,* 37, 165, 1986.
5. Everdeen, D. S., Kiefer, S., Willard, J. J., Muldoon, E. P., Dey, P. M., Li, X. B., and Lamport, D. T. A., Enzymic cross-linkage of monomeric extensin precursors in vitro, *Plant Physiol.,* 87, 616, 1988.
6. Grisebach, H., Lignins, in *Secondary Plant Products,* Conn, E. E., Ed., Academic Press, New York, 1981, 457–478.
7. Walter, M, H., Regulation of lignification in defense, in *Genes Involved in Plant Defense,* Boller, T. and Meins, F., Eds., Springer-Verlag, New York, 1992, 327–352.
8. Espelie, K. E. and Kolattukudy, P. E., Purification and characterization of an abscisic acid-inducible peroxidase associated with suberization in potato (*Solanum tuberosum*), *Arch. Biochem. Biophys.,* 240, 539, 1985.
9. Espelie, K. E., Franceschi, V. R., and Kolattukudy, P. E., Immunocytochemical localization and time course of appearance of an anionic peroxidase associated with suberization in wound-healing potato tuber tissue, *Plant Physiol.,* 87, 487, 1986.
10. Ride, J. P., Cell walls and other structural barriers in defense, in *Biochemical Plant Pathology,* Callow, J. A., Ed., Wiley-Interscience, New York, 1983, 215.
11. Venere, R. J., Wang, X., Dyer, J. H., and Zheng, L., Role of peroxidase in cotton resistant to bacterial blight: purification and immunological analysis of phospholipase D from castor bean endosperm, *Plant Sci. Lett.,* 306, 486, 1993.
12. Thordal-Christensen, H., Zhang, Z., Wei, Y., and Collinge, D. B., Subcellular localization of H2O2 in plants. H2O2 accumulation in papillae and hypersensitive response during the barley-powdery mildew interaction, *Plant J.,* 11, 1187, 1997.
13. Kolattukudy, P. E., Biopolyester membranes of plants: cutin and suberin, *Science,* 208, 990, 1980.
14. Pearce, R. B. and Rutherford, J., A wound-associated suberized barrier to the spread of decay in the sapwood of oak (*Quercus robur* L.), *Physiol. Plant Pathol.,* 19, 359, 1981.
15. Beardmore, J., Ride, J. P., and Granger, J. W., Cellular lignification as a factor in the hypersensitive resistance of wheat to stem rust, *Physiol. Plant Pathol.,* 22, 209, 1983.
16. Tiburzy, R., Untersuchungen zur bedeutung der hypersensitiven reaktion befallener weizenzellen als resistenzfaktor gegen *Puccinia graminis* f. sp. *tritici,* Ph.D. thesis, RWTH, Aachen, F.R.G., 1984.
17. Reisener, H. J., Tiburzy, R., Kogel, K. H., Moerscbacher, B., and Heck, B., Mechanism of resistance of wheat against stem rust in the Sr5/P5 interaction, in *Biology and Molecular Biology of Plant-Pathogen Interactions,* Bailey, J. A., Ed. NATO ASI Series, Springer, Berlin, 1986, 141.
18. Robb, J., Powell, D. A., and Street, P. F. S., Time course of wall-coating secretion in *Verticillium*-infected tomatoes, *Physiol. Plant Pathol.,* 31, 217, 1987.
19. Street, P. F. S., Robb, J., and Ellis, B. E., Secretion of vascular coating components by xylem parenchyma cells of tomatoes infected with *Verticillium albo-atrum, Protoplasma,* 132, 1, 1986.

20. Reimers, P. J. and Leach, J. E., Race-specific resistance to *Xanthomonas oryzae* pv. *oryzae* conferred by bacterial blight resistance gene *Xa-10* in rice *Oryzae sativa* involves accumulation of a lignin-like substance in host tissues, *Physiol. Mol. Plant Pathol.*, 38, 39, 1991.
21. Dixon, R. A. and Harrison, M. J., Activation, structure, and organization of genes involved in microbial defense in plants, *Adv. Genet.*, 28, 165, 1990.
22. Kawalleck, P., Schmelzer, E., Hahlbrock, K., Somosich, I. E., and Dangle, J. L., Two pathogen-responsive genes in parsley encode a tyrosine-rich hydroxyproline-rich glycoprotein (HPRG) and an anionic peroxidase, *Mol. Gen. Genet.*, 247, 444, 1995.
23. Reimers, P. J., Guo, A., and Leach, J. E., Increased activity of a cationic peroxidase associated with an incompatible interaction between *Xanthomonas oryzae* pv. *oryzae* and rice (*Oryza sativa*), *Plant Physiol.*, 99, 1044, 1992.
24. Young, S. A., Guo, A., Guikema, J. A, White, F. F., and Leach, J. A., Rice cationic peroxidase accumulates in xylem vessels during incompatible interactions with *Xanthomonas oryzae* pv. *oryzae*, *Plant Physiol.*, 107, 1333, 1995.
25. Flott, B. E., Moerschbacher, B. M., and Reisener, H., Peroxidase isoenzyme patterns of resistant and susceptible wheat leaves following stem rust infection, *New Phytol.*, 111, 413, 1989.
26. Kerby, K. and Somerville, S. C., Purification of an infection-related, extracellular peroxidase from barley, *Plant Physiol.*, 100, 397, 1992.
27. Kerby, K., and Somerville, S. C., Enhancement of specific intercellular peroxidases following inoculation of barley with *Erysiphe graminis* f. sp. *hordei*, *Physiol. Mol. Plant Pathol.*, 35, 323, 1989.
28. Dai, G. H., Nicole, M., Andary, C., Martinez, C., Bresson, E., Boher, B., Daniel, J. F., and Geiger, J. P., Flavonoids accumulate in cell walls, middle lamellae and callose-rich papillae during an incompatible interaction between *Xanthomonas campestris* pv. *malvacearum* and cotton, *Physiol. Mol. Plant Pathol.*, 49, 285, 1996.
29. McGhie, T. K., Masel, N. P., Maclean, D., Croft, B. J., and Smith, G. R., Biochemical responses of suspension-cultured sugarcane cells to an elicitor derived from the root pathogen *Pachymetra chaunorhiza*, *Aust. J. Plant Physiol.*, 24, 143, 1997.
30. Smith, J, A., Fulbright, D., W., and Hammerschmidt, R., Rapid induction of systemic resistance in cucumber by *Pseudomonas syringae* pv. *syringae*, *Physiol. Mol. Plant Pathol.*, 38, 223, 1991.
31. Hammerschmidt, R., Nuckles, E., and Kuc, J., Association of enhanced peroxidase activity with induced systemic resistance of cucumber to *Colletotrichum lagenarium*, *Physiol. Plant Pathol.*, 20, 73, 1982.
32. Smith, J. A. and Hammerschmidt, R., Comparative study of acidic peroxidases associated with induced resistance in cucumber, muskmelon and watermelon, *Physiol. Mol. Plant Pathol.*, 33, 255, 1990.
33. Rasmussen, J. B., Smith, J. A., Williams, S., Burkhart, W., Ward, E., Somerville, S. C., Ryals, J., and Hammerschmidt, R., cDNA cloning and systemic expression of acidic peroxidases associated with systemic acquired resistance to disease in cucumber, *Physiol. Mol. Plant Pathol.*, 46, 389, 1995.
34. Stermer, B. A. and Hammerschmidt, R., Heat shock induces resistance to *Cladosporium cucumerinum* and enhances peroxidase activity in cucumber, *Physiol. Plant Pathol.*, 25, 239, 1984.
35. Salzwedel, J. L. and Dazzo, F. B., pSym *nod* gene influence on elicitation of peroxidase activity from white clover and pea roots by rhizobia and their cell-free supernatants, *Mol. Plant-Microbe Interact.*, 6, 127, 1993.

36. Goodman, R. N. and Novacky, A. J., *The Hypersensitive Reaction in Plants to Pathogens,* APS Press, St. Paul, 1994.
37. Levine, A., Tenhaken, R., Dixon, R., and Lamb, C., H_2O_2 from the oxidative burst orchestrates the plant hypersensitive disease resistance response, *Cell*, 79, 583, 1994.
38. Triggs-Raine, B. L. and Loewen, P. C., Physical characterization of *katG*, encoding catalase HPI of *Escherichia coli, Gene,* 52, 121, 1987.
39. Loprasert, S., Negoro, S., and Okada, H., Cloning, nucleotide sequence, and expression in *Escherichia coli* of the *Bacillus stearothermophilus* peroxidase gene (*perA*), *J. Bacteriol.,* 171, 4871, 1989.
40. Loewen, P. C., EMBL data library, accession number X53001 (STKATG), 1990.
41. Welinder, K. G., Mauro, J. M., and Norskov-Lauritsen, L., Structure of plant and fungal peroxidases, *Biochem. Soc. Trans.,* 20, 337, 1992.
42. Finzel, B. C., Poulos, T. L., and Kraut, J., Crystal structure of yeast cytochrome c peroxidase refined at 1.7 Å resolution, *J. Biol. Chem.,* 259, 13027, 1984.
43. Welinder, K. G. and Norskov-Lauritsen, L., Structure of plant peroxidases — preliminary fitting into the molecular model of yeast cytochrome c peroxidase, in *Molecular and Physiological Aspects of Plant Peroxidases,* Greppin, H., Penel, C., and Gaspar, T., Eds., University of Geneva, Geneva, 1986, 61.
44. Smith, A. T., Du, P., and Loew, G. H., Homology modeling of horseradish peroxidase, in *Nuclear Magnetic Resonance of Paramagnetic Macromolecules,* Mar, G. N., Ed., Kluwer Academic Publishers, Netherlands, 1995, 75.
45. Morita, Y., Mikami, B., Yamashita, H., Lee, J. Y., Aibara, S., Sato, M., Katsube, Y., and Tanaka, N., Primary crystal structure of horseradish peroxidase isozyme E5, in *Biochemical, Molecular and Physiological Aspects of Plant Peroxidases,* Lobarzewski, J., Greppin, H., Penel, C., and Gaspar, T., Eds., University of Geneva, Geneva, 1991, 81.
46. Schuller, D. J., Ban, N., Huystee, R. B., McPherson, A., and Poulos, T. L., The crystal structure of peanut peroxidase, *Structure*, 4, 311, 1996.
47. Poupos, T. L., Edwards, S. L., Wariishi, H., and Gold, M. H., Crystallographic refinement of lignin peroxidase at 2 Å, *J. Biol. Chem.,* 268, 4429, 1993.
48. Petersen, J. W. F., Kadziola, A., and Larsen, S., Three-dimensional structure of a recombinant peroxidase from *Coprinus cinereus* at 2.6 Å resolution, *FEBS Lett.,* 339, 291, 1994.
49. Kunishima, N., Fukuyama, K., Matsubara, H., Hatanaka, H., Shibano, Y., and Amachi, T., Crystal structure of the fungal peroxidase from *Arthromyces ramosus* at 1.9 Å resolution, *J. Mol. Biol.,* 235, 331, 1994.
50. Welinder, K. G. and Gajhede, M., Structure and evolution of peroxidase, in *Plant Peroxidases: Biochemistry and Physiology,* Welinder, K. G., Rasmussen, S. K., Penel, C., and Greppin, H., Eds., University of Geneva, Geneva, 1993, 35.
51. Welinder, K. G., Plant peroxidases. Their primary, secondary and tertiary structures, and relation to cytochrome peroxidase, *Eur. J. Biochem.,* 151, 497, 1985.
52. Sakurada, J., Takahashi, S., and Hosoya, T., Nuclear magnetic resonance studies on the spatial relationship of aromatic donor molecules to the heme iron of horseradish peroxidase, *J. Biol.Chem.,* 261, 9657, 1986.
53. Ator, M. A. and Ortiz de Montellano, P. R., Mechanism-based (suicide) enzyme inactivation, *Enzymes*, 19, 213, 1990.
54. Veitch, N. C. and Williams, R. J. P., Two-dimensional ^{1}H-NMR studies of horseradish peroxidase C and its interaction with indole-3-propionic acid, *Eur. J. Biochem.,* 189, 351, 1990.

55. Veitch, N. C. and Williams, R. J. P., Two-dimensional proton nuclear magnetic resonance studies of plant peroxidase interactions with aromatic donor molecules, in *Biochemical, Molecular and Physiological Aspects of Plant Peroxidases,* Lobarzewski, J., Greppin, H., Penel, C., and Gaspar, T., Eds., University of Geneva, Geneva, 1991, 99.
56. Veitch, N. C., Williams, R. J. P., Bone, N. M., Burke, J. F., and Smith, A. T., Solution characterization by NMR spectroscopy of two horseradish peroxidase isoenzyme C mutants with alanine replacing either Phe142 or Phe143, *Eur. J. Biochem.,* 233, 650, 1995.
57. Veitch, N. C. and Williams, R. J. P., The use of methyl-substituted benzhydroxamic acids as structural probes of peroxidase substrate binding, *Eur. J. Biochem.,* 229, 629, 1995.
58. Kimura, M., A simple method for estimating evolutionary rates of base substitutions through comparative studies of nucleotide sequences, *J. Mol. Evol.,* 16, 111, 1980.
59. Saitou, N. and Nei, M., The neighboring-joining method: a new method for reconstructing phylogenetic trees, *Mol. Biol. Evol.,* 4, 406, 1987.
60. Studier, J. A. and Keppler, K. J., A note on the neighbor-joining algorithm of Saitou and Nei, *Mol. Biol. Evol.,* 5, 729, 1988.
61. Cook, D., Dreyer, D., Bonnet, D., Howell, M., Nony, E., and VandenBosch, K., Transient induction of a peroxidase gene in *Medicago truncatula* precedes infection by *Rhizobium meliloti, Plant Cell,* 7, 43, 1995.
62. Jackson, I. J., A reappraisal of non-consensus mRNA splice sites, *Nucleic Acids Res.,* 19, 3795, 1991.
63. Senapathy, P., Shapiro, M. B., and Harris, N. L., Splice junctions, branch point sites and exons: sequence statistics, identification, and applications to genome projects, *Methods Enzymol.,* 183, 252, 1990.
64. Lagrimini, L. M., Burkhart, W., Moyer, M., and Rothstein, S., Molecular cloning of complementary DNA encoding the lignin-forming peroxidase from tobacco: molecular analysis and tissue-specific expression, *Proc. Natl. Acad. Sci. U.S.A.,* 84, 7542, 1987.
65. Roberts, E. and Kolattukudy, P. E., Molecular cloning, nucleotide sequence, and abscisic acid induction of a suberization-associated highly anionic peroxidase, *Mol. Gen. Genet.,* 217, 223, 1989.
66. Schweizer, P., Hunziker, W., and Mosinger, E., cDNA cloning, in vitro transcription and partial sequence analysis of mRNAs from winter wheat (*Triticum aestivum* L.) with induced resistance to *Erysiphe graminis* f. sp. *tritici, Plant Mol. Biol.,* 12, 643, 1989.
67. Harrison, S. J., Curtis, M. D., McIntyre, C. L., Maclean, D. J., and Manners, J. M., Differential expression of peroxidase isogenes during the early stages of infection of the tropical forage legume *Stylosanthes humilis* by *Colletotrichum gloeosporioides, Mol. Plant-Microbe Interact.,* 8, 398, 1995.
68. Osakabe, K., Koyama, H., Kawai, S., Kayatama, Y., and Morohoshi, N., Molecular cloning of two tandemly arranged peroxidase genes from *Populus kitakamiensis* and their differential regulation in the stem, *Plant Mol. Biol.,* 28, 677, 1995.
69. Fujiyama, K., Takemura, H., Shibayama, S., Kobayashi, K., Choi, J.-K., Shinmyo, A., Takano, M., Yamada, Y., and Okada, H., Structure of the horseradish peroxidase isozyme c genes, *Eur. J. Biochem.,* 173, 681, 1988.
70. Baga, M., Chibbar, R. N., and Kartha, K. K., Molecular cloning and expression analysis of peroxidase genes from wheat, *Plant Mol. Biol.,* 29, 647, 1995.

71. Johansson, T. and Nyman, P. O., A cluster of genes encoding major isozymes of lignin peroxidase and manganese peroxidase from the white-rot fungus *Trametes versicolor, Gene,* 170, 31, 1996.
72. Dean, C., Van den Elzen, P., Tamaki, S., Dunsmuir, P., and Bedbrook, J., Linkage and homology analysis divides the wigt genes for the small subunit of petunia ribulose 1,5-bisphosphate carboxylase into three gene families, *Proc. Natl. Acad. Sci. U.S.A.,* 82, 4964, 1985.
73. Dean, C., Van den Elzen, P., Tamaki, S., Dunsmuir, P., and Bedbrock, J., Differential expression of the eight genes of the petunia ribulose bisphosphate carboxylase small subunit multi-gene family, *EMBO J.,* 4, 3055, 1985.
74. Ryder, T. B., Hedrick, S. A., Bell, J. N., Liang, X., Clouse, S. D., and Lamb, C. J., Organization and differential activation of a gene family encoding the plant defense enzyme CHS in *Phaseolus vulgaris, Mol. Gen. Genet.,* 210, 219, 1987.
75. Johansson, A., Rasmussen, S. K., Harthill, J. E., and Wellinder, K. G., cDNA, amino acid and carbohydrate sequence of barley seed-specific peroxidase BP 1, *Plant Mol. Biol.,* 18, 1151, 1992.
76. Gadea, J., Mayda, M. E., Conejero, V., and Vera, P., Characterization of defense-related genes ectopically expressed in viroid-infected tomato plants, *Mol. Plant-Microbe Interact.,* 9, 409, 1996.
77. Chittoor, J. M., Leach, J. E., and White, F. F., Differential induction of a peroxidase gene family during infection of rice by *Xanthomonas oryzae* pv. *oryzae, Mol. Plant-Microbe Interact.,* 10, 861, 1997.
78. Mohan, R. and Kolattukudy, P. E., Differential activation of expression of a suberization- associated anionic peroxidase gene in near-isogenic resistance and susceptible tomato lines by elicitors of *Verticillium albo-atrum, Plant Physiol.,* 92, 276, 1990.
79. Thordal-Christensen, H., Brandt, J., Cho, B. H., Rasmussen, S. K., Gregersen, P. L., Smedegaard-Petersen, V., and Collinge, D. B., cDNA cloning and characterization of two barley peroxidase transcripts induced differentially by the powdery mildew fungus *Erysiphe graminis, Physiol. Mol. Plant Pathol.,* 40, 395, 1992.
80. Rebmann, G., Hertig, C., Bull, J., Mauch, F., and Dudler, R., Cloning and sequencing of cDNAs encoding a pathogen-induced putative peroxidase of wheat (*Triticum aestivum* L.), *Plant Mol. Biol.,* 16, 329, 1991.
81. Reimmann, C., Ringli, C., and Dudler, R., Complementary DNA cloning and sequence analysis of a pathogen-induced putative peroxidase from rice, *Plant Physiol.,* 100, 1611, 1992.
82. Roberts, E., Kutchan, T., and Kolattukudy, P. E., Cloning and sequencing of cDNA for a highly anionic peroxidase from potato and the induction of its mRNA in suberizing potato tubers and tomato fruits, *Plant Mol. Biol.,* 11, 15, 1988.
83. Scott-Craig, J. S., Kerby, K. B., Stein, B. D., and Somerville, S. C., Expression of an extracellular peroxidase that is induced in barley (*Hordeum vulgare*) by the powdery mildew pathogen (*Erysiphe graminis* f. sp. *hordei*), *Physiol. Mol. Plant Pathol.,* 47, 407, 1995.
84. Curtis, M. D., Rae, A. L., Rusu, A. G., Harrison, S. J., and Manners, J. M., A peroxidase gene promoter induced by phytopathogens and methyl jasmonate in transgenic plants, *Mol. Plant-Microbe Interact.,* 10, 326, 1997.
85. Ishige, F., Mori, H., Yamazaki, K. and Imaseki, H., Identification of a basic glycoprotein induced by ethylene in primary leaves of azuki bean as a cationic peroxidase, *Plant Physiol.,* 101, 193, 1993.

86. Morgens, P. H., Callahan, A. M., Dunn, L. J., and Abeles, F. B., Isolation and sequencing of cDNA clones encoding ethylene-induced putative peroxidases from cucumber cotyledons, *Plant Mol. Biol.,* 14, 715, 1990.
87. Vera, P., Tornero, P., and Conejero, V., Cloning and expression analysis of a viroid-induced peroxidase from tomato plants, *Mol. Plant-Microbe Interact.,* 6, 790, 1993.
88. Chaloupkova, K. and Smart, C. C., The abscisic acid induction of a novel peroxidase is antagonized by cytokinin in *Spirodela polyrrhiza* L., *Plant Physiol.,* 105, 497, 1994.
89. Hertig, C., Rebmann, G., Bull, J., Mauch, F., and Dudler, R., Sequence and tissue-specific expression of a putative peroxidase gene from wheat (*Triticum aestivum* L.), *Plant Mol. Biol.,* 16, 171, 1991.
90. DeVetten, N. C. and Ferl, R. J., Transcriptional regulation of environmentally inducible genes in plants by an evolutionary conserved family of G-box binding factors, *Int. J. Biochem.,* 26, 1055, 1994.
91. Sherf, B. A., Bajar, A. M., and Kolattukudy, P. E., Abolition of an inducible highly anionic peroxidase activity in transgenic tomato, *Plant Physiol.,* 101, 201, 1993.
92. McIntyre, C. L., Bettenay, H. M., and Manners, J. M., Strategies for the suppression of peroxidase gene expression in tobacco. II. *In vivo* suppression of peroxidase activity in transgenic tobacco using ribozyme and antisense constructs, *Transgenic Res.,* 5, 263, 1996.
93. Ralph, J., MacKay, J. J., Hatfield, R. D., O'Malley, D. M., Whetten, R. W., and Sederoff, R. R., Abnormal lignin in a loblolly pine mutant, *Science*, 277, 235, 1997.
94. Zhu, Q., Maher, E. A., Masoud, S., Dixon, R. A., and Lamb, C. J., Enhanced protection against fungal attack by constitutive co-expression of chitinase and glucanase genes in transgenic tobacco, *Bio/Technology,* 12, 807, 1994.
95. Lagrimini, L. M., Vaughn, J., Erb, W. A., and Miller, S. A., Peroxidase overproduction in tomato: wound-induced polyphenol deposition and disease resistance, *HortScience*, 28, 218, 1993.
96. Kristensen, B. K., Brandt, J., Bojsen, K., Thordal-Christensen, H., Kerby, K. B., Collinge, D. B., Mikkelsen, J. D., and Rasmussen, S. K., Expression of a defense-related intercellular barley peroxidase in transgenic tobacco, *Plant Sci.,* 122, 173, 1997.
97. El-Turk, J., Asemota, O., Leymarie, J., Sallaud, C., Mesnage, S., Breda, C., Buffard, D., Kondorosi, A., and Esnault, R., Nucleotide sequences of four pathogen-induced alfalfa peroxidase-encoding cDNAs, *Gene*, 170, 213, 1996.
98. Abrahams, S. L., Hayes, C. M., and Watson, J. M., Organ-specific expression of three peroxidase-encoding cDNAs from lucerne (*Medicago sativa*), GenBank Accession L36156.
99. Buffard, D., Breda, C., van Huystee, R. B., Asemota, O., Pierre, M., Dang Hu, D. B., and Esnault, R., Molecular cloning of complementary DNAs encoding two cationic peroxidases from cultivated peanut cells, *Proc. Natl. Acad. Sci. U.S.A.,* 87, 8874, 1990.
100. Curtis, M. D., Nourse, J. P., and Manners, J. M., Nucleotide sequence of a cationic peroxidase gene from the tropical forage legume *Stylosanthes humilis, Plant Physiol.,* 108, 1303, 1995.
101. Mazza, G. and Welinder, K. G., Covalent structure of turnip peroxidase 7. Cyanogen bromide fragments, complete structure and comparison to horseradish peroxidase C, *Eur. J. Biochem.,* 108, 481, 1980.
102. Criqui, M. C., Plesse, B., Durr, A., Marbach, J., Parmentier, Y., Jamet, E., and Fleck, J., Characterization of genes expressed in mesophyll protoplasts of *Nicotiana sylvestris* before the re-initiation of the DNA replicational activity, *Mech. Dev.,* 38, 121, 1992.

103. Osakabe, K., Kawai, S., Katayama, Y., and Morohoshi, N., Nucleotide sequence for the genomic DNA encoding the anionic peroxidase gene from *Nicotiana tabacum,* GenBank Accession D11396.
104. Kawai, S., Matsumoto, Y., Kajita, S., Yamada, K., Katayama, Y., and Morohoshi, N., Nucleotide sequence for the genomic DNA encoding an anionic peroxidase gene from a hybrid poplar, *Populus kitakamiensis, Biosci. Biotechnol. Biochem.,* 57, 131, 1993.
105. Intapruk, C., Higashimura, N., Yamamoto, K., Okada, N., Shinmyo, A., and Takano, M., Nucleotide sequences of two genomic DNAs encoding peroxidases from *Arabidopsis thaliana, Gene,* 98, 237, 1991.
106. Shinmyo, A., Fujiyama, K., Kawaoka, A., and Intapruk, C., Structure and expression of peroxidase isozyme genes in horseradish and *Arabidopsis*, in *Plant Peroxidases: Biochemistry and Physiology,* Welinder, K. G., Rasmussen, S. K., Penel, C., and Greppin, H., Eds., University of Geneva, Geneva, 1993, 221.
107. Fujiyama, K., Takemura, H., Shinmyo, A., Okada, H., and Takano, M., Genomic DNA structure of two new horseradish-peroxidase-encoding genes, *Gene*, 89, 163, 1990.
108. Intapruk, C., Takano, M., and Shinmyo, A., Nucleotide sequence of a new cDNA for peroxidase from *Arabidopsis thaliana, Plant Physiol.,* 104, 285, 1994.
109. Kawaoka, A., Kawamoto, T., Ohta, H., Sekine, M., Takano, M., and Shinmyo, A., Wound-induced expression of horseradish peroxidase, *Plant Cell Reports,* 13, 149, 1994.
110. Sakuma, Y., Azuma, T., Kato, Y., Kojima, Y., and Miura, K., Poplar peroxidase gene responding to cytokinin treatment, GenBank Accession D83225.
111. Osakabe, K., Koyama, H., Kawai, S., Katayama, Y., and Morohoshi, N., Molecular cloning of two tandemly arranged peroxidase genes from *Populus kitakamiensis* and their differential regulation in the stem, *Plant Mol. Biol.*, 28, 677, 1995.
112. Theilade, B. and Rasmussen, S. K., Structure and chromosomal localization of the gene encoding barley seed peroxidase BP 2A, *Gene*, 118, 261, 1992.
113. Narita, H., Asaka, Y., Ikura, K., Matsumoto, S., and Sasaki, R., Isolation, characterization and expression of cationic peroxidase isozymes released into the medium of cultured tobacco cells, *Eur. J. Biochem.,* 228, 855, 1995.
114. Botella, M. A., Quesada, M. A., Hasegawa, P. M., and Valpuesta, V., Nucleotide sequences of two peroxidase genes from tomato (*Lycopersicon esculentum*), *Plant Physiol.,* 103, 665, 1993.
115. Ito, H., Kimizuka, F., Ohbayashi, A., Matsui, H. H., Honma, M., Shinmyo, A., Ohashi, Y., Caplan, A. B., and Rodriguez, R., Molecular cloning and characterization of two complementary DNAs encoding putative peroxidases from rice (*Oryza sativa* L.) shoots, *Plant Cell Rep.,* 13, 361, 1994.
116. Ritter, D., Allen, R. D., Trolinder, N., Wayne Huges, D., and Galau, G. A., Cotton cotyledon cDNA encoding a peroxidase, *Plant Physiol.,* 102, 1351, 1993.

9 Signal Transduction and Pathogen-Induced PR Gene Expression

Jian-Min Zhou

CONTENTS

9.1 Introduction 195
9.2 Signals and Putative Receptors that Activate PR Gene Expression 196
9.3 Pathogens Activate PR Genes by Different Pathways 197
9.4 Transcriptional Regulation of PR Gene Expression 198
9.5 Genetic Studies on PR Gene Expression 200
9.6 Conclusion 202
Acknowledgments 202
References 202

9.1 INTRODUCTION

Pathogenesis-related (PR) proteins are so named because of the intimate association of their expression with pathogen attacks (van Loon, Chapter 1). The pathogen-activated PR gene expression plays an important role in plant defense against pathogens. A good example is the coordinated activation of a variety of PR genes during the onset of systemic-acquired resistance in tobacco and *Arabidopsis*.[1,2] Since the discovery of PR proteins, regulation of PR gene expression has been a highly active research area. However, how pathogen-infection leads to PR gene expression in plants is still poorly understood. This is due, in part, to the complexity of environmental stimuli and developmental cues that can induce the expression of various PR genes. For example, PR genes can be induced not only by the infection of a variety of pathogens, but also through stimulation by phytohormones and an array of stress conditions such as wounding, osmotic stress, UV light, and oxidative stress.[3] In addition, certain tissue types such as roots, flowers, and cultured cells are highly active in PR gene expression presumably due to developmental stimuli. Furthermore, our efforts to unequivocally identify signal transduction components that link a particular external stimulus to the expression of PR genes has been hampered, until recently, by the lack of a genetic system. Nevertheless, clues to this complex problem are now beginning to emerge due, in part, to the rapid progress in the plant disease

0-8493-0697-3/99/$0.00+$.50

resistance field. Because signal transduction in plant defense responses is an overwhelmingly large subject that has been reviewed extensively,[4-7] this chapter focuses on signal transduction mechanisms directly related to PR gene regulation.

9.2 SIGNALS AND PUTATIVE RECEPTORS THAT ACTIVATE PR GENE EXPRESSION

A number of molecules derived from pathogens during plant–pathogen interactions can serve as elicitors for PR gene induction. Elicitors of pathogen origin include chitin fragments and glucans that are derived from fungal cell wall, extracellular glycoproteins/peptides from certain fungal species, oligosaccharides and harpins from bacteria, and Avr proteins derived from bacterial and fungal pathogens.[8] Thus in the event of plant–pathogen interactions, expression of PR genes in plants is often subjected to multiple stimuli. In contrast to the vast number of signals known to induce PR gene expression, no receptors have been unequivocally established for these signal molecules.

One well-characterized elicitor is the β-glucan elicitor (GE) released from *Phytophthora sojae* cell wall by a β-1,3-glucanase from soybean. GE was originally identified based on its ability to induce phytoalexin biosynthesis.[9] Recently, a high-affinity GE-binding protein (GEBP) has been purified from soybean and the corresponding cDNA isolated.[10] The amino acid sequence of GEBP does not match any proteins in the database with known function. Supporting a role of GEBP as a *bona fide* receptor of GE, an antiserum raised against GEBP partially inhibited the binding of GE to soybean membrane proteins and reduced the phytoalexin accumulation elicited by GE. However, transgenic tobacco plants carrying the *GEBP* gene failed to respond to GE. Thus, an *in vivo* function of GEBP is yet to be demonstrated.

Infection of parsley by the nonhost pathogen *Phytophthora sojae* causes ion channel openings, increased protein phosphorylation, expression of defense genes including PR genes, accumulation of phytoalexin, and hypersensitive cell death.[11,12] A 42 kDa glycoprotein secreted by this fungus has been purified and shown to possess the elicitor activity. Furthermore, a 13 amino acid oligopeptide sequence (Pep13) in this protein was found necessary and sufficient for the elicitor activity when assayed in cultured parsley cells.[12] A 25 amino acid oligopeptide (Pep25) encompassing the Pep13 sequence was also found to be a potent elicitor and more stable than Pep13.[13] A high-affinity, specific binding site for this elicitor has been identified on parsley plasma membranes, but it remains to be determined whether this binding site corresponds to a receptor for this peptide. The major challenge is to purify a high-affinity-binding protein for the elicitor molecule and establish its role in plant defense.

Another class of elicitors is the polypeptide encoded by pathogen avirulence (*avr*) genes. A pathogen containing a particular *avr* gene is recognized by the host plant that carries a corresponding resistance (*R*) gene and activates disease resistance mechanisms in the latter. Extraordinary progress has been made on this subject in the past decade and numerous *R* genes and *avr* genes have been isolated. The reader is referred to several excellent reviews on this topic.[6,14,15] The recognition of the *avr* gene by *R* gene normally activates a variety of defense responses, including increased PR gene expression, in the plant. One excellent example is the apoplast-located race-specific elicitor, an oligopeptide, encoded by the *avr9* gene from *Cladosporium fulvum*, a fungal pathogen that causes leaf mold disease in greenhouse-grown tomato plants. Injection of tomato intercellular washing fluid (IF) containing this peptide into tomato rapidly activated the transcription of glucanase and chitinase genes in plants carrying the cognate *R* gene *Cf9*.[16,17] Although a high-affinity Avr9-binding site has been detected in tomato and tobacco, it does not correspond to any known *Cf* resistance genes.[18] Thus although *Cf9* encodes a transmembrane protein with a putative extracellular ligand-binding domain, whether it is a receptor for Avr9 remains undetermined. Another race-specific fungal elicitor known to activate PR genes is the NIP1 protein from the barley pathogen *Rhynchosporium secalis*.[19]

The only evidence that an Avr protein directly interacts with an *R* gene product came from the study of bacterial speck disease in tomato. Tomato plants harboring the resistance gene *Pto*, which encodes a cytoplasmic kinase, are resistant to the bacterial pathogen *Pseudomonas syringae* pv *tomato* carrying the *avrPto* gene.[20] When expressed in yeast, the Pto and AvrPto proteins were capable of highly specific association. Alterations of amino acid sequences of Pto or AvrPto that blocked Pto–AvrPto association also abolished the resistance function in the plant.[21,22] The results suggest that Pto acts as a receptor for the AvrPto protein. It is now believed that many bacterial Avr proteins are secreted and act directly within host cells. Using an inducible gene expression system, McNellis et al.[23] directly expressed the bacterium-encoded AvrRpt2 protein in an *Arabidopsis* line that carries the cognate *R* gene *RPS2* and stimulated hypersensitive response (HR, a typical defense reaction) and "super induction" of the *PR-1* gene. The availability of such a system is invaluable for the elucidation of defense signaling mechanisms mediated by *avr-R* interaction.

9.3 PATHOGENS ACTIVATE PR GENES BY DIFFERENT PATHWAYS

Upon infection by pathogens, plants often exhibit increased production of reactive oxygen species (ROS), salicylic acid (SA), ethylene, and jasmonates.[4,5] These molecules can serve as secondary signals to activate plant defense. Recent results also indicate that nitric oxide is an important secondary signal during plant defense.[24,25] Many of these secondary signals are well-known inducers for PR gene expression. For example, SA induces PR genes (encoding mostly acidic PR proteins) that are normally activated during SAR, whereas ethylene and jasmonates are known to induce genes encoding proteinase inhibitors, defensin, thionin, and basic PR proteins.[1,3,26-29] However, in the majority of studies, it's uncertain if the observed PR gene induction by a particular pathogen requires secondary messengers. To complicate the matter, cross talks are common between signaling pathways mediated by these secondary messengers. For instance, SA is known to inhibit the biosynthesis of ethylene and jasmonates, and jasmonate-activated *Pin* (proteinase inhibitor) gene expression in tomato, whereas ethylene and jasmonates synergistically enhance the expression of *Osmotin*, *PR1-b* in tobacco, and *Pin2* in tomato.[26,27,30,31] Fortunately, a number of *Arabidopsis* mutants have been isolated that are defective either in the biosynthesis or sensing of ethylene, jasmonates, and SA.[32-34] The use of these mutants should be helpful in clarifying the roles of these secondary messengers in plant defense responses against pathogen attacks.

9.4 TRANSCRIPTIONAL REGULATION OF PR GENE EXPRESSION

Transcriptional regulation has been one of the most active areas in PR gene research. Pathogen-induced PR gene expression often occurs at the level of transcription. Through traditional promoter deletion analysis coupled with mutagenesis of putative regulatory elements, gain-of-function studies with synthetic promoters, and DNA-fingerprinting analysis, several *cis*-regulatory elements mediating pathogen-induced PR gene expression have been identified.[5,35,36] These include the W-box (consensus TTGACC or TGAC-[N]$_x$-GTCA), GCC box (consensus AGCCGCC), MRE-like sequence (consensus A[A/C]C[A/T]A[A/C]C), G-box (consensus CACGTG), SA-responsive element (SARE, with a consensus of TTCGACCTCC), and an 11 bp *cis*-element mediating elicitor-induced *PR-2* expression in parsley. Detailed below are the GCC box and W-box that have been studied extensively and have a wide implication in PR gene regulation.

The GCC box was originally identified as an ethylene responsive element in the promoter of a number of tobacco basic PR genes.[37,38] To date, more than 20 plant genes are known to contain the GCC box, mostly PR genes from solanaceous species.[39] Interestingly, other ethylene-responsive genes that function in fruit ripening and flower senescence do not appear to contain the GCC box. It may be that the GCC box is associated with defense responses mediated by ethylene. One exception to this is the GCC box in the promoter of *Arabidopsis HLS1* gene that is required for the formation of apical hook when etiolated seedlings are exposed to ethylene.[40] *HLS1* mRNA is also induced by ethylene, but the significance of the GCC box in this induction is not known. The GCC box is necessary and sufficient to confer ethylene-induced transcription of the tobacco *gln2* gene that encodes a β-1,3-glucanase.[38] A similar conclusion has been reached with tobacco PRB-1b and Osmotin promoters. At this point, it is not clear whether the GCC box also confers the inducibility of these genes by stimuli other than ethylene. It should be noted that a mutated GCC box not only failed to confer ethylene inducibility, but also exhibited diminished basal transcriptional activity when fused to a synthetic promoter.[38,41] In addition, a 140 bp fragment that contains the GCC box from the Osmotin promoter is necessary and sufficient to confer the responsiveness to most of the stimuli known to induce Osmotin.[42,43] Thus, it is possible that the GCC box could be a point of cross talk between various signal transduction pathways. However, the requirement of GCC box for PR gene induction by pathogens still lacks direct evidence.

Ohme-Takagi and Shinshi[38] screened a tobacco cDNA library with a radiolabeled GCC box DNA fragment in a South–western blot and isolated 4 cDNA clones encoding GCC box-binding proteins EREBP-1, EREBP-2, EREBP-3, and EREBP-4. These proteins share a central domain of 59 amino acids that is responsible for binding to the GCC box. *EREBP* transcripts are induced by ethephon, a compound known to release ethylene upon its degradation, suggesting that ethylene further induces the expression of *EREBP* genes. However, caution should be exercised because ethephon degradation also leads to the accumulation of acid that is known to activate certain genes.[44] Therefore, it is possible that the observed *EREBP* gene induction following ethephon application is caused by acid rather than ethylene. Most recently, Ecker's group reported strong evidence that an *Arabidopsis* EREBP homolog, AtERF1, acts directly downstream of EIN3, a nuclear component of the ethylene signaling pathway, and activates PR gene expression.[45,46] The *AtERF1* gene was found to be induced by ethylene in wild-type *Arabidopsis* but not in the *ein3* mutant. The EIN3 protein was capable of binding to the promoter sequence of *AtERF1,* suggesting that EIN3 functions as a transcription regulator of *AtERF1*. The AtERF1 protein binds to the GCC box and, when overexpressed in *Arabidopsis*, resulting in constitutive activation of a basic chitinase gene and *PDF1.2* (encoding a protein called defensin).

The *EREBP-1* gene was also found to be induced by *Pseudomonas* bacteria and SA, suggesting a role of this gene in plant defense.[39,47] A direct connection of the EREBP proteins with a disease-resistance pathway came from the study of the signaling pathway mediated by the tomato gene *Pto*. Using Pto as a "bait," Zhou et al.[39] identified several proteins that interact with Pto in a "yeast two-hybrid" assay. These proteins are collectively called Pto-interacting proteins (Pti). Three proteins, Pti4, Pti5, and Pti6, are highly similar to tobacco EREBPs at the amino acid level. Pti5 and Pti6 have been demonstrated to bind the GCC box of the tobacco *gln2* gene. Thus, it is likely that Pti4/5/6 and EREBPs act in the *R* gene pathway. Phosphorylation has been suggested to play an important role in the activation of PR gene expression during pathogen attacks.[48,49] The direct interaction of the Pto kinase with Pti4/5/6 reinforces this possibility. Indeed, *in vitro* phosphorylation of Pti4 by Pto has been demonstrated (Y. Gu and G. Martin, personal communication). It is possible that Pto can directly activate these transcription factors by phosphorylation. Interestingly, *Pti4* and *Pti5* display differential regulation during pathogen infections (V. Thara and J. Zhou, unpublished results). While *Pti5* transcripts are barely detectable prior to pathogen-inoculation, they accumulate rapidly upon the inoculation of *P. syringae tomato.* In contrast, the expression of *Pti4* was readily detected without inoculation and was further enhanced by *Pseudomonas* inoculation. *Pti4* was also strongly induced by mock inoculation with buffer, suggesting that *Pti4* is responsive to mechanical or osmotic stress. While these findings strongly suggest that the Pti proteins are transcription activators mediating pathogen-induced PR gene expression, *in vivo* experiments are necessary to confirm this possibility. Furthermore, it will be important to determine if the Pti proteins also mediate the ethylene-activated PR gene expression.

Another highly conserved *cis*-element is the W box. The W box is present in parsley *PR1-1* and *PR1-2* (both encoding the PR1 protein), tobacco *CHN50* (encoding a class I basic chitinase), asparagus *AoPR1* (encoding the PR10 protein), potato *PR-10a* (encoding the PR10 protein), and maize *PRms* (encoding the PR1 protein). Other pathogen inducible genes such as the potato glutathione S-transferase gene *prp1* and the grape phytoalexin synthesis gene *Vst1* also contain the W box in the promoter, suggesting a wider role for this element in pathogen-induced gene expression.[36] Promoter deletion analysis has demonstrated that the W box is required for Pep25-induced expression of the parsley *PR1-1* and *PR1-2* genes.[50] A tetramer of the W box fused to a 35S minimal promoter was able to confer the elicitor-induced expression in a transient assay indicating that the W box is sufficient to confer such induction. Similar results were obtained for the maize *PRms* gene using a crude elicitor prepared from fragmented mycelium of *Fusarium moniliforme*.[51]

By using a South–western screening, Rushton et al.[50] cloned three parsley cDNAs encoding W box-binding proteins. These proteins contain the consensus sequence, WRKYGQK, and are termed WRKY family proteins. The carboxy terminus of these proteins contains a zinc finger structure that may function in DNA-binding or protein–protein interaction. When transiently expressed in plant cells, the *Arabidopsis* WRKY protein ZAP1 can activate a promoter containing a W box, indicating that ZAP1 is capable of trans-activating W box-containing genes *in vivo*.[52] Similar to the pathogen-induced *EREBP-1* and *Pti5*, parsley *WRKY-1* and *WRKY-3* are transiently induced by the Pep25 elicitor.[50]

A major challenge in the future will be to address the *in vivo* function of the cloned transcription factors. Obviously, rigorous tests on transgenic plants with altered expression of the transcription factor genes are required to establish their roles in PR gene expression and defense responses. It should be noted that *EREBPs* and *WRKY* genes exist as large families. Whether different family members possess distinct or redundant functions awaits investigation. In addition, it is necessary to understand how different signals affect PR gene expression. For example, do different signals converge on the same transcription factor? Do these transcription factors interact? In *Arabidopsis*, the EREBP homolog AtEBP1 was found to interact with a basic-region leucine zipper transcription factor OBF4.[53] Answering these questions may enable us to better understand cross talks between different signaling pathways.

9.5 GENETIC STUDIES ON PR GENE EXPRESSION

In *Arabidopsis*, several PR genes, such as *PR-1, PR-2*, and *PR-5*, are known to be induced by pathogens and SA.[2,24] One of the most dramatic effects of SA on plants is the rapid induction of a large set of PR genes at very high levels. By using an SA-inducible promoter-*GUS* fusion construct (*BGL2::GUS*), Dong and her colleagues[54] have identified a number of interesting *Arabidopsis* mutant plants with altered PR gene expression. One of the mutants is *npr1* (nonexpressor of PR genes) in which both the *GUS* gene and endogenous *BGL2* (*PR2*) gene failed to respond to SA and pathogens.[55] The same mutant also has been identified by other screening schemes.[56-58] Subsequent characterization of this mutant indicated that the wild-type *NPR1* gene is necessary for the expression of a number of PR genes when infected by pathogens (bacteria and fungi) or treated with SA. Another class of mutants is called *cpr* (constitutive expressors of PR genes) mutants.[59-61] These mutants, including *cpr1, cpr5,* and *cpr6,* constitutively express PR genes in the absence of pathogen or exogenously applied SA.

The *NPR1* gene has been cloned, and it encodes a protein containing ankyrin repeats.[62,63] Ankyrin repeats are involved in protein–protein interactions and have been found, but not limited to, mammalian IκB and *Drosophila* Cactus proteins. Both animal proteins function by masking the nuclear localization sequence (NLS) of transcription factors such as NFκB, thus sequestering the latter in the cytoplasm.[64] Upon activation by external stimuli, the IκB, and likely Cactus, are phosphorylated and rapidly turned over by the ubiquitin/proteasome pathway, thus allowing nuclear-entry of the transcription factor NF-κB. Interestingly, the same mechanism is used both by mammals and *Drosophila* to activate defense gene expression which, similar to PR gene expression, leads to the accumulation of antimicrobial proteins.[6,65] While the presence of ankyrin repeats suggests that interactions with other proteins are involved in its function, NPR1 must act differently than IκB. First, while IκB negatively regulates gene transcription, NPR1 appears to be a positive regulator of PR gene expression. Loss of the NPR1 function in the *npr1* mutant resulted in the inability of PR gene induction rather than constitutive PR gene expression.[55] Conversely, overexpression of *NPR1* has enhanced pathogen-induced PR gene expression in *Arabidopsis*.[66] Furthermore, the NPR1 protein is targeted to the nucleus at the onset of SAR, suggesting that NPR1 acts in the nucleus to activate PR genes, contrasting the cytoplasmic location of IκB.[34]

Epistatic analysis has placed the *CPR5* and *CPR1* genes upstream of SA production, and the *NPR1* and *CPR6* downstream of SA production.[59,61] The wild-type *CPR5* gene functions by negatively regulating hypersensitive response, whereas the wild-type function of the *CPR1* gene is to negatively regulate SA biosynthesis. In addition to the classical PR genes (such as *PR1, PR2,* and *PR5*), *PDF1.2* (encoding a defensin) and *Thi2.1* (encoding a thionin) are also constitutively expressed in the *cpr5* and *cpr6* mutants. In contrast, the *cpr1* mutant accumulates only the classical PR genes but not *PDF1.2.* A *cpr5:npr1* double mutant accumulated mRNA for *PDF1.2* but not *PR1, PR2*, or *PR5*, indicating that the expression of *PDF1.2* is independent of NPR1 function.[59] Thus, the activation of *Arabidopsis* defense genes appears to follow two separate pathways: an NPR1-dependent pathway for *PR1, PR2,* and *PR5* expression, and an NPR1-independent pathway for *PDF1.2* expression and, possibly, *Thi2.1* expression. The NPR1-independent pathway is likely to involve jasmonates and ethylene; both are known to activate *PDF1.2* and *Thi2.1* expression. By constructing double mutants between the *cpr5* or *cpr6* mutants with *ein2,* an ethylene insensitive mutant, it was discovered that ethylene signaling is required for *PDF1.2* expression in both *cpr5* and *cpr6* mutants.[34] Several jasmonate insensitive mutants are available in *Arabidopsis*.[32] Double mutant analysis between the *cpr* mutants and jasmonate-insensitive mutants should aid in our understanding of the roles of *CPR* genes and jasmonates in plant defense responses. Genetic studies suggest that CPR6 may be responsible for the cross talk between the SA-mediated signaling pathway and the jasmonates/ethylene-mediated signaling pathway.[61] Thus, it will be critical to isolate the *CPR6* gene and determine the relationship between CPR6, NPR1, and components from the jasmonates and ethylene signaling pathways such as COI1, EIN2, EIN3, and AtERF1.[32,46,49,50,67]

9.6 CONCLUSION

The mechanisms underlying pathogen-induced PR gene expression remain largely unknown. The EREBP/Pti and WRKY proteins may play crucial roles in PR gene regulation. This possibility cannot be confirmed without evidence from reverse genetics. Should these proteins represent transcription factors that globally regulate PR gene expression, they may provide means to engineer plants with enhanced disease resistance. Genetic studies not only have identified important genes such as *NPR1* and *CPRs* in the signaling pathways leading to PR gene expression, they have also permitted various signaling pathways to be distinguished. Continued efforts combining both molecular and genetic approaches should greatly advance our knowledge in this area.

ACKNOWLEDGMENTS

The author is grateful to Scot Hulbert, Xiaoyan Tang, Krishnarajapuram Thara, and Subbaratnam Muthukrishnan for critical reading of the manuscript. J. Zhou was supported by a NSF grant MCB-98-08710 and Kansas Agricultural Experimental Station (Contribution No. 99-129-B).

NOTE: The *Arabidopsis PDF1.2* promoter has been reported.[68] The promoter confers pathogen and jasmonate responsiveness and contains a perfect GCC box. Together with the activation of *PDF1.2* expression by ERF1, the current data clearly demonstrate that the *PDF1.2* gene is a target for the EREBP/Pti class of transcription factors.

REFERENCES

1. Ward, E. R., Uknes, S. J., Williams, S. C., Dincher, S. S., Wiederhold, D. L., Alexander, D. C., Ahl-Goy, P., Metraux, J.-P., and Ryals J. A., Coordinate gene activity in response to agents that induce systemic acquired resistance, *Plant Cell*, 3, 1085, 1991.
2. Uknes, S., Mauch-Mani, B., Moyer, M., Potter, S., William, S., Dincher, S., Chandler, D., Slusarenko, A., Ward, E., and Ryals, J. A., Acquired resistance in *Arabidopsis*, *Plant Cell*, 4, 645, 1992.
3. Brederode, F. T., Linthorst, H. J. M., and Bol, J. F., Differential induction of acquired resistance and PR gene expression in tobacco by virus infection, ethephon treatment, UV light and wounding, *Plant Mol. Biol.*, 17, 1117, 1991.
4. Hammond-Kosack, K. E. and Jones, J. D. G., Resistance gene-dependent plant defense responses, *Plant Cell*, 8, 1773, 1996.
5. Yang, Y., Shah, J., and Klessig, D. F., Signal perception and transduction in plant defense responses, *Genes Dev.*, 11, 1621, 1997.
6. Baker, B., Zambryski, P., Staskawicz, B., and Dinesh-Kumar, S. P., Signaling in plant-microbe interactions, *Science,* 276, 726, 1997.
7. Innes, R. W., Genetic dissection of R gene signal transduction pathways, *Current Opin. Plant Biol.*, 1, 299, 1998.

8. Boller, T. and Felix, G., Olfaction in plants: specific perception of common microbial molecules, in *Biology of Plant-Microbe Interactions*, Stacey, G., Mullin, B., and Gresshoff, P., Eds., International Society for Molecular Plant-Microbe Interactions, St. Paul, Minnesota, 1996, 1.
9. Darvill, A. G. and Abersheim, P., Phytoalexins and their elicitors — a defense against microbial infection in plants, *Annu. Rev. Plant Physiol.*, 35, 243, 1984.
10. Umemoto, N., Kakitani, M., Iwamatsu, A., Yoshikawa, M., Yamaoka, N., and Ishida, I., The structure and function of a soybean β-glucan-elicitor-binding protein., *Proc. Natl. Acad. Sci. U.S.A.*, 94, 1029, 1997.
11. Jahnen, W. and Hahlbrock, K., Cellular localization of nonhost resistance reactions of parsley (*Petroselinum crispum*) to fungal infection, *Planta,* 173, 197, 1988.
12. Nurnberger, T., Nennsteil, D., Jabs, T., Sacks, W. R., Hahlbrock, K., and Scheel, D., High affinity binding of a fungal oligopeptide elicitor to parsley plasma membranes triggers multiple defense responses, *Cell,* 78, 449, 1994.
13. Ligterink, W., Kroj, T., zur Nieden, U., Hirt, H., and Scheel, D., Receptor-mediated activation of a MAP kinase in pathogen defense of plants, *Science*, 276, 2054, 1997.
14. Staskawicz, B. J., Ausubel, F. M., Baker, B. J., Ellis, J. G., and Jones, J. D. G., Molecular genetics of plant disease resistance, *Science*, 268, 661, 1995.
15. Bent, A., Plant disease resistance genes: function meets structure, *Plant Cell*, 8, 1757, 1996.
16. Ashfield, M., Hammond-Kosack, K. E., Harrison, K., and Jones, J. D. G., *Cf* gene-dependent induction of a β–1,3-glucanase promoter in tomato plants infected with *Cladosporium fulvum*, *Mol. Plant-Microbe Interact.*, 7, 645, 1994.
17. Wubben, J. P., Lawrence, C. B., and de Wit, P. J. G. M., Differential induction of chitinase and 1,3-β-glucanase gene expression in tomato by *Cladosporium fulvum* and its race-specific elicitors, *Physiol. Mol. Plant Pathol.*, 48, 105, 1996.
18. Kooman-Gersmann, M., Honee, G., Bonnema, G., and de Wit, P. J. G M., A high-affinity binding site for the Avr9 peptide elicitor of *Cladosporium fulvum* is present on plasma membranes of tomato and other solanaceous plants, *Plant Cell*, 8, 929, 1996.
19. Roche, M., Gierlich, A., Hermann, H., Hahn, M., Schmidt, B., Rosahl, S., and Knogge, W., The race-specific elicitor, NIP1, from the barley pathogen, *Rhynchosporium secalis*, determines avirulence on host plants of the *Rrs1* resistance genotype, *EMBO J.*, 14, 4168, 1995.
20. Martin, G. B., Brommonschenkel, S., Chunwongse, J., Frary, A., Ganal, M. W., Spivey, R., Wu, T., Earle, E. D., and Tanksley, S. D., Map-based cloning of a protein kinase gene conferring disease resistance in tomato, *Science*, 262, 1432, 1993.
21. Scolfield, S. R., Tobias, C. M., Rathjen, J. P., Chang, J. H., Lavelle, D. T., Michelmore, R. W., and Staskawicz, B. J., Molecular basis of gene-for-gene specificity in bacterial speck disease of tomato, *Science*, 274, 2063, 1996.
22. Tang, X., Frederick, R. D., Zhou, J., Halterman, D. A., Jia, Y., and Martin, G. B., Initiation of plant disease resistance by physical interaction of AvrPto and the Pto kinase, *Science,* 274, 2060, 1995.
23. McNellis, T. W., Mudgett, M. B., Li, K., Aoyama, T., Horvath, D., Chua, N.-H., and Staskawicz, B. J., Glucocorticoid-inducible expression of a bacterial avirulence gene in transgenic *Arabidopsis* induces hypersensitive cell death, *Plant J.,* 14, 247, 1998.
24. Delledonne, M., Xia, Y. J., Dixon, R. A., and Lamb, C. J., Nitric oxide functions as a signal in plant disease resistance, *Nature*, 394, 585, 1998.

25. Durner, J., Wendehenne, D., and Klessig, D. F., Defense gene induction in tobacco by nitric oxide, cyclic GMP and cyclic ADP ribose, *Proc. Natl. Acad. Sci. U.S.A.*, 95, 10328, 1998.
26. Farmer, E. E., Caldelari, D., Pearce, G., Walker-Simmons, M. K., and Ryan, C. A., Diethyldithiocarbamic acid inhibits the octadecanoid signaling pathway for the wounding induction of proteinase inhibitors in tomato leaves, *Plant Physiol.*, 106, 337, 1994.
27. O'Donnell, P. J., Calvert, C., Atzorn, R., Wasternack, C., Leyser, H. M. O., and Bowles, D. J., Ethylene as a signal mediating the wound response of tomato plants, *Science*, 274, 1914, 1996.
28. Penninckx, I. A. M. A., Eggermont, K., Terras, F. R. G., Thomma, B. P. H. J., de Samlanx, G. W., Bauchala, A., Metraux, J. P., Manners, J., and Broekaert, W. F., Pathogen-induced systemic activation of a plant defensin gene in *Arabidopsis* follows a salicylic acid-independent pathway, *Plant Cell,* 8, 2309, 1996.
29. Epple, P., Apel, K., and Bohlmann, H., An *Arabidopsis thaliana* thionin gene is inducible via a signal transduction pathway different from that for pathogenesis-related proteins, *Plant Physiol.*, 109, 813, 1995.
30. Doares, S. H., Narvaez-Vasquez, J., Conconi, A., and Ryan, C. A., Salicylic acid inhibits synthesis of proteinase inhibitors in tomato leaves induced by systemin and jasmonic acid, *Plant Physiol.*, 108, 1741, 1995.
31. Xu, Y., Chang, P. F. L., Liu, D., Narasimhan, M. L., Raghothanma, K. G., Hasegawa, P. M., and Bressan, R. A., Plant defense genes are synergistically induced by ethylene and methyl jasmonate, *Plant Cell*, 6, 1077, 1994.
32. Ecker, J. R., The ethylene signal-transduction pathway in plants, *Science*, 268, 667, 1995.
33. Creelman, R. A. and Mullet, J. E., Biosynthesis and action of jasmonates in plants, *Annu. Rev. Plant Physiol. Plant Mol. Biol.*, 48, 355, 1997.
34. Dong, X., SA, JA, ethylene, and disease resistance in plants, *Curr. Opin. Plant Biol.*, 1, 316, 1998.
35. Zhu, Q., Droge-Laser, W., Dixon, R. A., and Lamb, C. J., Transcriptional activation of plant defense genes, *Curr. Opin. Genet. Dev.*, 6, 624, 1996.
36. Rushton, P. J. and Somssich, I. E., Transcriptional control of plant genes responsive to pathogens, *Curr. Opin. Plant Biol.*, 1, 311, 1998.
37. Hart, C. M., Nagy, F., and Meins, F., Jr., A 61 bp enhancer element of the tobacco β-1,3-glucanase B gene interacts with one or more regulated nuclear proteins, *Plant Mol. Biol.,* 21, 121, 1993.
38. Ohme-Takagi, M. and Shinshi, H., Ethylene-inducible DNA-binding proteins that interact with an ethylene-responsive element, *Plant Cell*, 7, 173, 1995.
39. Zhou, J., Tang, T., and Martin, G., The Pto kinase conferring resistance to tomato bacterial speck disease interacts with proteins that bind a *cis*-element of pathogenesis-related genes, *EMBO J.,* 16, 3207, 1997.
40. Lehman, A., Black, R., and Ecker, J. R., *HOOKLESS1*, an ethylene response gene is required for differential cell elongation in the *Arabidopsis* hypocotyl, *Cell,* 85, 183, 1996.
41. Sessa, G., Meller, Y., and Fluhr, R., A GCC element and a G-box motif participate in ethylene-induced expression of the *PRB–1b* gene, *Plant Mol. Biol.*, 28, 145, 1995.
42. Liu, D., Narasimhan, M. L., Xu, Y., Raghothama, K. G., Hasegawa, P. M., and Bressan, R. A., Fine structure and function of the *osmotin* gene promoter, *Plant Mol. Biol.,* 29, 1015, 1995.

43. Raghothama, K. G., Kononowicz, A. K., Maggio, A., Narasimhan, M. L., Kononowicz, A. K., Wang, G., D'Urzo, M. P., Hasegawa, P. M., and Bressan, R. A., Tissue specific activation of the *osmotin* gene by ABA, C_2H_4 and NaCl involves the same promoter region, *Plant Mol. Biol.*, 34, 393, 1997.
44. Lawton, K. A., Potter, S. L., Uknes, S., and Ryals, J., Acquired resistance signal transduction in *Arabidopsis* is ethylene independent, *Plant Cell*, 6, 581, 1994.
45. Cao, Q. M., Rothenberg, M., Solano, R., Roman, G., Terzaghi, W., and Ecker, J. R., Activation of the ethylene gas response pathway in *Arabidopsis* by the nuclear protein ETHYLENE-INSENSITIVE3 and related proteins, *Cell*, 89, 1133, 1997.
46. Solano, R., Stepanova, A., Cao, Q. M., and Ecker, J. R., Nuclear events in ethylene signaling: a transcriptional cascade mediated by ETHYLENE-INSENSITIVE3 and ETHYLENE-RESPONSE-FACTOR1, *Genes & Development*, 12, 3703, 1998.
47. Horvath, D. M., Huang, D. J., and Chua, N. H., Four classes of salicylate-induced tobacco genes, *Mol. Plant-Microbe Interact.*, 11, 895, 1998.
48. Raz, V. and Fluhr, R., Ethylene signal is transduced via protein phosphorylation events in plants, *Plant Cell*, 5, 523, 1993.
49. Despres, C., Subramanian, R., Matton, D. P., and Brisson, N., The activation of the potato *PR-10a* gene requires phosphorylation of the nuclear factor PBF-1, *Plant Cell*, 7, 589, 1995.
50. Rushton, P. J., Torres, J. T., Parniske, M., Wernert, P., Hahlbrock, K., and Somssich, I. E., Interaction of elicitor-induced DNA-binding proteins with elicitor response elements in the promoters of parsley PR-1 genes, *EMBO J.*, 15, 5690, 1996.
51. Raventos, D., Jensen, A. B., Rask, M. B., Casacuberta, J. M., Mundy, J., and Segundo, B. S., A 20 bp *cis*-acting element is both necessary and sufficient to mediate elicitor response of a maize PRms gene, *Plant J.*, 7, 147, 1995.
52. de Parter, S., Greco, V., Pham, K., Memelink, J., and Kijne, J., Characterization of a zinc-dependent transcription activator from *Arabidopsis*, *Nucleic Acids Res.*, 24, 4624, 1996.
53. Buttner, M. and Singh, K. B., *Arabidopsis thaliana* ethylene-responsive element binding protein (AtEBP), an ethylene-inducible, GCC box-binding protein interacts with an ocs element binding protein, *Proc. Natl. Acad. Sci. U.S.A.*, 94, 5961, 1997.
54. Dong, X., Mindrinos, M., Davis, K. R., and Ausubel, F. M., Induction of *Arabidopsis* defense genes by virulent and avirulent *Pseudomonas syringae* strains and by a cloned avirulence gene, *Plant Cell*, 3, 61, 1991.
55. Cao, H., Bowling, S. A., Gordon, A. S., and Dong, X., Characterization of an *Arabidopsis* mutant that is nonresponsive to inducers of systemic acquired resistance, *Plant Cell*, 6, 1583, 1994.
56. Delaney, T. P., Friedrich, L., and Ryals, J. A., *Arabidopsis* signal transduction mutant defective in chemically and biologically induced disease resistance, *Proc. Natl. Acad. Sci. U.S.A.*, 92, 6602, 1995.
57. Shah, J., Tsui, F., and Klessig, D. F., Characterization of a salicylic acid-insensitive mutant (*sai1*) of *Arabidopsis thaliana*, identified in a selective screen utilizing the SA-inducible expression of the *tms2* gene, *Mol. Plant-Microbe Interact.*, 10, 69, 1997.
58. Hammond-Kosack, K. E. and Jones, J. D. G., Resistance gene-dependent plant defense responses, *Plant Cell*, 8, 1773, 1996.
59. Bowling, S. A., Guo, A, Cao, H., Gordon, A. S., Klessig, D. F., and Dong, X., A mutation in *Arabidopsis* that leads to constitutive expression of systemic acquired resistance, *Plant Cell*, 6, 1845, 1994.

60. Bowling, S. A., Clarke, J. D., Liu, Y., Klessig, D. F., and Dong, X., The *cpr5* mutant of *Arabidopsis* express both NPR1-dependent and NPR1-independent resistance, *Plant Cell*, 9, 1573, 1997.
61. Clarke, J. D., Liu, Y., Klessig, D. F., and Dong, X., Uncoupling PR gene expression from NPR1 and bacterial resistance: characterization of the dominant *Arabidopsis cpr6-1* mutant, *Plant Cell*, 10, 557, 1998.
62. Cao, H., Glazebrook, J., Clarke, J. D., Volko, S., and Dong, X., The *Arabidopsis NPR1* gene that controls systemic acquired resistance encodes a novel protein containing ankyrin repeats, *Cell*, 88, 57, 1997.
63. Ryals, J., Weymann, K., Lawton, K., Friedrich, L., Ellis, D., Steiner, J. Y., Johnson, J., Delaney, T. P., Jesse, T., Vos, P., and Uknes, S., The *Arabidopsis NIM1* protein shows homology to the mammalian transcription factor inhibitor IκB, *Plant Cell*, 9, 425, 1997.
64. Krappmann, D., Wulczn, F. G., and Scheidereit, C., Different mechanisms control signal-induced degradation and basal turnover of the NF-κB inhibitor IκB *in vivo*, *EMBO J*, 15, 6717, 1996.
65. Hoffmann, J. A. and Reichhart, J. M., *Drosophila* immunity, *Trends Cell Biol.*, 7, 309, 1997.
66. Cao, H., Li, X., and Dong, X., Generation of broad-spectrum disease resistance by overexpression of an essential regulatory gene in systemic acquired resistance, *Proc. Natl. Acad. Sci. U.S.A.*, 95, 6531, 1998.
67. Xie, D. X., Feys, B. F., James, S., Nieto-Rostro, M., and Turner, J. G., COI1: an *Arabidopsis* gene required for jasmonate-regulated defense and fertility, *Science*, 280, 1091, 1998.
68. Manners, J. M., Penninckx, I. A. M. A., Vermaere, K., Kazan, K., Brown, R. L., Morgan, A., Mclean, D. J., Curtis, M. D., Cammue, B. P. A., and Broekaert, W. F., The promoter of the plant defensin gene *PDF1.2* from *Arabidopsis* is systemically activated by fungal pathogens and responds to methyl jasmonate but not to salicylic acid, *Plant Mol. Biol.*, 38, 1071, 1998.

10 The Role of Thionins in the Resistance of Plants

Holger Bohlmann

CONTENTS

10.1 Introduction 207
10.2 Distribution of Thionins 208
10.3 Structure of Thionins 211
10.4 Toxicity 212
10.5 Function 214
10.6 Processing and Localization 215
10.7 Barley Leaf Thionins 218
10.8 Arabidopsis Thionins 220
10.9 Genetic Engineering 222
References 222

10.1 INTRODUCTION

Plants are steadily threatened by a variety of pests and pathogens against which they have evolved different defense mechanisms. Preformed physical barriers[1] and toxic metabolites[2] are two parts of the defense. Other defense reactions are set in action only after attack by pathogens or pests. These include the formation of phytoalexins,[3] pathogenesis-related (PR) proteins,[4] papillae,[5] a strengthening of cell walls,[6] and the hypersensitive response (HR), which is thought to function by isolating and killing microorganisms in a small region of the plant by localized cell death.[7,8] Following an HR, plants often develop a systemic resistance (SAR) against subsequent pathogen attacks. Salicylate plays a key rôle in SAR,[9] although it is still debated if it is the transmitted signal. Resistance coincides with the expression of certain PR-proteins.[10]

A different inducible defense mechanism is triggered by wounding, which occurs, for instance, through the activities of grazing insects or their larvae. This leads to the expression of, among others, proteinase inhibitors which are thought to act on enzymes in the digestive tract of insects. The signal transduction pathway leading to the induction of proteinase inhibitors has been studied intensively in tomato.[11,12] Systemin, an 18 amino acid oligopeptide,[13] and jasmonate[14] have been identified as key components of this system.

0-8493-0697-3/99/$0.00+$.50

The defense system of plants (and animals) also includes a variety of relatively small, basic, cysteine-rich polypeptides with *in vitro* antimicrobial activities (for reviews, see references 15 and 16). Among these are lipid transfer proteins,[17] plant defensins,[18] and thionins (for previous reviews, see references 19 through 22). Thionins are the subject of this review. After a general review of thionins, I will discuss work on barley leaf thionins and Arabidopsis thionins which strongly supports the view that thionins are defense proteins.

10.2 DISTRIBUTION OF THIONINS

The toxic activity of thionins was apparently first described by Jago and Jago in 1885 (cited in Ohtani et al.[23]), who reported that wheat flour contains a substance that is toxic to yeast. Balls and Hale[24] found that it could be extracted with petroleum ether. The polypeptide responsible for the toxicity was isolated by Balls et al.[25] and named purothionin. It was later shown to consist of three different isoforms whose sequences have been determined (Figure 10.1). Purothionins, as with the majority of thionins, are basic and cysteine-rich polypeptides with a molecular weight of about 5 kD. Related polypeptides, all with eight conserved cysteine residues, have been isolated and sequenced from the endosperm of oat and barley (see Figure 10.1) and are also present in other species of the *Poaceae*.[38-43] The occurrence of thionins in maize[44,45] is doubtful, because a polypeptide with similar properties, but clearly not homologous to thionins, has been isolated from maize grains.[46] Recently, seed-specific thionins have been discovered in wheat and *Aegilops squarrosa,* which are neutral and lack the cysteine residues 2 and 8.[33,34] All thionins mentioned thus far are localized in the endosperm of cereals. Leaf-specific thionins have been discovered in barley in addition to the seed-specific hordothionins[35,36] and will be discussed later. Furthermore, five different cDNAs for thionins were recently cloned from another monocotyledonous plant species, *Tulipa gesneriana.*[37] One of the thionins encoded by the cDNAs has an unusual pattern of cysteines (see Figure 10.1), and their acidic domains are also unique in having only three cysteine residues.

Thionins are also present in dicot species. One large group of thionins (Figure 10.2) has been found in different species of mistletoes. The toxic effects of viscotoxins toward mammals have been investigated in detail[62-65] because of the medical use of mistletoe extracts in the treatment of cancer for instance,[66] but it should be noted that mistletoes contain other toxic compounds in addition to the viscotoxins, for instance lectins.[67] The crude thionin extract from mistletoe leaves was first isolated by Winterfeld and Bijl.[68] This crude mixture was separated into three distinct viscotoxins by Samuelsson[69] and Samuelsson and Pettersson,[70] and later into four.[71] Several viscotoxins from different species have been sequenced (see Figure 10.2). However, mistakes occurred in the determination of the C-terminal amino acids, as has also been discussed by Vernon et al.[47] The correct sequences for viscotoxin A3 and B have been deduced from cDNA sequences,[51] and the C-terminus of all viscotoxins in Figure 10.2 has been corrected accordingly. Using a PCR-based approach, Schrader-Fischer and Apel[48] isolated DNA fragments for two other viscotoxins from *V. album.* In contrast to the classical viscotoxins which are characterized by six cysteine residues, these newly identified viscotoxins contained eight cysteine

Position:	1 10 20 30 40	
Purothionin α1	KSCCRSTLGRNCYNLCRARG---AQKLCAGVCRCKISSGLSCPKGFPK	23, 26, 27
Purothioninα2	KSCCRSTLGRNCYNLCRSRG---AQKLCSTVCRCKLTSGLSCPKGFPK	26
Purothionin β	KSCCKSTLGRNCYNLCRARG---AQKLCANVCRCKLTSGLSCPKDFPK	23, 27, 28
Avenothionin α	KSCCRNTLGRNCYNLCRSRG---APKLCATVCRCKISSGLSCPKDFPK	29
Avenothionin β	KSCCKNTLGRNCYNLCRARG---APKLCSTVCRCKLTSGLSCPKDFPK	29
Hordothionin α	KSCCRSTLGRNCYNLCRVRG---AQKLCAGVCRCKLTSSGKCPTGFPK	30, 31
Hordothionin β	KSCCRSTLGRNCYNLCRVRG---AQKLCANACRCKLTSGLKCPSSFPK	32
pTTH20,TthV2,AthV1	VDCGANPFKVACFNSCLLGPS--TVFQCADFCACRLPAG---------	33, 34
DB4	KSCCKDTLARNCYNTCHFAGG--SRPVCAGACRCKIISGPKCPSDYPK	35
DC4	KSCCKDTLARNCYNTCRFAGG--SRPVCAGACRCKIISGPKCPSDYPK	35
DG3	KSCCKNTTGRNCYNACRFAGG--SRPVCATACGCKIISGPTCPRDYPK	35
pKG1348	KSCCKNTTGRNCYNACRFAGG--SRPVCATACGCKIISGPTCPRDYPK	36
TgThi1-1	KSCCRTTAARNCYNVCR-LGGT-PQTLCARTCDCIHITTGNCPRSHPK	37
TgThi1-2	KSCCRNTTARNCYNVCR-LPGT-PRPVCAATCDCKIISSGKCPPGYEK	37
TgThi1-3	KSCCRNTTARNCYNVCR-LPGT-PRPVCAATCDCKIISSGKCPPGYEK	37
TgThi1-4	KSCCESTAARNCYNVCR-FPGT-PRPVCAATCGCKIITGTKCPPDYPK	37
TgThi4-1	KSCFPSTAAKYCYNACR-LPGCRPETICAARCGCKIISSGNCPPGYEN	37

FIGURE 10.1 Sequences of thionins from monocotyledonous plants.

Position:	1 . . . . 10 . . . . 20 . . . . 30 40	
Pyrularia Toxin	K S C C R N T W A R N C Y N V C - R L P G T I S R E I C A K K C D C K I I S G T T C P S D Y P K	47
Thi1Va1	K I C C R A P A G K K C Y N L C - T A L L - - S S E T C A N T C Y C K D V S G E T C P A D Y P A	48
Thi1Va2	K S C C R N T T G R N C Y N A C - R V P G - T P R P V C A N T C D C K I I S G S K C P A D Y P R	48
Viscotoxin A2	K S C C P N T T G R N I Y N T C - R F G G G - S R E V C A S L S G C K I I S A S T C P S D Y P K	49
Viscotoxin A3	K S C C P N T T G R N I Y N A C - R L T G A - P R P T C A K L S G C K I I S G S T C P S D Y P K	50, 51
Viscotoxin B	K S C C P N T T G R N I Y N T C - R L G G G - S R E R C A S L S G C K I I S A S T C P S D Y P K	51, 52
Viscotoxin 1-PS	K S C C P N T T G R N I Y N T C - R F G G G - S R E V C A R I S G C K I I S A S T C P S D Y P K	53
Phoratoxin A	K S C C P T T T A R N I Y N T C - R F G G G - S R P V C A K L S G C K I I S G T K C D S G W N H	54
Phoratoxin B	K S C C P T T T A R N I Y N T C - R F G G G - S R P I C A K L S G C K I I S G T K C D S G W N H	55
Ligatoxin	K S C C P S T T A R N I Y N T C - R L T G T - S R P T C A S L S G C K I I S G S T C D S G W N H	56
Denclatoxin B	K S C C P T T A A R N G Y N I C - R L P G T - P R P V C A A L S G C K I I S G T G C P P G Y R H	57
Crambin 1	T T C C P S I V A R S N F N V C - R L P G T - P E A L C A T Y T G C I I I P G A T C P G D Y A N	58, 59
Crambin 2	T T C C P S I V A R S N F N V C - R L P G T - S E A I C A T Y T G C I I I P G A T C P G D Y A N	58, 59
THI2.1	K I C C P S N Q A R N G Y S V C - R I R F - - S K G R C M Q V S G C Q N S - - D T C P R G W V N	60
THI2.2	K I C C P T K D D R S V Y F V C - M L S V - S S Q F Y C L L K S K C K N T S Q T I C P P G Y T N	60
Rs-thionin	K V C C P S K E A R L A F Y V C N R T K A T - - - A T C A Q L N G C I I V P E T I C P S G Y . .	61

FIGURE 10.2 Sequences of thionins from dicotyledonous plants.

residues, and their transcripts have been found in seeds only. Recently, Schaller et al.[72] have also isolated further viscotoxin variants from the leaves of *V. album* subspecies. Their sequences have not been reported yet, but they are probably different from the eight-cysteine viscotoxins reported by Schrader-Fischer and Apel[48] whose transcripts were found in seeds.

Several thionins with six cysteine residues have been described from species of the family Brassicaceae (Figure 10.2). These include crambin, with six cysteine residues from the seeds of *Crambe abyssinica.*[73] Crambin is neutral and has no known toxic activities.[74,75] As has been found for other thionins, crambin has been shown to be actually a mixture of two different isoforms.[59] Schrader-Fischer and Apel[76] found, in addition to crambin, several novel and highly variable thionins in *Crambe abyssinica* using a PCR-based approach. A thionin from radish was found to have weak antimicrobial activity.[61] The sequences from two different thionins of *Arabidopsis thaliana* have been deduced from cDNA clones.[60]

Another thionin has been isolated from *Pyrularia pubera.*[47] Again, preliminary results[47] indicate that this plant also most likely contains additional thionin variants in other tissues. In addition, Daley and Theriot[77] demonstrated the existence of proteins with characteristics similar to thionins in tomato (*Lycopersicum esculentum,* Fam. *Solanaceae*), mango (*Mangifera indica,* Fam. *Anacardiaceae*), papaya (*Carica papaya,* Fam. *Caricaceae*), and walnut (*Juglans regia,* Fam. *Juglandaceae*), but their sequences have not been determined. At the moment, it is not known if all plants contain thionin genes or if their appearance in the plant kingdom is only sporadic.

Thionins were classified as five different types by García-Olmedo et al.[19] They considered endosperm thionins and barley leaf thionins as different types. These are now combined into one class together with *Pyrularia* thionin. Based on the number of cysteine residues and the pattern of disulfide bridges, thionins are now grouped in the following classes:[78]

- Class I: Thionins with eight cysteines as in purothionins and hordothionins, barley leaf thionins, and *Pyrularia* thionin.
- Class II: Without cysteines no. 3 and 6, for instance crambin and viscotoxins from leaves.
- Class III: Without cysteines no. 2 and 8, originally described as type V.[33,34]
- Class IV: Without cysteine no. 2. New cysteine between no. 4 and 5. Thionin 4-1 from *Tulipa.*[37]

10.3 STRUCTURE OF THIONINS

The three-dimensional structure of crambin has been determined in detail (e.g., see references 79 through 82) because it can easily be crystallized.[73,83] Crambin has thus become a model for studying the three-dimensional structure of proteins. Subsequently, similar studies have also been undertaken with several other thionins: purothionins (e.g., see references 84 through 87), hordothionin,[88,89] viscotoxins,[84,90,91] and phoratoxin (e.g., see reference 92). In all cases a compact L-shaped structure has been revealed which is stabilized by three or four disulfide bridges.[93-95,58] The long arm of the L is formed by two α-helices, and the short arm by two short antiparallel β-sheets. The last, approximately 10 amino acids form a loop-like structure which is partly locked through a disulfide bridge between Cys3 and the last cysteine residue. Thionins are amphipathic; hydrophobic residues are mainly located at the outer surface of the long arm of the L, whereas hydrophilic residues are found primarily at the inner surface of the L and at the outer surface of the corner of the L.

Recent studies have indicated a possible phospholipid binding site in thionins.[86,87,92] Interaction of crambin, phoratoxin, and purothionins with phospholipids has been directly demonstrated by NMR experiments.[92] Furthermore, both phosphate and glycerol, as well as acetate, were found in purothionin structures. Phosphate and acetate were not used in the crystallization medium, but had been used during purification of the proteins. The phosphate ion must be bound very tightly to survive the purification procedure. Addition of phosphate also enhanced the crystallization of β-purothionin.[86] The phosphate ion is bound through Lys1 and Pro44 in purothionins[86,87] and through Lys1 and Arg10 in Phoratoxin-A.[92] Binding of the phosphate ion stabilizes dimer formation, perhaps also in solution.[92] Glycerol is bound through Ser2, Tyr13, Arg17, and Gln22/Lys45 in α_1-/β-purothionin.[86,87] Together these results provide a link between the structure of the thionins and their toxicity.

10.4 TOXICITY

A variety of toxic activities have been described for most thionins. Soon after the first isolation of purothionin from wheat flour, its toxic effects against yeast,[25] bacteria and fungi,[96] and mammals[97] were discovered. The latter authors also showed that purothionin was toxic to mammals only when injected intraperitoneally or intravenously but had no effect when eaten. Viscotoxins are also only toxic if administered intraperitoneally or intravenously but are not toxic when eaten.[98] Similar effects on mammals have since been reported for other thionins.[99,100] The LD_{50} for intraperitoneal application of thionins is in the range of 1 to 10 mg per kg body weight. Several endosperm thionins have also been shown to be toxic to insects.[101]

Toxicity against yeasts[96,102-108] and bacteria[17,96,102,109-113] and fungi other than yeasts[17,61,110,112-116] is well documented. In addition to these effects on whole organisms, several reports describe toxic activities of thionins on cultured cells which result in leakage of cytoplasmic ions and enzymes.[47,65,100,117-119] Hemolytic activity of thionins has also been reported.[63,120]

The common denominator for the toxic effects of thionins *in vivo* seems to be a destruction of membranes. Several reports have demonstrated a leakage of low-molecular-weight compounds and ions from different cells following treatment with thionins.[65,106,117,121] Formation of pores in the cytoplasmic membrane has been observed, and thionins have been proposed to display homology to the cysteine-rich domain of mammalian pore-forming proteins.[122] However, this homology does not include the first two cysteine residues and the first tyrosine residue of the thionins, which are important for function.[108] Membrane permeability following thionin treatment has also been studied by preloading the cells with nonmetabolizable compounds. $^{86}Rb^+$ was used with hamster BHK-21 cells[65] and ^{14}C-isoaminobutyric was used with *Neurospora crassa*.[121] An efflux of phosphate ions[106] can result in ATP hydrolysis.[123] Depletion of ATP results in cell death and can be used to measure the toxicity of thionins.[124] However, Vernon and Rogers[125] have shown that hemolytic activity of *Pyrularia* thionin is stimulated by phosphate ions. An influx of Ca^{2+} has been observed[121,126] and could play a role in the toxicity of thionins against filamentous fungi which extend by tip growth. This process is thought to be regulated by Ca^{2+} ions[127] and would obviously be disturbed by toxins such as thionins which disturb the permeability of the membrane for Ca^{2+} ions. Whether or not a binding of thionins to calmodulin which has been described *in vitro*,[128] plays any role in toxicity, is still unknown.

Two mechanisms have been proposed as possible explanations for the lytic effect of thionins on membranes. The amphipathic structure of the thionins indicates that the toxicity might be exerted by a direct, detergent-like interaction with the lipid bilayers of biological membranes. The hydrophobic face of the thionins could interact with the hydrophobic aliphatic chains of the membrane lipids, while the positively charged basic amino acids could interact with the negatively charged phosphate groups of the phospholipids. These intermolecular salt bridges cannot be formed by crambin, which has only two basic residues (Arg10 and Arg17), both of which are blocked through intramolecular bonds.[84] This might explain that up to now no toxic activities have been found for this thionin.

Recent work on *Pyrularia* thionin (reviewed in reference 129) led Vernon and co-workers to propose a different model, which includes the specific binding to a membrane receptor, after showing that *Pyrularia* thionin has specific binding sites on the surface of erythrocytes. The receptor is probably a specific phospholipid.[125,130] This model is further substantiated by recent findings of a putative phospholipid binding site in the three-dimensional structure of purothionins.[86,87] Furthermore, as has been discussed before, endosperm thionins can be isolated as lipid–protein complexes,[25,39,131,132] indicating a tight binding to lipids. The binding site for *Pyrularia* thionin seems to be the same as those of the cardiotoxins, small (about 60 amino acids), basic proteins with eight cysteine residues, from the venoms of cobras and related snakes.[133,134] Iodinated *Pyrularia* thionin, which is no longer toxic,[126] competitively inhibits hemolysis both by the native *Pyrularia* thionin and by cardiotoxin, whereas the hemolytic activity of melittin from bee venom was not inhibited.[135] In addition, both toxins are inhibited by extracellular calcium ions and are stimulated by external phosphate ions.[125] That higher concentrations (5 m*M*) of calcium ions and other divalent metal ions can abolish the toxic activity of thionins has been known for quite some time.[105] The binding of *Pyrularia* thionin to the membrane leads to the stimulation of an internal phospholipase, probably PLA_2.[75,126] *Pyrularia* thionin itself has no phospholipase activity.[75,120] This has also been demonstrated for cardiotoxins,[136] but snake venom contains, in addition, phospholipases which might act synergistically with the cardiotoxins.[137] Such a synergistic effect has also been demonstrated for viscotoxin B and phospholipase A.[63] Other effects mediated by *Pyrularia* thionin on different mammalian cell cultures have been described and can also be related to the membrane-destructive action of this thionin. Shaw et al.[119] presented evidence that the cytotoxic effect of *Pyrularia* thionin on spleen lymphocytes and hepatocytes from rats may be mediated by the activation of phospholipase A_2 and a subsequent influx of calcium ions. In mouse P388 cell, the concentration of cytosolic calcium ions also increased after application of *Pyrularia* thionin,[126] and an endogenous phospholipase A_2 was also activated in NIH 3T3 cells.[75] In S49 lymphoma cells, *Pyrularia* thionin enhanced adenylate cyclase activity.[138] Adenylate cyclase plays an important role in signal transduction pathways of mammalian cells by catalyzing the formation of the second messenger cAMP from ATP, but Huang et al.[138] concluded that *Pyrularia* thionin had only an indirect effect on adenylate cyclase through its membrane activity. Taken together, there is good evidence that the mechanism by which *Pyrularia* thionin exerts its toxicity starts with the binding to a membrane receptor, probably a specific phospholipid, and that cardiotoxins use a similar mechanism involving the same receptor.

The tyrosine residue in position 13 not only seems to be crucial for the toxicity of the *Pyrularia* thionin based on the iodination studies, but also for the other toxic thionins as well since this residue and the region around it (positions 10 through 14) are conserved in all toxic thionins (see Figures 10.1 and 10.2). The importance of the tyrosine residue in purothionin has been demonstrated — also using iodination — by Wada et al.,[108] supporting the model of Vernon and co-workers of specific binding sites on the membrane for *Pyrularia* thionin and perhaps for other thionins. Recent improvements in the three-dimensional structure of purothionins also indicate that Tyr13 is probably part of a phospholipid binding site.[86,87] In a simple detergent-like interaction, the crucial role for this tyrosine residue would be hard to imagine. However, crambin, which has no known toxic activities, has also been shown to bind phospholipids.[92] On the other hand, all toxic thionins have a conserved basic nature and amphipathic structure, which suggests that the second step after the specific binding might still be a detergent-like interaction leading to the destruction of membranes and all of the well-known effects of the toxic thionins. The recently reported production of recombinant crambin[139] opens the way to further study of structure–function relationship using mutated proteins. Another approach for such studies would be to chemically synthesize thionins and test their activity, as has been demonstrated by Rao et al.[140] for α-hordothionin.

10.5 FUNCTION

The biological function of thionins has been discussed for a long time. In addition to the toxic effects on cells and whole organisms, several inhibitory effects on *in vitro* systems have been reported. Purothionin can be reduced by the thioredoxin system of wheat seeds and can in turn activate fructose-1,6-biphosphatase or inhibit ribonucleotide reductase.[141,142] Similar redox reactions have also been demonstrated for the interaction of thionins with the reporter enzymes β-glucuronidase and neomycin phosphotransferase *in vitro* and in transiently transformed tobacco protoplasts.[143-146] Viscotoxins can bind to DNA and protect it against thermal denaturation.[147] Thionins have also been shown to inhibit cell-free protein synthesis.[148,149] However, many reports on *in vitro* effects did not take into account that plant cells are highly compartmentalized and that components which can be easily brought together *in vitro* are separated *in vivo*. Thus, care should be taken in deducing a biological function from *in vitro* effects.

One short report claims that purothionin has α-amylase-inhibiting activity,[150] but the inhibition was only partial. To my knowledge, a thorough investigation of possible enzyme-inhibiting activities of thionins has not been done. Different families of enzyme inhibitors are, similar to thionins, rather small proteins with several disulfide bridges and pronounced stability.[151] Furthermore, some thionin genes have been shown to be inducible by methyl jasmonate.[60,152] Jasmonate is involved in the induction of proteinase inhibitors in tomato[153] and other species. Therefore, an enzyme-inhibitory function of at least certain thionins cannot be excluded. Members with α-amylase inhibitory activities have been found in the plant defensin family[154] and among lipid transfer proteins,[155] while the majority of proteins in both families has *in vitro* antimicrobial activity.

Work by Schrader-Fischer and Apel[156] has demonstrated a decomposition of viscotoxins in senescent mistletoe leaves. This indicates that the plant might be reutilizing the sulfur of the thionins, although it does not prove a primary function of viscotoxins as storage proteins. A storage protein function, especially for sulfur, might also be envisaged for the seed thionin crambin which has no known toxic activities.[74,75] However, its nitrogen content is lower than that of the basic and toxic thionins. These apparent contradictions could be resolved by considering that many proteins might have dual functions. This might also apply for many seed-borne enzyme inhibitors and for thionins.[157]

A report by Fernandez de Caleya et al.[109] pointed in another direction. The authors found that purothionins are toxic to phytopathogenic bacteria. Subsequently, thionins from the leaves or the endosperm of barley have also been shown to be toxic to plant pathogenic bacteria and fungi.[17,111] Furthermore, it has been shown that expression of a hordothionin in transgenic tobacco plants can lead to enhanced resistance against a phytopathogenic bacterium.[158] However, in similar experiments, Florack et al.[159] did not find enhanced resistance in transgenic tobacco lines expressing a synthetic α-hordothionin. Maddox et al.[160] have expressed α-hordothionin in transgenic tobacco as well, but nothing has been reported yet about the resistance of these plants. The overexpression of the barley leaf thionin DB4 in transgenic tobacco plants also did not give enhanced resistance (Mollenhauer and Apel, unpublished results). However, other work on barley leaf thionins supports a role for thionins in plant defense and will be discussed later. Recently, *Arabidopsis* has been used as an experimental system. High-level expression of a viscotoxin and overexpression of the endogenous *Arabidopsis* thionins resulted in significantly enhanced resistance against *Plasmodiophora brassicae* and *Fusarium oxysporum* f sp *matthiolae,*[161] respectively. This and other work on Arabidopsis thionins strongly supports the view that thionins are defense proteins and will also be discussed separately.

10.6 PROCESSING AND LOCALIZATION

Thionins have been isolated as the 5 kD mature proteins from different plant species but are synthesized as much larger precursors.[162] Cloning of different thionin cDNAs (as well as genomic clones) confirmed the original observation and revealed that these mature thionins are first synthesized as preproproteins (Figure 10.3). All

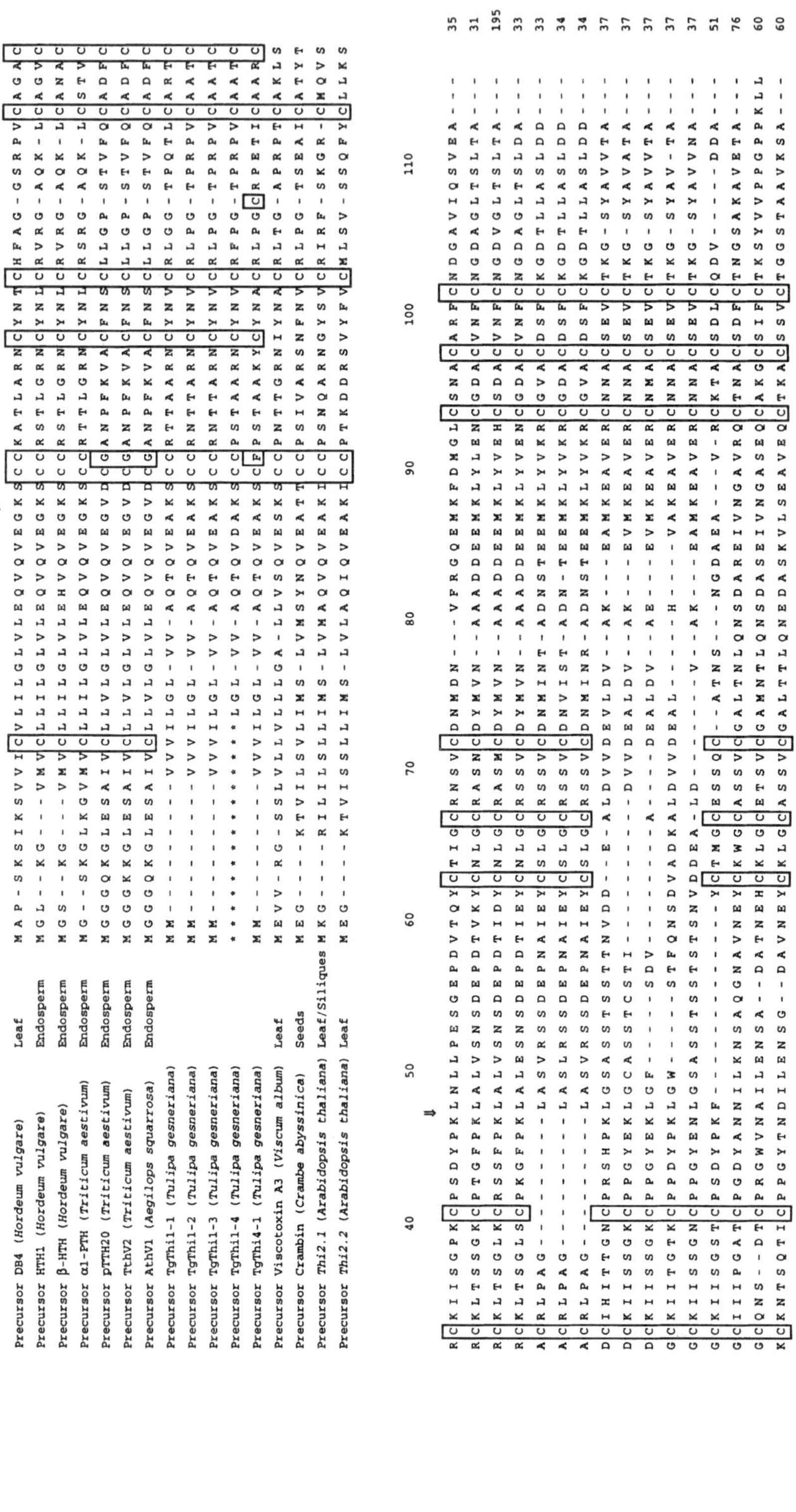

FIGURE 10.3 Sequences of thionin preproteins.

precursors have a typical N-terminal signal sequence with a conserved signal sequence cleavage site[163] which directs the proprotein into the endoplasmic reticulum (ER). The proprotein consists of the thionin itself and a C-terminal acidic domain.

Thionins have been demonstrated in cell walls, vacuoles, and protein bodies. Barley leaf thionins are found in the vacuole and the cell wall,[114,164] and hordothionins are localized in the periphery of protein bodies.[132,165] Viscotoxins have been detected in protein storage vacuoles in the leaves.[48] The localization of viscotoxins in seeds is not totally clear. There, viscotoxins were detected in protein storage vacuoles but also to a large extent in the cytoplasm. It is not known whether the latter were sequestered in the ER.[48] How the plant protects itself from the thionins in these compartments is not known. One possibility is that the thionins are complexed by other cellular compounds. Endosperm thionins, for instance, have repeatedly been isolated as lipid–protein complexes.[25,39,131,132] These lipid-thionin complexes, however, also could be an artifact due to the isolation procedure. On the other hand, studies of the three-dimensional structure (see above) of different thionins have revealed that they most likely contain a phospholipid binding site,[86,87,92] and a reduction of the antimicrobial activity of purothionin by phospholipids has been demonstrated.[102]

The acidic domains have six cysteine residues (Figure 10.3) which are highly conserved, as are the cysteine residues in the thionin, and are mostly acidic (as the name says). An exception are the acidic domains deduced from the *Tulipa* thionin precursors which have only three cysteine residues.[37] In the case of the neutral crambin, the acidic domain too is neutral.[76] This has also been observed for the class III endosperm thionins.[33,34] A protein corresponding to the acidic domain has, to my knowledge, never been isolated from plants. To date, there is no experimental information available about possible functions of the acidic domain, but it is clearly not dispensable, as shown by the high conservation of the cysteine residues, even in the case of viscotoxin precursors which have several deletions in the acidic domain.[51] Furthermore, Florack et al.[159] found that expression of α-hordothionin in transgenic tobacco plants without the acidic domain resulted in significantly lower levels of the mature thionin.

One possible function might be that the acidic domain contains information to guide the thionin through the secretory pathway to its final destination. Furthermore, the acidic domain might function as an intramolecular chaperone to assist in the folding of the thionin. Such a function has been demonstrated for the bovine pancreatic trypsin inhibitor, which is synthesized with a 13 amino acid pro region. This pro region contains a cysteine residue and has been shown to facilitate proper folding of the mature protein *in vitro*.[166] Another function of the acidic domain might be to neutralize the basic thionin and thereby protect the cell against its own toxin. A toxic effect of barley leaf thionins has been demonstrated against tobacco protoplasts[164] and against barley protoplasts.[167] Several other small, basic, cysteine-rich polypeptides with antimicrobial activities are also synthesized as preproproteins with acidic domains. Defensins from mammals and insects, for example, have an acidic domain between the signal sequence and the mature defensin.[168,169] A plant defensin from tobacco,[170] originally described as a thionin, is also produced with a C-terminal acidic domain, but this domain has no cysteine residues. Not much is known to date about the final processing of the thionin proprotein and about what happens to the acidic domain. As a first step in studying this process, Romero et al.[171] have recently isolated a proteinase from barley which was able to process *in vitro* synthesized thionin precursors.

10.7 BARLEY LEAF THIONINS

Experimental evidence which has been obtained for barley leaf thionins supports the view of thionins being defense proteins. Cultivated barley (*Hordeum vulgare* L.) and also many wild *Hordeum* species contain a large multigene family of closely related genes for leaf thionins[35,36,114,172,173] in addition to a small gene family for endosperm specific hordothionins.[31] Mature leaf thionins have been found in the vacuole[164] and in the cell wall[114] of barley leafs. Especially high concentrations of leaf thionins are present in the outer cell walls of epidermal cells,[174] which is in agreement with a defense function against pathogens.

A prominent feature of leaf thionin genes in barley is their light responsiveness. Etiolated barley seedlings have a very high level of leaf thionin transcripts, which drops drastically after illumination, probably mediated by two photoreceptors, phytochrome and a blue-light-absorbing photoreceptor.[174] Blue light leads to a much stronger decline of thionin mRNAs compared to the red-light effect, but only at high light intensities. Such high light intensities are not reached under normal light conditions in the meristematic zone at the leaf basis, which is covered by the sheath of the preceding leaves. Thionin synthesis is therefore possible in these young developing cells, and since thionins are very stable,[47,98] their concentration declines only slowly once these cells are no longer shaded by the leaf sheath but exposed to full light, which leads to a drastic decline in the thionin mRNA level. This expression of barley leaf thionin genes is not influenced by the plastid compartment[175] but is under circadian regulation[176] with maximal mRNA levels during the dark period. Thus, barley seedlings growing in a day/night cycle can accumulate a basic level of thionins.

Barley leaf thionins and hordothionins have repeatedly been shown to have antimicrobial activities *in vitro* against different phytopathogenic bacteria and fungi.[111-114,116] This *in vitro* toxicity and the cellular distribution of barley leaf thionins suggested a role in plant defense. Abiotic and biotic inducers have therefore been tested for effectiveness in leaf thionins. Indeed, this has been demonstrated for different chemical inducers: heavy metals,[177] jasmonic acid,[152] and 2,6-dichloroisonicotinic acid (INA).[178] Conflicting reports are available for salicylic acid. Kogel et al.[179] reported an induction by salicylate, whereas García-Olmedo et al.[17] found no induction by salicylate. All of these inducers have been shown to activate defense reactions in other plant species.[3,180-183] Work by Andresen et al.[152] has revealed that leaf thionins are identical with the so-called JIP6, a 6 kD jasmonate-inducible protein described by Weidhase et al.[184] While it has been shown that induction of proteinase inhibitors in tomato can be induced by volatile methyl jasmonate, much higher concentrations are needed in barley. This different responsiveness might be due to an impeded uptake of volatile methyl jasmonate in uninjured barley plants, or jasmonate might not be the final effector in barley. The jasmonate responsiveness of leaf thionin genes might also be related to the induction by abscisic acid[17,185] and drought.[17] In this regard, it should also be tested if wounding also induces barley leaf thionin genes.

Fungal and bacterial pathogens of barley have also been shown to be effective inducers of leaf thionins. Infection by *Septoria nodorum* results in strong expression of leaf thionins in barley coleoptiles.[186,187] Induction of leaf thionin genes was also demonstrated after inoculation with *Rhynchosporium secalis* and *Pseudomonas syringae*[17] and in roots after infection with *Drechslera graminea.*[188] Powdery mildew infection of barley leaves leads to a transient rise in the leaf thionin transcript level. The maximum of the leaf thionin transcript level was found at 44 h after inoculation in the cultivar "Emir," with both a virulent and an avirulent powdery mildew race.[114] In the cultivars "Midas," "Pallas," and "Siri," which all carry the *Mla6* resistance gene, different maxima of the leaf thionin transcript level were observed.[189,190] A first maximum was found at about 6 hours after inoculation in the cultivar "Midas,"[190] but at about 12 hours in the cultivars "Pallas" and "Siri."[189] The reason for these differences is not known. The accumulation of thionins after mildew infection was also investigated by immunogold labeling using an antibody obtained against a fusion protein of β-galactosidase and the leaf thionin precursor DB4.[35] Resistance against powdery mildew infection can result from the ability of epidermal cells to produce cell wall appositions (papillae) at the point of invasion.[5] After inoculation with the mildew race C17, leaf thionins were detected in the papillae of the resistant barley cultivar "Stamm 41." The cell wall surrounding the infection site showed normal levels of thionins. In contrast, thionins could not be detected in papillae of the susceptible cultivar "Peruvian." Only reduced levels, compared to cell walls of uninfected cells, were found in the cell wall surrounding the infection site. Leaves of the adult-plant resistant cultivar "Osiris" showed a differential response. Primary leaves are susceptible and had a low thionin level, as in the susceptible cultivar "Peruvian." The fifth leaves, on the other hand, are resistant to this powdery mildew race. Here, thionins were detectable in the papillae just as in the resistant cultivar "Stamm 41."[191,192] These results demonstrate that thionins are newly synthesized following powdery mildew infection and incorporated into the papillae. Furthermore, it has to be concluded from these results that the powdery mildew fungus is able to protect itself against the leaf thionins in a compatible interaction by somehow destroying or masking the thionins so that they are no longer recognized by the antibody. A similar mechanism might also be effective in the interaction with *Pyrenophora (Drechslera) teres,* which is inhibited by barley leaf thionins *in vitro*[114] but is a pathogen of barley *in vivo.*

Results similar to those obtained with powdery mildew were also obtained for *Puccinia hordei.*[193] Furthermore, cross-reactivity of the antibody against barley leaf thionin with wheat has been found.[193,194] However, occurrence of leaf thionins in wheat as suggested by these reports has not been confirmed yet by isolation and sequencing of the proteins or by DNA clones. Purothionins have not been detected in parts of the wheat plant other than the seeds.[43]

Barley leaf thionins are encoded by a large multigene family with as many as 50 genes.[114] Leaf thionin genes have been mapped to the *Lth* Locus at the distal end of the satellite of chromosome 6.[114,191,196,197] A complex leaf thionin multigene family has also been detected in several other wild barley species. Other species, for instance *H. bulbosum* and *H. marinum,* had only very few fragments cross-hybridizing on Southern blots with a barley leaf thionin probe.[172] Whether the latter species have indeed only very few leaf thionin genes, or whether these genes are so divergent that they do not hybridize with the cDNA probe from *H. vulgare,* is currently unknown. In *H. vulgare* cv Carina, the known DNA clones[35,114,198] can be grouped into two closely related subfamilies with microheterogeneity at the DNA and protein levels in each subfamily. The cDNA clones which have been isolated from *H. murinum,*[172] on the other hand, are much more divergent. Thus, it is possible that part of the diversity of leaf thionin genes present in certain wild barley species was lost during breeding of cultivated barley. However, Pecchioni et al.[199] found that the leaf thionin cDNA DB4[35] revealed complex hybridization patterns on Southern blots and polymorphisms among different barley varieties, indicating that there is still a high diversity for leaf thionin genes in cultivated barley.

Sequencing of *H. murinum* leaf thionin cDNAs has revealed that the thionin domain shows a higher variability than the signal sequence or the acidic domain.[172] The same variability has been found for viscotoxins.[48] Castagnaro et al.[33] described a similar phenomenon for a class III endosperm thionin from wheat. However, it has turned out that in this case very closely related thionins can be found in wheat and *Aegilops.*[34] The reason for the high variability of the thionin domain could be a selective pressure from pathogens which try to overcome the toxicity of thionins. Another example for a polymorphism of polypeptide toxins can be seen in the case of the cone shell neurotoxins,[200] this time as a consequence of the interplay between predator and prey.

In addition to leaves, where they have originally been discovered,[35,36] leaf thionins are also found in other parts of barley plants. Expression of leaf thionin genes in coleoptiles after induction by *Septoria nodorum* has been described.[186,187] Steinmüller et al.[201] showed that roots of *H. vulgare* cv Carina have very low levels of leaf-thionin-related transcripts. Valè et al.[188] reported thionins in roots after infection with *Drechslera graminea.* A root-specific cDNA has been isolated from *H. murinum* (Bohl and Apel, unpublished results). Thus, leaf thionin genes are expressed in many parts of barley plants, and the large multigene family could at least in part result from specialization of single genes for expression in certain tissues or in response to certain external stimuli, as has been shown for other multigene families.[202]

10.8 ARABIDOPSIS THIONINS

We have recently identified two thionin genes in *Arabidopsis thaliana.* This opened the way to use the advantages of *Arabidopsis* as an experimental system also for the study of thionins. The two genes were found to be regulated differently.[60] The *Thi2.2* gene showed a low constitutive expression in seedlings and rosettes and may be under circadian regulation. The *Thi2.1* gene on the other hand, was found to be inducible by jasmonate, silver nitrate, and necrotrophic fungi but did not respond to salicylate or ethephon, indicating a signal transduction pathway different from that for PR proteins. Expression of the gene in seedlings could not be detected on northern blots but was very strong in flowers and siliques. Some of the pathogen-inducible PR proteins have also been shown to be constitutively expressed in tobacco flowers.[203-206] Other antimicrobial peptides, such as plant defensins, are constitutively expressed in tobacco flowers[170] and *Arabidopsis* siliques as well.[207,208]

The highest transcript level of the *Thi2.1* gene was found after spray inoculation of seedlings grown on MS agar plates with a spore suspension of *Fusarium oxysporum* f sp *matthiolae*. The ecotype Col-2, which has been used in most expression studies, was found to be susceptible to this fungal strain. However, it is usually assumed that defense-related genes are expressed at higher levels in incompatible interactions compared with compatible interactions. We have therefore tested several other Arabidopsis ecotypes for their resistance against *F. oxysporum* f sp *matthiolae*. Ecotypes Ws and Ler were also susceptible, but ecotypes Mt-0 and Uk-4 turned out to be resistant (Epple, Vignutelli, Apel, Bohlmann, manuscript submitted). The transcript level for the *Thi2.1* gene after inoculation was found to correlate with resistance, being 5 to 10 times higher in the resistant than in the susceptible ecotypes. This effect was not due to gene amplification as shown by Southern blot analysis.

To further test the possible function of the THI2.1 thionin in resistance against *F. oxysporum* f sp *matthiolae*, transgenic lines of the susceptible ecotype Col-2 were generated which expressed a high constitutive THI2.1 level.[161] Several of these lines showed significantly enhanced resistance against *F. oxysporum* f sp *matthiolae*. Similar results were also obtained with transgenic lines over-expressing the THI2.2 thionin. First, the transgenic lines had less loss of chlorophyll after inoculation. Second, the transgenic seedlings supported much less fungal growth on the cotyledons as shown by directly staining fungal hyphae. In addition, fungal hyphae growing on the transgenic lines often showed a hyperbranching effect which was never found on the susceptible ecotype Col-2. As has been discussed before, fungal hyphae extend by tip growth, a process which is thought to be regulated by Ca^{2+} ions.[127] It is evident that this mechanism would be disturbed by toxins such as thionins, which alter the permeability of the membrane for Ca^{2+} ions. Similar effects on fungal growth have been reported for certain fungicides[209,210] and for plant defensins *in vitro*.[18,115] Recently, such an effect has also been demonstrated for *Fusarium culmorum* treated *in vitro* with a purified thionin from radish tubers.[61]

Induction of the *Thi2.1* gene by MeJ indicated that the *Thi2.1* gene might be regulated via the octadecanoid pathway, which has been intensively studied for the regulation of vegetative storage proteins of soybean[211] and for the induction of proteinase inhibitors of potato and tomato.[12,153] Further analysis of transgenic *Thi2.1* promoter-*uidA* seedlings showed that the promoter is also inducible by wounding, both locally and systemically. Such an induction pattern is also produced if *F. oxysporum* f sp *matthiolae* spores are inoculated as a single drop on a cotyledon (Vignutelli, Wasternack, Apel, Bohlmann, manuscript submitted). The systemic induction in all cases is strongest in the apical meristem and the youngest leaves. This would protect the most valuable parts of the young plants against further infections, again supporting the view that thionins are defense proteins.

In conclusion, our work has provided strong evidence for a role of the thionin genes *Thi2.1* and *Thi2.2* in the resistance of Arabidopsis. In the case of the *Thi2.1* gene, the systemic induction might be regulated via the octadecanoid pathway, perhaps coordinately with genes for other antimicrobial peptides such as plant defensins.[60,207,208] One might speculate that at least in *Arabidopsis,* the salicylate-dependent SAR pathway, which regulates the synthesis of PR proteins, controls mainly biotrophic fungi, while a jasmonate-dependent pathway for antimicrobial peptides is also needed for resistance against necrotrophic fungi.

10.9 GENETIC ENGINEERING

During the past several years, strong evidence has been obtained for a resistance function of at least some thionins. As discussed before, overexpression of different thionins has resulted in enhanced resistance against bacterial and fungal pathogens. So far, however, no protection of economically important crop plants has been reported. To achieve this, it may be necessary to express combinations of antimicrobial peptides, for instance thionins and plant defensins which exert their activity via different toxic mechanisms.[121] Similar effects have also been found by combining chitinases and glucanases.[212] Instead of introducing different artificial genes into different crop plants, enhanced resistance has also been achieved by inducing the plant's own defenses with chemicals such as INA[180] or benzothiadiazole (BTH).[213] It is thought that these chemicals induce SAR via the production of PR proteins. However, resistance induced by BTH, for instance, is not effective against necrotrophic pathogens such as *Alternaria alternata* and *Botrytis cinerea.*[213] It is tempting to speculate that the salicylate-dependent SAR pathway, which leads to the expression of PR-proteins, controls mainly biotrophic pathogens, but that a jasmonate-dependent pathway for antimicrobial peptides is needed in addition for resistance against necrotrophic pathogens.[161,207] In line with this, MeJ has been shown to induce resistance against different fungal pathogens in several plant species,[214-216] but the resistance mechanism is not known in any of these cases. It might therefore be interesting to look for chemicals which could induce this jasmonate-dependent pathway, and therefore the expression of antimicrobial peptides, without inducing the negative effects, such as growth inhibition or senescence, that are a consequence of jasmonate treatments.

REFERENCES

1. Ride, J. P., Non-host resistance to fungi, in *Mechanisms of Resistance to Plant Diseases,* Fraser, R. S. S., Ed., Martinus Nijhoff/Dr. W. Junk, Dordrecht, The Netherlands, 1985, 29.
2. Harborne, J. B., *Introduction to Ecological Biochemistry,* Academic Press, London, 1988.
3. Dixon, R. A., The phytoalexin response: elicitation, signalling and control of host gene expression, *Biol. Rev.,* 61, 239, 1986.
4. Linthorst, H. J. M., Pathogenesis-related proteins of plants, *Crit. Rev. Plant Sci.,* 10, 123, 1991.
5. Aist, J. R., Structural responses as resistance mechanisms, in *The Dynamics of Host Defense,* Bailey, J. A., Ed., Academic Press, New York, 1983, 33.
6. Brisson, L. F., Tenhaken, R., and Lamb, C., Functions of oxidative cross-linking of cell wall structural proteins in plant disease resistance, *Plant Cell,* 6, 1703, 1994.

7. Dangl, J. L., Dietrich, R. A., and Richberg, M. H., Death don't have no mercy: cell death programs in plant-microbe interactions, *Plant Cell,* 8, 1793, 1996.
8. Mittler, R. and Lam, E., Sacrifice in the face of foes: pathogen-induced programmed cell death in plants, *Trends Microbiol.,* 4, 10, 1996.
9. Gaffney, T., Friedrich, L., Vernooij, B., Negrotto, D., Nye, G. Uknes, S., Ward, E., Kessmann, H., and Ryals, J., Requirement of salicylic acid for the induction of systemic acquired resistance, *Science*, 261, 754, 1993.
10. Hunt, M. D. and Ryals, J. A., Systemic acquired resistance signal transduction, *Crit. Rev. Plant Sci.,* 15, 583, 1996.
11. Doares, S. H., Syrovets, T., Weiler, E. W., and Ryan, C., Oligogalacturonides and chitosan activate plant defensive genes through the octadecanoid pathway, *Proc. Natl. Acad. Sci. U.S.A.,* 92, 4095, 1995.
12. Peña-Cortés, H., Fisahn, J., and Willmitzer, L., Signals involved in wound-induced proteinase inhibitor II gene expression in tomato and potato plants, *Proc. Natl. Acad. Sci. U.S.A.,* 92, 4106, 1995.
13. Pearce, G., Strydom, D., Johnson, S., and Ryan, C., A polypeptide from tomato leaves activates the expression of proteinase inhibitor genes, *Science*, 253, 895, 1991.
14. Farmer, E. E. and Ryan, C., Interplant communication. Airborne methyl jasmonate induces synthesis of wound-inducible proteinase inhibitors in plant leaves, *Proc. Natl. Acad. Sci. U.S.A.,* 87, 7713, 1990.
15. Hancock, R. E. W., Falla, T., and Brown, M., Cationic bactericidal peptides, *Adv. Microb. Physiol.,* 37, 135, 1995.
16. Nicolas, P. and Mor, A., Peptides as weapons against microorganisms in the chemical defense system of vertebrates, *Annu. Rev. Microbiol.,* 49, 277, 1995.
17. García-Olmedo, F., Molina, A., Segura, A., and Moreno, M., The defensive role of nonspecific lipid-transfer proteins in plants, *Trends Microbiol.,* 3, 72, 1995.
18. Terras, F. R. G., Eggermont, K., Kovaleva, V., Raikhel, N. V., Osborn, R. W., Kester, A., Rees, S. B. Torrekens, S., Van Leuven, F., Vanderleyden, J., Cammue, B. P. A., and Broekaert, W. F., Small cysteine-rich antifungal proteins from radish: their role in host defense, *Plant Cell,* 7, 573, 1995.
19. García-Olmedo, F., Rodriguez-Palenzuela, P., Hernandez-Lucas, C., Ponz, F., Marana, C., Carmona, J. J., Lopez-Fando, J., Fernandez, J. A., and Carbonero, P., The thionins: a protein family that includes purothionins, viscotoxins and crambins, *Oxf. Surv. Plant Mol. Cell Biol.,* 6, 31, 1989.
20. Bohlmann, H. and Apel, K., Thionins, *Annu. Rev. Plant Phys. Plant Mol. Biol.,* 42, 227, 1991.
21. Bohlmann, H., The role of thionins in plant protection, *Crit. Rev. Plant Sci.,* 13, 1, 1994.
22. Florack, D. E. A. and Stiekema, W. J., Thionins: properties, possible biological roles, and mechanisms of action, *Plant Mol. Biol.,* 26, 25, 1994.
23. Ohtani, S., Okada, T., Yoshizumi, H., and Kagamiyama, H., Complete primary structures of two subunits of purothionin A, a lethal protein for brewer's yeast from wheat flour, *J. Biochem.,* 82, 53, 1977.
24. Balls, A. K. and Hale, W. S., A sulphur-bearing constituent of the petroleum ether extract of wheat flour (preliminary report), *Cereal Chem.,* 17, 43, 1940.
25. Balls, A. K., Hale, W. S., and Harris, T. H., A crystalline protein obtained from a lipoprotein of wheat flour, *Cereal Chem.,* 19, 79, 1942.
26. Jones, B. L. and Mak, A. S., Amino acid sequences of the two α-purothionins of hexaploid wheat, *Cereal Chem.,* 54, 11, 1977.

27. Ohtani, S., Okada, T., Kagamiyama, H., and Yoshizumi, H., The amino acid sequence of purothionin A, a lethal toxic protein for brewer's yeasts from wheat, *Agric. Biol. Chem.,* 39, 2269, 1975.
28. Mak, A. S. and Jones, B. L., The amino acid sequence of wheat β-purothionin, *Can. J. Biochem.,* 22, 835, 1976.
29. Bekes, F. and Lasztity, R., Isolation and determination of amino acid sequence of avenothionin, a new purothionin analogue from oat, *Cereal Chem.,* 58, 360, 1981.
30. Ponz, F., Paz-Ares, J., Hernandez-Lucas, C., García-Olmedo, F., and Carbonero, P., Cloning and nucleotide sequence of a cDNA encoding the precursor of the barley toxin α-hordothionin, *Eur. J. Biochem.,* 156, 131, 1986.
31. Rodriguez-Palenzuela, P., Pintor-Toro, J. A., Carbonero, P., and García-Olmedo, F., Nucleotide sequence and endosperm-specific expression of the structural gene for the toxin a-hordothionin in barley (*Hordeum vulgare* L.), *Gene*, 70, 271, 1988.
32. Hernandez-Lucas, C., Royo, J., Paz-Ares, J., Poz, F., García-Olmedo, F., and Carbonero, P., Polyadenylation site heterogeneity in mRNA encoding the precursor of the barley toxin β-hordothionin, *FEBS Lett.,* 200, 103, 1986.
33. Castagnaro, A., Marana, C., Carbonero, P., and García-Olmedo, F., Extreme divergence of a novel wheat thionin generated by a mutational burst specifically affecting the mature protein domain of the precursor, *J. Mol. Biol.,* 224, 1003, 1992.
34. Castagnaro, A., Segura, A. and García-Olmedo, F., High conservation among sequences encoding type-V thionins in wheat and *Aegilops, Plant Physiol.,* 107, 1475, 1995.
35. Bohlmann, H. and Apel, K., Isolation and characterization of cDNAs coding for leaf-specific thionins closely related to the endosperm-specific hordothionin of barley (*Hordeum vulgare* L.), *Mol. Gen. Genet.,* 207, 446, 1987.
36. Gausing, K., Thionin genes specifically expressed in barley leaves, *Planta*, 171, 241, 1987.
37. De Boer, D. A. D., *Tulipa gesneriana* thionin cDNAs, EMBL database, 1994.
38. Carbonero, P. and García-Olmedo, F., Purothionins in *Aegilops-Triticum* spp., *Experientia*, 25, 1110, 1969.
39. Redman, D. G. and Fisher, N., Purothionin analogues from barley flour, *J. Sci. Food Agric.,* 20, 427, 1969.
40. Fernandez de Caleya, R., Hernandez-Lucas, C., Carbonero, P., and García-Olmedo, F., Gene expression in alloploids: genetic control of lipopurothionins in wheat, *Genetics*, 83, 687, 1976.
41. Hernandez-Lucas, C., Carbonero, P., and García-Olmedo, F., Identification and purification of a purothionin homologue from rye (*Secale cereale* L.), *J. Agric. Food Chem.,* 26, 794, 1978.
42. Jones, B. L., Lookhart, G. L., Mak, A. S., and Cooper, D. B., Sequences of purothionins and their inheritance in diploid, tetraploid, and hexaploid wheats, *J. Hered.,* 73, 143, 1982.
43. Wada, K., Localization of purothionins and genome expression in wheat plants, *Plant Cell Physiol.,* 23, 1357, 1982.
44. Jones, B. L. and Cooper, D. B., Purification and characterization of a corn (*Zea mays*) protein similar to purothionins, *J. Agric. Food Chem.,* 28, 904, 1980.
45. Wada, K. and Buchanan, B. B., Purothionins and related proteins: seed proteins with thioredoxin activity, in *Thioredoxins. Structure and Function,* Gadal, P., Ed., International Conference CNRS-NASA, Berkeley, CA, 1981, 213.

46. Duvick, J. P., Rood, T., Rao, A. G., and Marshak D. R., Purification and characterization of a novel antimicrobial peptide from maize (*Zea mays* L.) kernels, *J. Biol. Chem.*, 267, 18814, 1992.
47. Vernon, L. P., Evett, G. E., Zeikus, R. D., and Gray, W. R., A toxic thionin from *Pyrularia pubera*: purification, properties, and amino acid sequence, *Arch. Biochem. Biophys.*, 238, 18, 1985.
48. Schrader-Fischer, G. and Apel, K., cDNA-derived identification of novel thionin precursors in *Viscum album* that contain highly divergent thionin domains but conserved signal and acidic polypeptide domains, *Plant Mol. Biol.*, 23, 1233, 1993.
49. Olson, T. and Samuelsson, G., The amino acid sequence of viscotoxin A2 from the European mistletoe (*Viscum album* L., Loranthaceae), *Acta Chem. Scand.*, 26, 585, 1972.
50. Samuelsson, G., Seger, L., and Olson, T., The amino acid sequence of oxidized viscotoxin A3 from the European mistletoe (*Viscum album* L., Loranthaceae), *Acta Chem. Scand.*, 22, 2624, 1968.
51. Schrader, G. and Apel, K., Isolation and characterization of cDNAs encoding viscotoxins of mistletoe (*Viscum album*), *Eur. J. Biochem.*, 198, 549, 1991.
52. Samuelsson, G. and Pettersson, B., The amino acid sequence of viscotoxin B from the European mistletoe (*Viscum album* L., Loranthaceae), *Eur. J. Biochem.*, 21, 86, 1971.
53. Samuelsson, G. and Jayawardene, A. L., Isolation and characterization of viscotoxin 1Ps from *Viscum album* L. ssp. *austriacum* (Wiesb.) Vollmann, growing on *Pinus silvestris, Acta Pharmacol. Suecica,* 11, 175, 1974.
54. Mellstrand, S. T. and Samuelsson, G., Phoratoxin, a toxic protein from the mistletoe *Phoradendron tomentosum* subsp. *macrophyllum* (Loranthaceae). The amino acid sequence, *Acta Pharmacol. Suecica,* 11, 347, 1974.
55. Thunberg, E., Phoratoxin B, a toxic protein from the mistletoe *Phoradendron tomentosum* subsp. *macrophyllum, Acta Pharmacol. Suecica,* 20, 115, 1983.
56. Thunberg, E. and Samuelsson, G., The amino acid sequence of ligatoxin A, from the mistletoe *Phoradendron liga, Acta Pharmacol. Suecica,* 19, 447, 1982.
57. Samuelsson, G. and Pettersson, B., Toxic proteins from the mistletoe *Dendrophtora clavata.* II. The amino acid sequence of denclatoxin B, *Acta Pharmacol. Suecica,* 14, 245, 1977.
58. Teeter, M. M., Mazer, J. A., and L'Italien, J. J., Primary structure of the hydrophobic plant protein crambin, *Biochemistry,* 20, 5437, 1981.
59. Vermeulen, J. A. W. H., Lamerichs, R. M. J. N., Berliner, L., De Marco, A., Llinas, M., Boelens, R., Alleman, U., and Kaptein, R., ^{1}H-NMR characterization of two crambin species, *FEBS Lett.*, 219, 426, 1987.
60. Epple, P., Apel, K., and Bohlmann, H., An *Arabidopsis thaliana* thionin gene is inducible via a signal transduction pathway different from that for pathogenesis-related proteins, *Plant Physiol.*, 109, 813, 1995.
61. Terras, F. R. G., Torrekens, S., Van Leuven, F., and Broekaert, W. F., A six-cysteine type thionin from the radish storage organ displays weak *in vitro* antifungal activity against *Fusarium culmorum, Plant Physiol. Biochem.*, 34, 599, 1996.
62. Rosell, S. and Samuelsson, G., Effect of mistletoe viscotoxin and phoratoxin on blood circulation, *Toxicon*, 4, 107, 1966.
63. Lankisch, P. G. and Vogt, W., Potentiation of haemolysis by the combined action of phospholipase A and a basic peptide containing S–S-bonds (viscotoxin B), *Experientia*, 27, 122, 1971.

64. Andersson, K. E. and Johannsson, M., Effects of viscotoxin on rabbit heart and aorta, and on frog skeletal muscle, *Eur. J. Pharm.*, 23, 223, 1973.
65. Carrasco, L., Vazques, D., and Hernandez-Lucas, C., Thionins: plant peptides that modify membrane permeability in cultured mammalian cells, *Eur. J. Biochem.*, 116, 185, 1981.
66. Selawry, O. S., Vester, F., and Mai, W., Zur kenntnis der inhaltsstoffe von *Viscum album*. II. tumorhemmende inhaltsstoffe, *Hoppe-Seyler's Z. Physiol. Chem.*, 324, 262, 1961.
67. Buessing, A., Suzart, K., and Bergmann, J., Induction of apoptosis in human lymphocytes treated with *Viscum album* L. is mediated by the mistletoe lectins, *Cancer Lett.*, 99, 59, 1996.
68. Winterfeld, K. and Bijl, L. H., Viscotoxin, ein neuer inhaltsstoff der mistel (*Viscum album* L.), *Liebigs Ann. Chem.*, 561, 107, 1948.
69. Samuelsson, G., Chromatography of viscotoxin and oxidized viscotoxin on phosphate cellulose, *Acta Chem. Scand.*, 20, 1546, 1966.
70. Samuelsson, G. and Pettersson, B., Separation of viscotoxins from the European mistletoe, *Viscum album* L. (Loranthaceae) by chromatography on sulfoethyl sephadex, *Acta Chem. Scand.*, 24, 2751, 1970.
71. Konopa, J., Woynarowski, J. M., and Lewandowska-Gumieniak, M., Isolation of viscotoxins. Cytotoxic basic polypeptides from *Viscum album* L., *Hoppe-Seyler's Z. Physiol. Chem.*, 361, 1525, 1980.
72. Schaller, G., Urech, K., and Giannattasio, M., Cytotoxicity of different viscotoxins and extracts from the European subspecies of *Viscum album* L., *Phytotherapy Res.*, 10, 473, 1996.
73. Van Etten, C. H., Nielsen, H. C., and Peters, J. E., A crystalline polypeptide from the seed of *Crambe abyssinica, Phytochemistry,* 4, 467, 1965.
74. Van Etten, C. H., Gagne W. E., Robbins, D., Booth, A. N., Daxenbichler, M. E., and Wolff, I. A., Biological evaluation of *Crambe* seed meals and derived products by rat feeding, *Cereal Chem.*, 46, 145, 1969.
75. Angerhofer, C. K., Shier, W. T., and Vernon, L. P., Phospholipase activation in the cytotoxic mechanism of thionin purified from nuts of *Pyrularia pubera, Toxicon,* 28, 547, 1990.
76. Schrader-Fischer, G. and Apel, K., Organ-specific expression of highly divergent thionin variants that are distinct from the seed-specific crambin in the crucifer *Crambe abyssinica, Mol. Gen. Genet.*, 245, 380, 1994.
77. Daley, L. S. and Theriot, L. J., Electrophoretic analysis, redox activity, and other characteristics of proteins similar to purothionins from tomato (*Lycopersicum esculenta*), mango (*Mangifera indica*), papaya (*Carica papaya*), and walnut (*Juglans regia*), *J. Agric. Food Chem.*, 35, 680, 1987.
78. Bohlmann, H., Apel, K., and García-Olmedo, F., Thionins, *Plant Mol. Biol. Rep.*, 12, S75, 1994.
79. Hendrickson, W. A. and Teeter, M. M., Structure of the hydrophobic protein crambin determined directly from the anomalous scattering of sulfur, *Nature*, 290, 107, 1981.
80. Teeter, M. M., Water structure of a hydrophobic protein at atomic resolution: pentagon rings of water molecules in crystals of crambin, *Proc. Natl. Acad. Sci. U.S.A.*, 81, 6014, 1984.
81. Stec, B., Zhou, R., and Teeter, M. M., Full-matrix refinement of the protein crambin at 0.83 Å and 130 K, *Acta Cryst.*, D51, 663, 1995.

82. Weeks, C. M., Hauptman, H. A., Smith, G. D., Blessing, R. H., Teeter, M. M., and Miller, R., Crambin: a direct solution for a 400-atom structure, *Acta Cryst.,* D51, 33, 1995.
83. Teeter, M. M. and Hendrickson, W. A., Highly ordered crystals of the plant seed protein crambin, *J. Mol. Biol.,* 127, 219, 1979.
84. Whitlow, M. and Teeter, M. M., Energy minimization for tertiary structure prediction of homologous proteins: (1-purothionin and viscotoxin A3 models from crambin, *J. Biochem. Struct. Dynam.,* 2, 831, 1985.
85. Rao, U. and Teeter, M. M., Improvement of turn structure prediction by molecular dynamics: a case study of α_1-purothionin, *Protein Eng.,* 6, 837, 1993.
86. Rao, U., Stec, B., and Teeter, M. M., Refinement of purothionins reveals solute particles important for lattice formation and toxicity. Part 1: α_1-purothionin revisited, *Acta Cryst.,* D51, 904, 1995.
87. Stec, B., Rao, U., and Teeter, M. M., Refinement of purothionins reveals solute particles important for lattice formation and toxicity. Part 2: Structure of β-purothionin at 1.7 Å resolution, *Acta Cryst.,* D51, 914, 1995.
88. Lecomte, J. T. J., Jones, B. L., and Llinas, M., Proton magnetic resonance studies of barley and wheat thionins: structural homology with crambin., *Biochemistry,* 21, 4843, 1987.
89. Han, K. H., Park, K. H., Yoo, H. J., Cha, H., Suh, S. W., Thomas, F., Moon, T. S., and Kim, S. M., Determination of the three-dimensional structure of hordothionin-α by nuclear magnetic resonance, *Biochem J.,* 313, 885, 1996.
90. Lecomte, J. T. J., Kaplan, D., Llinas, M., Thunberg, E., and Samuelsson, G., Proton magnetic resonance characterization of phoratoxins and homologous proteins related to crambin, *Biochemistry,* 26, 1187, 1987.
91. Teeter, M. M. and Whitlow, M., Test of circular dichroism (CD) methods for crambin and CD-assisted secondary structure prediction of its homologous toxins, *Proteins: Struct. Funct. Genet.,* 4, 262, 1988.
92. Markman, O., Rao, U., Lewis, K., Heffron, G. J., Stec, B., and Teeter, M. M., Mode of phospholipid binding to the membrane active plant toxin phoratoxin-A, in *New Developments in Lipid-Protein Interactions and Receptor Function,* Wirtz, K. W. A., Ed., Plenum Press, New York, 1993, 263.
93. Mellstrand, S. T. and Samuelsson, G., Phoratoxin, a toxic protein from the mistletoe *Phoradendron tomentosum* subsp. *macrophyllum* (Loranthaceae). The disulfide bonds, *Acta Pharmacol. Suecica,* 11, 367, 1974.
94. Olson, T. and Samuelsson, G., The disulfide bonds of viscotoxin A_2 from the European mistletoe (*Viscum album* L., Loranthaceae), *Acta Pharmacol. Suecica,* 11, 381, 1974.
95. Hase, T., Matsubara, H., and Yoshizumi, H., Disulfide bonds of purothionin, a lethal toxin for yeasts, *J. Biochem.,* 83, 1671, 1978.
96. Stuart, L. S and Harris, T. H., Bactericidal and fungicidal properties of a crystalline protein isolated from unbleached wheat flour, *Cereal Chem.,* 19, 288, 1942.
97. Coulson, E. J., Harris, T. H., and Axelrod, B., Effect on small laboratory animals of the injection of the crystalline hydrochloride of a sulfur protein from wheat flour, *Cereal Chem.,* 19, 301, 1942.
98. Samuelsson, G., Mistletoe toxins, *Syst. Zool.,* 22, 566, 1974.
99. Samuelsson, G., Pettersson, B., and Ljunggren, A., Toxic proteins from the mistletoe *Dendrophtora clavata.* I. Isolation and properties of denclatoxin A and denclatoxin B, *Acta Pharmacol. Suecica,* 14, 237, 1977.

100. Evett, G. E., Donaldson, D. M., and Vernon, L. P., Biological properties of *Pyrularia* thionin prepared from nuts of *Pyrularia pubera*, *Toxicon*, 24, 622, 1986.
101. Kramer, K. J., Klassen, L. W., Jones, B. L., Speirs, R. D., and Kammer, A. E., Toxicity of purothionin and its homologues to the tobacco hornworm, *Manduca sexta* (L.) (Lepidoptera: Sphingidae), *Toxicol. Appl. Pharmacol.*, 48, 179, 1979.
102. Woolley, D. W. and Krampitz, L. O., Reversal by phosphatides of the antimicrobial action of a crystalline protein from wheat, *J. Biol. Chem.*, 146, 273, 1942.
103. Nose, Y. and Ichikawa, M., Studies on the effects of flour extract on baker's yeast, *J. Ferment. Technol.*, 46, 915, 1968.
104. Okada, T. and Yoshizumi, H., A lethal toxic substance for brewing yeast in wheat and barley. Part II. Isolation and some properties of toxic principle, *Agr. Biol. Chem.*, 34, 1089, 1970.
105. Okada, T., Yoshizumi, H., and Terashima, Y., A lethal toxic substance for brewing yeast in wheat and barley. Part I. Assay of toxicity on various grains, and sensitivity of various yeast strains, *Agr. Biol. Chem.*, 34, 1084, 1970.
106. Okada, T. and Yoshizumi, H., The mode of action of toxic protein in wheat and barley on brewing yeast, *Agr. Biol. Chem.*, 37, 2289, 1973.
107. Hernandez-Lucas, C., Fernandez de Caleya, R., and Carbonero, P., Inhibition of brewer's yeasts by wheat purothionins, *Appl. Microbiol.*, 28, 165, 1974.
108. Wada, K., Ozaki, Y., Matsubara, H., and Yoshizumi, H., Studies on purothionin by chemical modifications, *J. Biochem.*, 91, 257, 1982.
109. Fernandez de Caleya, R., Gonzalez-Pascual, B., García-Olmedo, F., and Carbonero, P., Susceptibility of phytopathogenic bacteria to wheat purothionins in vitro, *Appl. Microbiol.*, 23, 998, 1972.
110. Cammue, B. P. A., De Bolle, M. F. C., Terras, F. R. G., Proost, P., Van Damme, J., Rees, S. B., Vanderleyden, J., and Broekaert, W. F., Isolation and characterization of a novel class of plant antimicrobial peptides from *Mirabilis jalapa* L. seeds, *J. Biol. Chem.*, 267, 2228, 1992.
111. Florack, D. E. A., Visser, B., De Vries, P. H. M., Van Vuurde, J. W. L., and Stiekema, W. J., Analysis of the toxicity of purothionins and hordothionins for plant pathogenic bacteria, *Neth. J. Pl. Path.*, 99, 259, 1993.
112. Molina, A., Ahl-Goy, P., Fraile, A., Sanchez-Monge, R., and García-Olmedo, F., Inhibition of bacterial and fungal plant pathogens by thionins of types I and II, *Plant Science*, 92, 169, 1993.
113. Molina, A., Segura, A., and García-Olmedo, F., Lipid transfer proteins (nsLTPs) from barley and maize leaves are potent inhibitors of bacterial and fungal plant pathogens, *FEBS Lett.*, 316, 119, 1993.
114. Bohlmann, H., Clausen, S., Behnke, S., Giese, H., Hiller, C., Schrader, G., Barkholt, V., and Apel, K., Leaf-thionins of barley — a novel class of cell wall proteins toxic to plant-pathogenic fungi and possibly involved in the defense mechanism of plants, *EMBO J.*, 7, 1559, 1988.
115. Terras, F. R. G., Schoofs, H. M. E., De Bolle, M. F. C., Van Leuven, R., Rees, S. B., Vanderleyden, J., Cammue, B. P. A., and Broekaert, W. F., Analysis of two novel classes of plant antifungal proteins from radish (*Raphanus sativus* L.) seeds, *J. Biol. Chem.*, 267, 15302, 1992.
116. Terras, F. R. G., Schoofs, H. M. E., Thevissen, K., Osborn, R. W., Vanderleyden, J., Cammue, B. P. A., and Broekaert, W. F., Synergistic enhancement of the antifungal activity of wheat and barley thionins by radish and oilseed rape 2S albumins and by barley trypsin inhibitors, *Plant Physiol.*, 103, 1311, 1993.

117. Kashimoto, T., Sakakibara, R., Huynh, Q. K., Wada, H., and Yoshizumi, H., The effect of purothionin on bovine adrenal medullary cells, *Res. Commun. Chem. Pathol. Pharmacol.,* 26, 221, 1979.
118. Nakanishi, T., Yoshizumi, H., Tahara, S., Hakura, A., and Toyoshima, K., Cytotoxicity of purothionin-A on various animal cells, *Gann.,* 70, 323, 1979.
119. Shaw, K. P., Kauffman, F. C., Vernon, L. P., and Albuquerque, E. X., Cytotoxic mechanism of *Pyrularia* thionin, *Soc. Neurosc. Abstr.,* 13, 930, 1987.
120. Osorio e Castro, V. R., Van Kuiken, B A., Vernon, L P., and Albuquerque, E. X., Action of a thionin isolated from nuts of *Pyrularia pubera* on human erythrocytes, *Toxicon*, 27, 501, 1989.
121. Thevissen, K., Ghazi, A., De Samblanx, G. W., Brownlee, C., Osborn, R. W., and Broekaert, W. F., Fungal membrane responses induced by plant defensins and thionins, *J. Biol. Chem.,* 271, 15018, 1996.
122. Oka, T., Murata, Y., Nakanishi, T., Yoshizumi, H., Hayashida, H., Ohtsuki, Y., and Toyoshima, K., Similarity, in molecular structure and function, between the plant toxin purothionin and the mammalian pore-forming proteins, *Mol. Biol. Evol.,* 9, 707, 1992.
123. Guihard, G., Bénédetti, H., Besnard, M., and Letellier, L., Phosphate efflux through the channels formed by colicins and phage T5 in *Escherichia coli* cells is responsible for the fall in cytoplasmic ATP, *J. Biol. Chem.,* 268, 17775, 1993.
124. Jones, B. L., Lookhart, G. L., and Johnson, D. E., Improved separation and toxicity analysis methods for purothionins, *Cereal Chem.,* 62, 327, 1985.
125. Vernon, L. P. and Rogers, A., Effect of calcium and phosphate ions on hemolysis induced by *Pyrularia* thionin and *Naja naja kaouthia* cardiotoxin, *Toxicon*, 30, 701, 1992.
126. Evans, J., Wang, Y., Shaw, K. P., and Vernon, L. P., Cellular responses to *Pyrularia* thionin are mediated by Ca^{2+} influx and phospholipase A_2 activation and are inhibited by thionin tyrosine iodination, *Proc. Natl. Acad. Sci. U.S.A.,* 86, 5849, 1989.
127. Jackson, S. L. and Heath, I. B., Roles of calcium ions in hyphal tip growth, *Microbiol. Rev.,* 57, 367, 1993.
128. Rao, U., Teeter, M. M., Erickson-Viitanen, S., and DeGrado, W. F., Calmodulin binding to α_1-purothionin: solution binding and modeling of the complex, *Proteins: Struct. Funct. Genet.,* 14, 127, 1992.
129. Vernon, L. P., *Pyrularia* thionin: physical properties, biological responses and comparison to other thionins and cardiotoxins, *J. Toxicol. Toxin Rev.,* 11, 169, 1992.
130. Gasanov, S. E., Vernon, L. P., and Aripov, T. F., Modification of phospholipid membrane structure by the plant toxic peptide *Pyrularia* thionin, *Arch. Biochem. Biophys.,* 301, 367, 1993.
131. Hernandez-Lucas, C., Fernandez de Caleya, R., Carbonero, P., and García-Olmedo, F., Reconstitution of petroleum ether soluble wheat lipopurothionin by binding of digalactosyl diglyceride to the chloroform-soluble form, *J. Agric. Food Chem.,* 25, 1287, 1977.
132. Carbonero, P., García-Olmedo, F., and Hernandez-Lucas, C., External association of hordothionin with protein bodies in mature barley, *J. Agric. Food Chem.,* 28, 399, 1980.
133. Harvey, A. L., Cardiotoxins from cobra venoms: possible mechanisms of action, *J. Toxicol. Toxin Rev.,* 4, 41, 1985.
134. Harvey, A. L., Cardiotoxins from cobra venoms, in *Handbook of Natural Toxins,* Vol. 5. Reptile Venoms and Toxins, Tu, A. T., Ed., Marcel-Decker, New York, 1991, 85.

135. Osorio e Castro, V. R. and Vernon, L. P., Hemolytic activity of thionin from *Pyrularia pubera* nuts and snake venom toxins of *Naja naja* species: *Pyrularia* thionin and snake venom cardiotoxin compete for the same membrane site, *Toxicon*, 27:511–517, 1989.
136. Jiang, M. S., Fletcher, J. E., and Smith, L. A., Factors influencing the hemolysis of human erythrocytes by cardiotoxins from *Naja naja kaouthia* and *Naja naja atra* venoms and a phospholipase A_2 with cardiotoxin-like activities from *Bungarus fasciatus* venom, *Toxicon*, 27, 47, 1989.
137. Louw, A. I. and Visser, L., The synergism of cardiotoxin and phospholipase A_2 in hemolysis, *Biochim. Biophys. Acta,* 163, 512, 1989.
138. Huang, W., Vernon, L. P., and Bell, J. D., Enhancement of adenylate cyclase activity in S49 lymphoma cell membranes by the toxin thionin from *Pyrularia pubera, Toxicon,* 32, 789, 1994.
139. Lobb, L., Stec, B., Kantrowitz, E. K., Yamano, A., Stojanoff, V., Markman, O., and Teeter, M. M., Expression, purification and characterization of recombinant crambin, *Prot. Engineering,* 9, 1233, 1996.
140. Rao, A. G., Hassan, M., and Hempel, J. C., Structure-function validation of high lysine analogs of α-hordothionin designed by protein modeling, *Protein Engineering,* 7, 1485, 1994.
141. Wada, K. and Buchanan, B. B., Purothionin: a seed protein with thioredoxin activity, *FEBS Lett.,* 124, 237, 1981.
142. Johnson, T. C., Wada, K., Buchanan, B. B., and Holmgren, A., Reduction of purothionin by the wheat seed thioredoxin system, *Plant Physiol.,* 85, 446, 1987.
143. Diaz, I., Carmona, M. J., and García-Olmedo, F., Effects of thionins on β-glucuronidase in vitro and in plant protoplasts, *FEBS Lett.,* 296, 279, 1992.
144. García-Olmedo, F., Piñeiro, M., and Diaz, I., Dances to a redox tune, *Plant Mol. Biol.,* 26, 11, 1994.
145. Piñeiro, M., García-Olmedo, F., and Diaz, I., Redox modulation of the expression of bacterial genes encoding cysteine-rich proteins in plant protoplasts, *Proc. Natl. Acad. Sci. U.S.A.,* 91, 3867, 1994.
146. Piñeiro, M., Diaz, I., Rodriguez-Palenzuela, P., Titarenko, E., and García-Olmedo, F., Selective disulfide linkage of plant thionins with other protein, *FEBS Lett.,* 369, 239, 1995.
147. Woynarowski, J. M. and Konopa, J., Interaction between DNA and viscotoxins, cytotoxic basic polypeptides from *Viscum album* L., *Hoppe-Seyler's Z. Physiol. Chem.,* 361, 1535, 1980.
148. García-Olmedo, F., Carbonero, P., Hernandez-Lucas, C., Paz-Ares, J., Ponz, F., Vicente, O., and Sierra, J. M., Inhibition of eukaryotic cell-free protein synthesis by thionins from wheat endosperm, *Biochim. Biophys. Acta,* 740, 52, 1983.
149. Brümmer, J., Thole, H., and Kloppstech, K., Hordothionins inhibit protein synthesis at the level of initiation in the wheat-germ system, *Eur. J. Biochem.,* 219, 425, 1994.
150. Jones, B. L. and Meredith, P., Inactivation of α-amylase activity by purothionins, *Cereal Chem.,* 59, 321, 1982.
151. Richardson, M., Seed storage proteins: the enzyme inhibitors, *Meth. Plant Biochem.,* 5, 259, 1991.
152. Andresen, I., Becker, W., Schlüter, K., Burges, J., Parthier, B., and Apel, K., The identification of leaf thionin as one of the main jasmonate-induced proteins of barley (*Hordeum vulgare*), *Plant Mol. Biol.,* 19, 193, 1992.

153. Bergey, D. R., Howe, G. A., and Ryan, C. A., Polypeptide signaling for plant defensive genes exhibits analogies to defense signaling in animals, *Proc. Natl. Acad. Sci. U.S.A.*, 93, 12053, 1996.
154. Bloch, C., Jr. and Richardson, M., A new family of small (5 kDa) protein inhibitors of insect α-amylases from seeds of sorghum (*Sorghum bicolor* (L) Moench) have sequence homologies with wheat γ-purothionins, *FEBS Lett.*, 279, 101, 1991.
155. Campos, F. A. P. and Richardson, M., The complete amino acid sequence of the bifunctional α-amylase/trypsin inhibitor from seeds of ragi (Indian finger millet; *Eleusine coracana* Goertn.), *FEBS Lett.*, 152, 300, 1984.
156. Schrader-Fischer, G. and Apel, K., The anticyclic timing of leaf senescence in the parasitic plant *Viscum album* is closely correlated with the selective degradation of sulphur-rich viscotoxins, *Plant Physiol.*, 101, 745, 1993.
157. Bohlmann, H., Significance of sulfur-rich proteins in seeds and leaves, in *Sulfur Nutrition and Assimilation in Higher Plants,* DeKok, L. J., Stulen, I., Rennenberg, H., Brunold, C., and Rauser, W. E., Eds., Academic Publishing, The Hague, 1993, 211.
158. Carmona, M. J., Molina, A., Fernandez, J. A., Lopez-Fando, J. M., and García-Olmedo, F., Expression of the alpha-thionin gene from barley in tobacco confers enhanced resistance to bacterial pathogens, *Plant J.*, 3, 457, 1993.
159. Florack, D. E. A., Dirkse, W. G., Visser, B., Heidekamp, F., and Stiekema, W. J., Expression of biologically active hordothionin in tobacco. Effects of pre- and pro-sequences at the amino and carboxyl termini of the hordothionin precursor on mature protein expression and sorting, *Plant Mol. Biol.*, 24, 83, 1994.
160. Maddox, J., Duvick, J., Hastings, C., Rouch, H., Ross, M., Lane, S., and Daywalt, M., Cloning of a barley gene, α-hordothionin, and expression in transgenic tobacco, *J. Cell Biochem.*, Suppl 16F, 217, 1992.
161. Epple, P., Apel, K., and Bohlmann, H., Overexpression of an endogenous thionin enhances resistance of *Arabidopsis* against Fusarium oxysporum, *Plant Cell,* 9, 509, 1997.
162. Ponz, F., Paz-Ares, J., Hernandez-Lucas, C., Carbonero, P., and García-Olmedo, F., Synthesis and processing of thionin precursors in developing endosperm from barley (*Hordeum vulgare* L.), *EMBO J.*, 2, 1035, 1983.
163. Von Heijne, G., A new method for predicting signal sequence cleavage sites, *Nucl. Acids Res.*, 14, 4683, 1986.
164. Reimann-Philipp, U., Schrader, G., Martinoia, E., Barkholt, V., and Apel, K., Intracellular thionins of barley. A second group of leaf thionins closely related to but distinct from cell wall-bound thionins, *J. Biol. Chem.*, 264, 8978, 1989.
165. Carmona, M. J., Hernandez Lucas, C., San Martin, C., Gonzales, P., and García-Olmedo, F., Subcellular localization of type I thionins in the endosperms of wheat and barley, *Protoplasma*, 173, 1, 1993.
166. Weissman, J. S. and Kim, P. S., The pro region of BPTI facilitates folding, *Cell*, 71, 841, 1992.
167. Lee, D. W., Apel, K., and Binding, H., The formation of leaf thionins and their effect on the viability of isolated protoplasts of barley (*Hordeum vulgare* L.), *J. Plant Physiol.*, 145, 131, 1995.
168. Daher, K. A., Lehrer, R. I., Ganz, T., and Kronenberg, M., Isolation and characterization of human defensin cDNA clones, *Proc. Natl. Acad. Sci. U.S.A.*, 85, 7327, 1988.
169. Dimarcq, J. L., Zachary, D., Hoffmann, J. A., Hoffmann, D., and Reichhart, J. M., Insect immunity: expression of the two major inducible antibacterial peptides, defensin and diptericin, in *Phormia terranovae, EMBO J.*, 9, 2507, 1990.

170. Gu, Q., Kawata, E. E., Morse, M. J., Wu, H. M., and Cheung, A. Y., A flower-specific cDNA encoding a novel thionin in tobacco, *Mol. Gen. Genet.,* 234, 89, 1992.
171. Romero, A., Alamillo, J. M., and García-Olmedo, F., Processing of thionin precursors in barley leaves by a vacuolar proteinase, *Eur. J. Biochem.,* 243, 202, 1997.
172. Bunge, S., Wolters, J., and Apel, K., A comparison of leaf thionin sequences of barley cultivars and wild barley species, *Mol. Gen. Genetics,* 231, 460, 1992.
173. Bohl, S. and Apel, K., A novel fingerprint method for analyzing the expression of complex multigene families of very low transcript abundance, *Plant J.,* 3, 887, 1993.
174. Reimann-Philipp, U., Behnke, S., Batschauer, A., Schäfer, E., and Apel, K., The effect of light on the biosynthesis of leaf-specific thionins in barley, *Hordeum vulgare, Eur. J. Biochem.,* 182, 283, 1989.
175. Hess, W. R., Müller, A., Nagy, F., and Börner, T., Ribosome-deficient plastids affect transcription of light-induced nuclear genes: genetic evidence for a plastid-derived signal, *Mol. Gen. Genet.,* 242, 305, 1994.
176. Beator, J., Pötter, E., and Kloppstech, K., The effect of heat shock on morphogenesis in barley, *Plant Physiol.,* 100, 1780, 1992.
177. Fischer, R., Behnke, S., and Apel, K., The effect of chemical stress on the polypeptide composition of the intercellular fluid of barley leaves, *Planta,* 178, 61, 1989.
178. Wasternack, C., Atzorn, R., Jarosch, B., and Kogel, K. H., Induction of a thionin, the jasmonate-induced 6 kDa protein of barley by 2,6-dichloroisonicotinic acid, *J. Phytopathology,* 140, 280, 1994.
179. Kogel, K. H., Ortel, B., Jarosch, B., Atzorn, R., Schiffer, R., and Wasternack, C., Resistance in barley against the powdery mildew fungus (*Erysiphe graminis f. sp. hordei*) is not associated with enhanced levels of endogenous jasmonates, *Eur. J. Plant Pathol.,* 101, 319, 1995.
180. Métraux, J. P., Ahl-Goy, P., and Staub, T., Induced resistance in cucumber in response to 2,6-dichlorisonicotinic acid and pathogens, in *Advances in Molecular Genetics of Plant-Microbe Interactions,* Hennecke, H. and Verma, D. P. S., Eds., Kluwer Academic Publishers, Dordrecht, The Netherlands, 1991, 432.
181. Gundlach, H., Müller, M. J., Kutchan, T. M., and Aenk, M. H., Jasmonic acid is a signal transducer in elicitor-induced plant cell cultures, *Proc. Natl. Acad. Sci. U.S.A.,* 89, 2389, 1992.
182. Uknes, S., Mauch-Mani, B., Moyer, M., Potter, S., Williams, S., Dincher, S., Chandler, D., Slusarenko, A., Ward, E., and Ryals, J., Acquired resistance in Arabidopsis, *Plant Cell,* 4, 645, 1992.
183. Dempsey, D' M. A. and Klessig, D. F., Salicylic acid, active oxygen species and systemic acquired resistance in plants, *Trends Cell Biol.,* 4, 334, 1994.
184. Weidhase, R. A., Kramell, H., Lehmann, J., Liebisch, H. W., Lerbs, W., and Parthier, B., Methyljasmonate-induced changes in the polypeptide pattern of senescing barley leaf segments, *Plant Sci.,* 51, 177, 1987.
185. Molina, A. and García-Olmedo, F., Developmental and pathogen-induced expression of three barley genes encoding lipid transfer proteins, *Plant J.,* 4, 983, 1993.
186. Titarenko, E., Hargreaves, J., Keon, J., and Gurr, S. J., Defense-related gene expression in barley coleoptile cells following infection by *Septoria nodorum,* in *Mechanisms of Plant Defense Responses,* Fritig, B. and Legrand, M., Eds., Kluwer Academic Publishers, Dordrecht, The Netherlands, 1993, 308.
187. Stevens, C., Titarenko, E., Hargreaves, J. A., and Gurr, S. J., Defense-related gene activation during an incompatible interaction between *Stagonospora* (*Septoria*) *nodorum* and barley (*Hordeum vulgare* L.) coleoptile cells, *Plant Mol. Biol.,* 31, 741, 1996.

188. Valè, G. P., Torrigiani, E., Gatti, A., Delogu, G., Porta-Puglia, A., Vannacci, G., and Cattivelli, L., Activation of genes in barley roots in response to infection by two *Drechslera graminea* isolates, *Physiol. Mol. Plant Pathol.*, 44, 207, 1994.
189. Boyd, L. A., Smith, P. H., and Brown, J. K. M., Molecular and cellular expression of quantitative resistance in barley to powdery mildew, *Physiol. Mol. Plant Pathol.*, 45, 47, 1994.
190. Boyd, L. A., Smith, P. H., Green, R. M., and Brown, J. K. M., The relationship between the expression of defense-related genes and mildew development in barley, *Mol. Plant-Microbe Interact.*, 7, 401, 1994.
191. Ebrahim-Nesbat, F., Behnke, S., Kleinhofs, A., and Apel, K., Cultivar-related differences in the distribution of cell-wall bound thionins in compatible and incompatible interactions between barley and powdery mildew, *Planta*, 179, 203, 1989.
192. Ebrahim-Nesbat, F., Bohl, S., and Heitefuss, R., Thionin in cell walls and papillae of barley in compatible and incompatible interactions with *Erysiphe graminis f. sp. hordei, Physiol. Mol. Plant Pathol.*, 43, 343, 1993.
193. Ebrahim-Nesbat, F., Rohringer, R., and Heitefuss, R., Effect of rust infection on cell walls of barley and wheat; immunocytochemistry using anti-barley thionin as a probe, *J. Phytopathol.*, 141, 38, 1994.
194. Ebrahim-Nesbat, F., Apel, K., and Heitefuss, R., Immunochemical evidence for the presence of barley thionins in rachis and stalk of wheat: decrease of antibody binding at *Fusarium culmorum* penetration sites, *J. Phytopathol.*, 131, 259, 1991.
195. Rasmussen, S. and Rasmussen, C. E., Full-length cDNA clone encoding barley toxin β-hordothionin shows a signal peptide 27 amino acid residues long, *Plant Mol. Biol.*, 21, 579, 1993.
196. Kleinhofs, A., Kilian, A., Saghai Maroof, M. A., Biyashev, R. M., Hayes, P., Chen, F. Q., Lapitan, N., Fenwick, A., Blake, T. K., Kanazin, V., Ananiev, E., Dahleen, L., Kudrna, D., Bollinger, J., Knapp, S. J., Liu, B., Sorrells, M., Heun, M., Franchowiak, J. D., Hoffman, D., Skadsen, R., and Steffenson, B. J., A molecular, isozyme and morphological map of the barley (*Hordeum vulgare*) genome, *Theor. Appl. Genet.*, 86, 705, 1993.
197. Pedersen, C., Giese, H., and Linde-Laursen, I., Towards an integration of the physical and the genetic chromosome maps of barley by in situ hybridization, *Hereditas*, 123, 77, 1995.
198. Holtorf, S., Apel, K., and Bohlmann, H., Specific and different expression patterns of two members of the leaf thionin multigene family of barley in transgenic tobacco, *Plant Sci.*, 111, 27, 1995.
199. Pecchioni, N., Stanca, A. M., Terzi, V., and Cattivelli, L., RFLP analysis of highly polymorphic loci in barley, *Theor. Appl. Genet.*, 85, 926, 1993.
200. Olivera, B. M., Rivier, J., Clark, C., Ramilo, C. A., Corpuz, G. P., Abogadie, F. C., Mena, E. E., Woodward, S. R., Hillyard, D. R., and Cruz, L. J., Diversity of *Conus* neuropeptides, *Science*, 249, 257, 1990.
201. Steinmüller, K., Batschauer, A., and Apel, K., Tissue-specific and light-dependent changes of chromatin organization in barley (*Hordeum vulgare*), *Eur. J. Biochem.*, 158, 519, 1986.
202. Smith, H., Signal perception, differential expression within multigene families and the molecular basis of phenotypic plasticity, *Plant Cell Environ.*, 13, 585, 1990.
203. Lotan, T., Ori, N., and Fluhr, R., Pathogenesis-related proteins are developmentally regulated in tobacco flowers, *Plant Cell*, 1, 881, 1989.

204. Neale, A. D., Wahleithner, J. A., Lund, M., Bonnett, H. T., Kelly, A., Meeks-Wagner, D. R., Peacock, W. J., and Dennis, E. S., Chitinase, β-1,3-glucanase, osmotin, and extensin are expressed in tobacco explants during flower formation, *Plant Cell,* 2, 673, 1990.
205. Ori, N., Sessa, G., Lotan, T., Himmelhoch, S., and Fluhr, R., A major stylar matrix polypeptide (sp41) is a member of the pathogenesis-related proteins superclass, *EMBO J.,* 9, 3429, 1990.
206. Coté, F., Cutt, J. R., Asselin, A., and Klessig, D. F., Pathogenesis-related acidic beta-1,3-glucanase genes of tobacco are regulated by both stress and developmental signals, *Mol. Plant-Microbe Interact.,* 4, 173, 1991.
207. Penninckx, I. A. M. A., Eggermont, K., Terras, F. R. G., Thomma, B. P. H. J., De Samblanx, G. W., Buchala, A., Métraux, J. P., Manners, J. M., and Broekaert, W., Pathogen-induced systemic activation of a plant defensin gene in Arabidopsis follows a salicylic acid-independent pathway, *Plant Cell,* 8, 2309, 1996.
208. Epple, P., Apel, K., and Bohlmann, H., ESTs reveal a multigene family for plant defensins in *Arabidopsis thaliana, FEBS Lett.,* 400, 168, 1997.
209. Robson, G. D., Kuhn, P. J., and Trinci, A. P. J., Effect of validamycin A on the inositol content and branching of *Rhizoctonia cerealis* and other fungi, *J. Gen. Microbiol.,* 135, 739, 1989.
210. Wiebe, M. G., Robson, G. D., and Trinci, A. P. J., Edifenphos (Hinosan) reduces hyphal extension, hyphal growth unit length and phosphatidylcholine content of *Fusarium graminearum* A3/5, but has no effect on specific growth rate, *J. Gen. Microbiol.,* 136, 979, 1990.
211. Creelman, R. A. and Mullet, J. E., Jasmonic acid distribution and action in plants: regulation during development and response to biotic and abiotic stress, *Proc. Natl. Acad. Sci. U.S.A.,* 92, 4114, 1995.
212. Jongedijk, E., Tigelaar, H., van Roekel, J. S. C., Bres-Vloemans, S. A., Dekker, I., van den Elzen, P. J. M., Cornelissen, B. J. C., and Melchers, L. S., Synergistic activity of chitinases and β-1,3-glucanases enhances fungal resistance in transgenic tomato plants, *Euphytica*, 85, 173, 1995.
213. Friedrich, L., Lawton, K., Ruess, W., Masner, P., Specker, N., Gut Rella, M., Meier, B., Dincher, S., Staub, T., Uknes, S., Métraux, J. P., Kessmann, H., and Ryals, J., A benzothiadiazole derivative induces systemic acquired resistance in tobacco, *Plant J.,* 10, 61, 1996.
214. Mitchell, A. F. and Walters, D. R., Systemic protection in barley against powdery mildew infection using methyl jasmonate, *Aspects Appl. Biol.,* 42, 323, 1995.
215. Li, J., Zingen-Sell, I., and Buchenauer, H., Induction of resistance of cotton plants to *Verticillium* wilt and of tomato plants to *Fusarium* wilt by 3-aminobutyric acid and methyl jasmonate, *J. Plant Dis. Protect.,* 103, 288, 1996.
216. Cohen, Y., Gisi, U., and Niderman, T., Local and systemic protection against *Phytophthora infestans* induced in potato and tomato plants by jasmonic acid and jasmonic methyl ester, *Phytopathology,* 83, 1054, 1993.

11 Ribosome Inactivating Proteins: Structure, Function, and Engineering

Anders Bøgh Jensen, Robert Leah, Bushra Chaudhry, and John Mundy

CONTENTS

11.1 Introduction 235
11.2 Structural Classification of RIPs 236
11.3 Physiological Roles and Regulation of RIPs 237
11.4 Role of RIPs in Plant–Viral Interactions 239
11.5 Role of RIPs in Plant–Fungal Interactions 240
11.6 Engineering Chimeric, Targeted RIPs 242
11.7 Perspectives 242
References 243

11.1 INTRODUCTION

Ribosome-inactivating proteins (RIPs) are a widely distributed group of *N*-glycolytic enzymes that modify ribosomes by cleaving a specific adenine residue in a highly conserved sequence of 28S rRNA (Figure 11.1; references 1 and 2). This irreversible modification makes the ribosome unable to bind elongation factor 1a and therefore blocks translation. Many structurally and functionally related RIPs from plants and bacteria exhibit specific phylogenetic activity.[3] Thus, some RIPs inactivate host conspecific ribosomes, while others exhibit toxicity toward ribosomes from distantly related species, including animals and fungi.[4,5] In cases where a RIP may inactivate conspecific ribosomes, the protein is usually synthesized as an inactive precursor or is targeted out of the cytoplasm, usually extracellularly. Because of their potential toxicities, two major functions are postulated for RIPs. First, RIPs which inactivate conspecific ribosomes may function as inducible or developmentally programmed suicide genes. Second, RIPs which inactivate phylogenetically distant ribosomes may have defensive functions in protecting plants from herbivores or pathogens. The potential utility of either role has led to work to elucidate their function and to exploit their varied toxicities. Thus, work has been done on the potential of RIPs to increase the protection of plants against pathogens.[6-8] Similarly, RIPs have been

0-8493-0697-3/99/$0.00+$.50

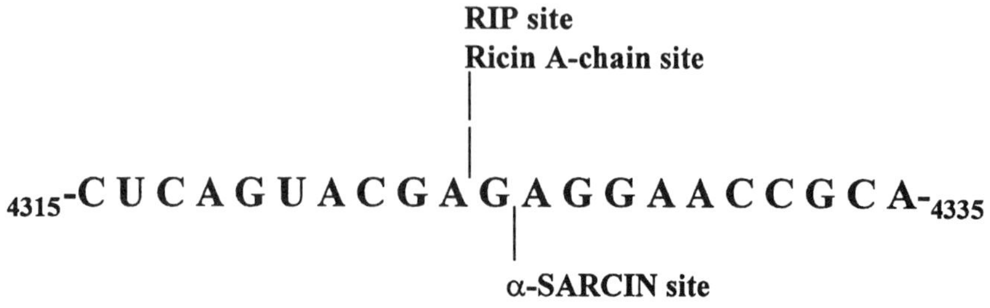

FIGURE 11.1 RIP, Ricin, and α-Sarcin inactivate ribosomes by specific cleavage of 28S rRNA. Ricin A-chain and many Type-1 RIPs act on eukaryotic ribosomes by hydrolytic cleavage of the *N*-glycosidic bond at residue A-4324 of the 28S rRNA, located close to the site of action of α-Sarcin from *Aspergillus giganteus.*

clinically assessed as anti-HIV agents in AIDS patients[9] and as toxic components of immunotoxins targeted at malignant cells.[10,11]

11.2 STRUCTURAL CLASSIFICATION OF RIPs

Since the isolation and characterization of ricin by Lord and co-workers,[12-13] many structurally and functionally related RIPs have been identified in a variety of plant species. On the basis of structural diversity, plant RIPs have been classified into two or more major groups.[14,15] Type-I RIPs, the most numerous group containing pokeweed antiviral protein (PAP), trichosmythin, barley RIP 30, and gelonin, are all synthesized as single-chain enzymes of about 30kD.[16] These may accumulate in starchy endosperm storage cells (RIP 30),[19] or be secreted from vegetative or seed cells (PAP).[7] In contrast, trichosanthin and α-trichosanthin accumulate in root cells.[17]

Type-II RIPs, such as prototypical ricin from castor bean endosperm, are synthesized as larger precursors which accumulate in protein bodies and are processed to a 30kD, RIP A-chain linked to a 30kD lectin B-chain through a single disulfide bond.[20] Upon maturation, the ricin C-terminal B-chain acts as a galactose-specific lectin, which facilitates uptake by cells displaying galactose on their surfaces, of the toxic A-chain, thereby causing cell death.[21] Similarly to Type-I RIPs, Type-II RIPs are also found in seed, vegetative cells, and in roots.[3] Type-I RIPs, which do not possess the B-chain binding domain, do not bind readily to cells and consequently have a relatively low mammalian cytotoxicity compared to Type-II RIPs.[4] However, Type-I RIPs can enter cells by fluid-phase endocytosis.[4]

Recently, a third RIP type has been characterized which apparently contains two cytoplasmically localized but quite different members from maize and barley.[22-25] The maize Type-III RIP accumulates in the kernel as a 34 kD inactive precursor which, upon germination, may be processed into a two-chain, active RIP by removal of an internal peptide from within the catalytic domain.[23] Similarly, the barley JIP60 protein, which has been shown to possess ribosome inactivating activity, is synthesized as a precursor that requires removal of a similar peptide for activation.[24] A primary distinction between Type-III and Types-I or -II RIPs is therefore that the former require removal of a peptide from the catalytic domain for activation.[1] JIP60, however, possesses a large C-terminal domain of unknown function which apparently also must be processed and/or removed for activation.[24]

The sequence similarities between RIP catalytic RNA-*N*-glycosidase domains are remarkably high. This is not only recognized by the alignment of the primary sequences but also reflected in the superimposition of three-dimensional model structures based on the known 3D structures. Interestingly, all RIPs share conserved 3D structures of the active cleft centered around the conserved catalytic amino acid residue E_{177} (numbering according to Ricin A) along with the invariant Y_{80}, Y_{123}, and W_{211}, which have been shown to be essential for substrate binding (see also the star in Figure 11.2).[26] Although RIPs from monocotyledonous species (barley, wheat, and maize) share a high similarity to all the other RIPs, as outlined in a Pileup (Figure 11.2), they seem to form a distinct group which diversified significantly earlier during evolution. Furthermore, until now it is only within this group of species that the distinct, Type-III RIPs have been identified.

Different RIPs have been isolated from many different plant species covering 13 families, including both monocotyledons and dicotyledons. RIP-producing plants are diverse, being found in crop plants including cereals, tropical trees, desert succulents, spring annuals, and weeds. Some families contain many RIP-producing members, and RIPs are frequently produced in amounts up to 10% of total, specific tissue protein, which may be far in excess of the amounts required to inhibit protein synthesis. RIPs have been detected in most plant organs and tissues including endosperm, fruit, roots, leaves, bark, and latex.[3] Most RIPs show developmentally regulated, organ-specific expression. However, several examples are known of inducibly expressed RIPs, including the jasmonate responsive JIP60 of barley and the beetins which are induced by viral infection and show responsiveness to salicylic acid and hydrogen peroxide.[24,27]

11.3 PHYSIOLOGICAL ROLES AND REGULATION OF RIPs

As noted in the Introduction, the potential toxicities of RIPs may be compatible with their having at least two physiological roles in plants: defense against pathogens and/or mediation of inducible or developmentally programmed suicide (programmed cell death, PCD).[28] Evidence for either role is partial and controversial at best. For example, although expression of Type-I RIP transgenes in plants has been shown to increase antiviral and antifungal protection, there is no direct evidence that this is their normal physiological role. Thus, several plants which express high levels of Type-I RIPs, that as transgenes increase the protection of other plants to pathogens, are themselves susceptible to virus infection. The endogenous, antiviral activity of these RIPs is therefore difficult to determine biochemically or molecularly.[7] For the very toxic Type-II lectins, such as ricin and abrin, which are found in large amounts in the endosperm, protective roles in deterring foraging birds and insects from eating the seeds have been postulated but not proven.[29]

The only evidence suggesting that most RIPs have roles in suicide are correlative: most organs which accumulate high levels of RIPs are either storage organs which

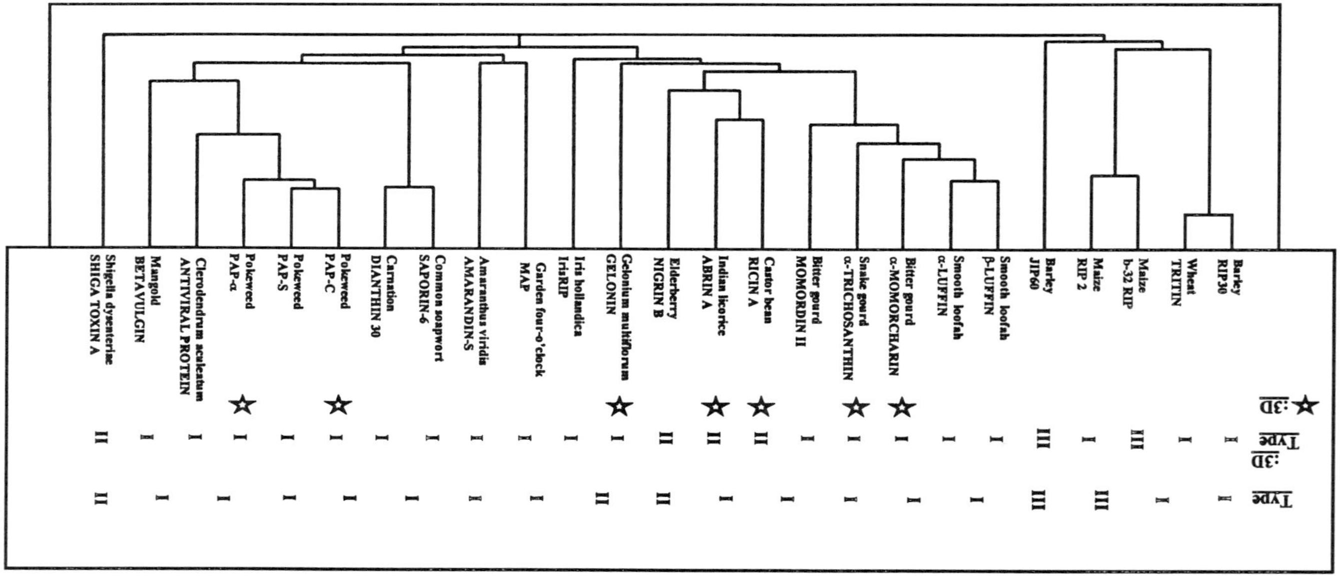

FIGURE 11.2 Phylogram derived from a Pileup comparison of RIP catalytic domain sequences. The RIP pileup is based on nonredundant and annotated Swiss-Prot (Release: 34) and GenBank (Update:103.0) accessions and use of the GCG (Ver.9.1) Software Package. No partial peptide sequences and only a single representative of known RIP isoforms are included. The accessions are (from top): RIP30: P22244; Tritin: D13795; B-32 protein: P25892; RIP 2: L26305; JIP60: X66376; β-Luffin: P22851; α-Luffin: Q00465; α-Momorcharin: P16094; α-Trichosanthin: P09989; Momordin II: P29339; Ricin A: A24041; Abrin A: S16022; Negrin B: U41299; Gelonin: P33186; IrisRIP: U78039; MAP: P21326; Amarandin-S: U70215; Saporin-6: P20656; Dianthin 30: P24476; PAP-C: P10297; PAP-S: P23339; PAP-a: Q03464; Antiviral protein: X96474; Betavulgin: X85967; Shiga Toxin A: P10149. The star indicates RIPs for which a three-dimensional structure has been determined. The RIP type (I, II, or III) for each accession is also indicated.

do not develop further, or organs which senesce rapidly. For example, the barley seed RIP 30, which accumulates to high levels in endosperm storage cells late in embryo development, could be involved in the programmed senescence of this storage organ at seed maturity.[19] Likewise, the vegetatively expressed, jasmonate-responsive JIP60 protein, which apparently is not secreted and accumulates in the cytosol, could have a role in the induction of leaf senescence by inhibiting conspecific translation after the protein has been processed into an active RIP.[24] The expression and presumed proteolytic activation of the JIP60 precursor accompanies the development of these senescence symptoms shown in Figure 11.3. However, neither JIP60 nor RIP 30 inhibit conspecific ribosomes *in vitro,* and if they do mediate barley PCD, they might do so only at high cellular concentrations. Therefore, there is no direct evidence linking the expression of most active RIPs with developmentally programmed or inducible plant PCD. However, as outlined below, there is good evidence that the pokeweed RIP, PAP, may function as a specialized suicide factor.

11.4 ROLE OF RIPs IN PLANT–VIRAL INTERACTIONS

Single chain RIPs, such as pokeweed antiviral protein, PAP, have been shown to possess antiviral activities toward viruses of both plant and animal origin.[7,30] Three distinct isoforms of PAP accumulate in the cell wall matrix of leaves and seeds and

FIGURE 11.3 Effect of methyl jasmonate on barley leaf segments. The phytohormone methyl jasmonate (5 μM) induces senescence symptoms in barley leaf segments over a period of 24 h. During this time, the Type-III RIP, JIP60, accumulates. It is unclear whether JIP60, induced under this artificial senescence, is ectopically expressed, or whether its accumulation reflects a role for the protein in defense, programmed cell death, or another physiological response.

can readily be obtained from water-macerated leaf tissue. Exogenous application of small amounts of PAP to the surface of plant leaves completely prevents mechanical transmission of unrelated viruses to several different host plants.[31] More important, expression of a PAP cDNA in transgenic tobacco and potato plants, which are thought to lack endogenous RIPs, showed resistance to a range of viruses.[7] In contrast to exogenously applied PAP, resistance was effective against both mechanical and aphid transmission of several potato viruses. However, the transgenic plants grew poorly and showed mottled and stunted phenotypes making them less useful for agricultural exploitation. Although the mechanisms of PAP-mediated resistance is not yet understood, one possibility is that PAP enters the host cells along with the virus and prevents translation of viral RNA by inactivating conspecific ribosomes. In this scenario, PAP is a secreted suicide factor whose re-entry into host cells is potentiated by cellular permeabilization by the pathogen. This is supported by comparing *in vitro* rRNA depurination activities of different RIPs on yeast and tobacco leaf ribosomes, as substrates.[32] All of the RIPs were active on yeast ribosomes whereas PAP and dianthin 32 were highly active on tobacco ribosomes. The same two proteins were also highly effective in inhibiting the formation of local lesions on tobacco leaves caused by tobacco mosaic virus. In contrast, tritin and barley RIP showed no depurination activities on tobacco ribosomes and they were ineffective in inhibiting virus spread. This supports the hypothesis that the antiviral activity of some RIPs, like PAP, work via inactivation of conspecific ribosomes.[32]

11.5 ROLE OF RIPs IN PLANT–FUNGAL INTERACTIONS

Pathogenesis related (PR) proteins have been shown to inhibit fungal growth *in vitro*[19,33] and to enhance protection against pathogens by over-expression in transgenic plants.[6,8] Moreover, combinations of PR proteins, such as the barley single-chain RIP 30, seed chitinase CHI 26, and (1-3)-b-glucanase BGL32, exhibited synergistic inhibition of the growth rate of *Trichoderma reesei*, a barley pathogen, and *Fusarium sporotrichioides,* a barley seed rot, in *in vitro*, microtiter plate assays. Mycelial growth of *T. reesei* was inhibited by the individual purified proteins from 20 to 90% by adding 1.5 μg/well. In contrast, growth was inhibited more than 95% when all three proteins were added at 0.25 μg/well. This synergistic inhibition was even more pronounced for *F. sporotrichioides*; moreover, the soil-borne pathogen *Rhizoctonia solani*, which infects a range of plant species, and *Botrytis cinerea* responded with a similar inhibition of growth when exposed to CHI 26, BGL 32, and RIP 30.[19] These results suggest that single-chain RIPs, like RIP 30, can be toxic to fungal cells. Furthermore, the synergistic inhibition of fungal growth by mixtures of CHI 26, BGL 32, and RIP 30 suggest that the toxicity of RIP 30 is enhanced when hyphal cell walls are permeabilized by the hydrolytic action of CHI 26 and BGL 32, resulting in an increased uptake of RIP 30 into the fungal cytoplasm (Figure 11.4).

These results prompted the production of transgenic tobacco plants expressing barley RIP 30 under control of the wound-inducible promoter of the potato *wun1* gene. These plants accumulated RIP mRNA and protein upon infection by *R. solani*. Moreover, *wun1*-RIP transgenic plants grown in soil inoculated with *R. solani*

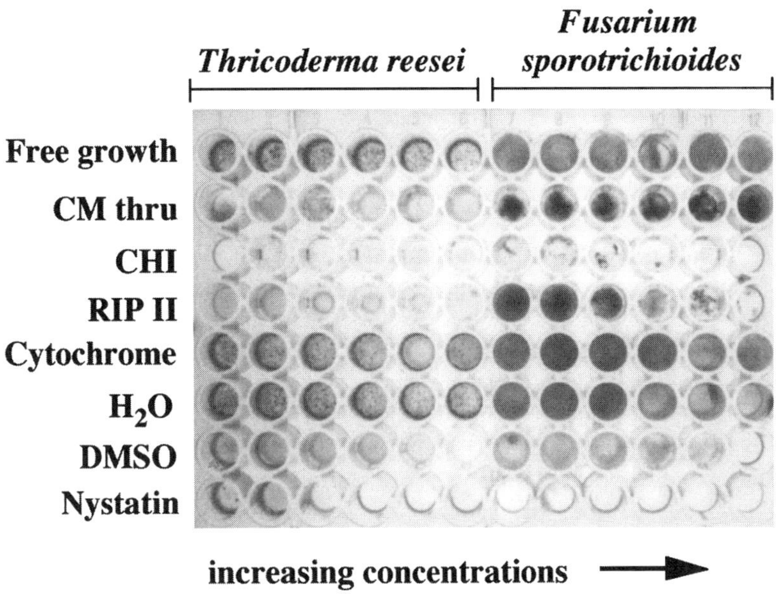

FIGURE 11.4 Purified barley RIP30 and chitinase inhibit fungal growth in a microtiter assay. Spores of two fungi, *T. reesei* and *F. sporotrichoides,* were added to potato dextrose media and allowed to germinate and grow for 4 days (free growth). Increasing amounts of a partially purified, unbound fraction (CM thru) of an ion exchange chromatography containing both chitinase and RIP 30 partially inhibited the growth of *T. reesei.* Increasing amounts of chitinase (CHI) and RIP30 (RIP II, isoform 2) further purified by FPLC inhibited the growth of both fungi. Negative controls include increasing amounts of Cytochrome C (same basic pI as chitinase and RIP30 isoform 2) and water. Positive controls include DMSO and nystatin, a fungal antibiotic. Quantitation of such assays are described in Leah et al.[19]

exhibited increased tolerance toward the pathogen.[6] To examine whether combinatorial expression of several antifungal proteins could lead to enhanced quantitative resistance, high-level expression of barley RIP 30 alone or in combination with barley CHI 26 and BGL 32 was obtained in transgenic tobacco plants using the strong constitutive 35S promoter. A construct was also tested containing a plant secretory peptide fused to RIP 30. Using a semiquantitative disease reduction assay based on a five-class, disease-severity-scale, a correlation between transgene RIP protein levels and disease reduction was observed.[8] In addition, plants expressing the signal peptide RIP form secreted the protein and showed a higher level of protection than transgenics expressing the cytosolic form of the RIP. Finally, combinatorial ectopic expression of the barley seed RIP 30/CHI 26 and RIP 30/BGL 32 from the 35S promoter in cotransformed tobacco plants exhibited significantly enhanced protection against fungal attack when compared with the protection levels obtained with corresponding isogenic lines expressing a single barley transgene to a similar level.[8] Clearly then, ectopic expression of some RIP forms can enhance protection for some fungal pathogens.

11.6 ENGINEERING CHIMERIC, TARGETED RIPs

Type-I RIPs which lack a lectin B-chain are generally less cytotoxic than Type-II molecules. Therefore, Type-I RIPs are thought not to be targeted to specific cells, but to enter cells by fluid-phase endocytosis. A model for improving the efficacy of toxin action comes from work on immunotoxin cells.[10-11] In general, these involve chimeric molecules, reminiscent of Type-II RIPs, in which a general toxin is fused to a single-chain antibody to potentiate cell-specific, toxin targeting. Following this paradigm, we are attempting to target the Type-I barley RIP 30 to fungal cells. This may be possible by constructing chimeras containing chitin-binding domains and plant secretion peptides. Chitin, an aminoglucoside polymer, is a major structural polysaccharide of fungal cell walls and is often exposed in growing hyphal tips.[34-35] Plants synthesize various chitin-binding proteins, such as Class 1 chitinases, wheat germ agglutinin (WGA), and other lectins.[36] All known chitin-binding proteins contain a conserved sequence of 30 to 43 amino acid residues, called the hevein domain, with several cysteines and glycines at conserved positions. Chimeric constructs were made by fusing PCR amplified chitin-binding domains from various genes to the *N*-terminus of the barley RIP 30 cDNA (Figure 11.5).

In initial experiments, these chimeric proteins were produced by *in vitro* transcription and translation in wheat germ, and their ability to bind chitin was tested by batch adsorption to chitin followed by SDS-PAGE. As expected, RIP 30 and signal peptide PRIP 30 did not bind chitin, while three chimeric proteins containing hevein domains did. To examine the ability of these proteins to inhibit translation on mammalian ribosome, *in vitro* protein synthesis assays with reticulocyte lysate were performed. All proteins, both RIP 30 and chimeras, were found to inhibit subsequent translation. These results indicate that the chimeric RIPs are enzymatically active and bind chitin. To begin to test the effects of the chimeras in plants, the chimeric RIP constructs were transferred to a vector carrying the 35S promoter for expression in transgenic *Arabidopsis*. Preliminary results show that high-level accumulation of the fusion proteins is not associated with phenotypic abnormalities. Antifungal assays will be pursued to examine the abilities of these transgenes to enhance the protection of the transgenic plants toward fungal pathogens.

11.7 PERSPECTIVES

Aspects of the biology of RIPs which require further attention include attempts to optimize their use as defensive transgenes, which depends on a better understanding of their *in vivo* functions in host plants. Future work on the endogenous functions of RIPs would benefit from genetic approaches in plants in which null or knockout RIP alleles are available — for example, B32. This kind of study could be extended to *Arabidopsis,* which has been reported to express a RIP,[37] although no clear RIP orthologous sequences have yet been uncovered in the Arabidopsis Genome Sequencing Program (http://genome-www.stanford.edu/Arabidopsis/).

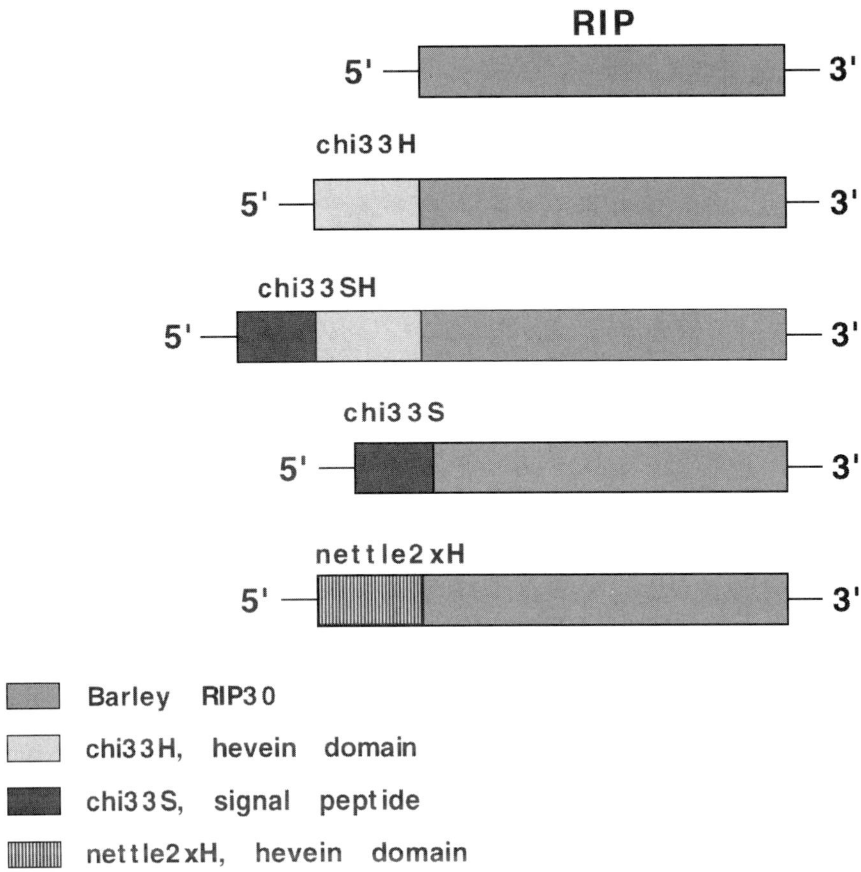

FIGURE 11.5 Chimeric RIP constructs. Transgene constructs made by fusing the barley chitinase 33 hevein domain and/or signal peptide,[38] or the nettle lectin double hevein domain[36] in frame to the *N*-terminus of the barley RIP 30.

REFERENCES

1. Endo, Y., Mitesui, K., Motezuki, M., and Tsurugi, K., The mechanism of action of ricin and related toxic lectins on eukaryotic ribosomes: the site and the characteristics of the modification in 28S ribosomal RNA caused by the toxins, *J. Biol. Chem.*, 262, 5908, 1987.
2. Endo, Y., Tsurugi, K., and Ebert, R. F., The mechanism of action of a barley toxin: a Type I ribosome inactivating protein with RNA *N*-glycosidase activity, *Biochim. Biophys. Acta*, 954, 224, 1988.
3. Hartley, M. R., Chaddock, J. A., and Bonness, M. S., The structure and function of ribosome-inactivating proteins, *Trends Plant Sci.*, 8, 254, 1996.
4. Stirpe, F., Barbieri, L., Battelli, M. G., Soria, M., and Lappi, D. A., Ribosome-inactivating proteins from plants: present status and future prospects, *Bio/Technology*, 10, 405, 1992.

5. Stirpe, F. and Huges, R. C., Specificity of ribosome inactivating proteins with RNA *N*-glycosidase activity, *Biochem. J.,* 262, 1001, 1992.
6. Logemann, J., Jach, G., Tommerup, H., Mundy, J., and Schell, J., Expression of a barley ribosome-inactivating protein leads to increased fungal protection in transgenic tobacco plants, *Bio/Technology,* 10, 305, 1991.
7. Lodge, J. K., Kaniewski, W. K., and Tumer, N. E., Broad-spectrum virus resistance in transgenic plants expressing pokeweed antiviral protein, *Proc. Natl. Acad. Sci. U.S.A.*, 90, 7089, 1993.
8. Jach, G., Gornhardt, B., Mundy, J., Logemann, J., Pinsdorf, E., Leah, R., Schell, J., and Maas, C., Enhanced quantitative resistance against fungal disease by combinatorial expression of different barley antifungal proteins in transgenic tobacco, *Plant J.,* 8, 97, 1995.
9. Barbieri, L., Battelli, M. G., and Stirpe, F., Ribosome-inactivating proteins from plants, *Biochim. Biophys. Acta,* 1154, 237, 1993.
10. Uhr, J. W., Immunotoxins: an overview, *Semin. Cell Biol.,* 2, 1, 1991.
11. Ovadia, M., Hager, C. C., and Oeltmann, T. N., An antimelanoma-barley ribosome inactivating protein conjugate is cytotoxic to melanoma cells *in vitro, Anticancer Research,* 10, 671, 1990.
12. Lamb, F. I., Roberts, L. M., and Lord, J. M., Nucleotide sequence of cloned complementary DNA coding for preproricin, *Eur. J. Biochem.*, 148, 265, 1985.
13. Roberts, L. M., Lamb, F. I., Pappins, D. J. C., and Lord, J. M., The primary sequence of *Ricinus communis* agglutinin: comparison with ricin, *J. Biol. Chem.*, 260, 15682, 1985.
14. Stirpe, F. and Barbieri, L., Ribosome-inactivating proteins up to date, *FEBS Lett.*, 195, 1, 1986.
15. Mundy, J., Leah, R., and Boston, R., Genes encoding ribosome inactivating proteins, *Plant Mol. Biol. Rep.,* 12, 60, 1994.
16. Maraganore, J. M., Joseph, M., and Bailey, M. C., Purification and characterization of trichosanthin: homology to the ricin A chain and implications as to mechanism of abortifacient activity, *J. Biol. Chem.*, 262, 11628, 1987.
17. Yeung, H. W., Li, W. W., Feng Z., Barbieri, L., and Stirpe, F., Trichosanthin, alpha-momorcharin and beta-momorcharin: identity of abortifacient and ribosome-inactivating proteins, *Int. J. Pept. Prot. Res.*, 31, 179, 1988.
18. Nolan, P. A., Garrison, D. A., and Better, M., Cloning and expression of a gene encoding gelonin, a ribosome-inactivating protein from *Gelonium multiflorum, Gene,* 134, 223, 1993.
19. Leah, R., Tommerup, H., Svendsen, I. J., and Mundy, J., Biochemical and molecular characterization of three seed proteins with anti-fungal properties, *J. Biol. Chem.*, 266, 564, 1990.
20. Lord, J. M., Synthesis and intracellular transport of lectin and storage protein precursors in endosperm from castor bean, *Eur. J. Biochem.*, 146, 403, 1985.
21. Tregear, J. W. and Roberts, L. M., The lectin gene family of *Ricnus coounis*: cloning of a functional ricin lectin and three lectin pseudogenes. *Plant Mol. Biol.*, 18, 515, 1992.
22. Bass, H. W., Webster, C., O'Brian, G. R., Roberts, J. K. M., and Boston, R. S., A maize ribosome-inactivating protein is controlled by the transcriptional activator opaque-2, *Plant Cell,* 4, 225, 1992.
23. Walsh, T. A., Morgan, A. E., and Hey, T.D., Characterization and molecular cloning of a proenzyme form of a ribosome-inactivating protein from maize, *J. Biol. Chem.*, 266, 23422, 1991.

24. Chaudhry, B., Muller-Uri, F., Cameron-Mills, G. S., Simpson, D., Skriver, K., and Mundy, J., The barley 60 kDa jasmonate-induced protein (JIP60) is a novel ribosome-inactivating protein, *Plant J.*, 6, 815, 1994.
25. Reinbothe, S., Reinbothe, C., Lehmann, J., Becker, W., Apel, K., and Parthier, B., JIP60, a methyl jasmonate-induced ribosome-inactivating protein involved in plant stress reactions, *Proc. Natl. Acad. Sci. U.S.A.*, 91, 7012, 1994.
26. Chen, J. K., Hung, C. H., Liaw, Y. C., and Lin, J. Y., Identification of amino acid residues of abrin-a A chain is essential for catalysis and reassociation with abrin-a B chain by site-directed mutagenesis, *Protein Eng.*, 10, 827, 1997.
27. Gibes, T., Iglesias, C. T. R., and Ferreras, J. M., RIP for viruses, *Nature*, 379, 777, 1996.
28. Greenberg, J. T., Programmed cell death in plant-pathogen interactions, *Annu. Rev. Plant Physiol. Plant Mol. Biol.*, 48, 525, 1997.
29. Olsnes, S. and Pihl, A., Toxic lectins and related proteins, in *Molecular Action of Toxins and Viruses*, Cohen, P. and van Heyringen, A., Eds., Elsevier, London, 1982, 121.
30. Hartley, M. R. and Lord, J. M., Ricin and related cytotoxin proteins, in *Biosynthesis and Manipulation of Plant Products*, Grierson, D., Ed., Chapman and Hall, Glasgow, 1993, 210.
31. Chen, Z. C., White, R. F., Antoniw J. F., and Lin, Q., Effect of pokeweed antiviral protein on the infection of plant viruses, *Plant Pathol.*, 40, 612, 1991.
32. Taylor, S., Massiah, A., Lomonossaff, G., Roberts, L. M., Lord, J. M., and Hartley, M., Correlation between the activities of five ribosome-inactivating proteins in depurination of tobacco ribosomes and inhibition of tobacco mosaic virus infection, *Plant J.*, 5, 827, 1994.
33. Roberts, W. K. and Selitrennikoff, C. P., Isolation and partial characterization of two antifungal protein from barley, *Biochem. Biophys. Acta*, 880, 161, 1986.
34. Boller, T. A., A method for the study of fungal growth inhibition by plant proteins, *FEMS Microbiol. Lett.*, 57, 61, 1990.
35. Bartnicki-Garciam, S., Cell wall chemistry, morphogenesis and taxonomy of fungi, *Annu. Rev. Microbiol.*, 87, 108, 1968.
36. Raikhel, N. V. and Lee, H. I., Structure and function of chitin-binding proteins, *Annu. Rev. Plant Physiol. Plant Mol. Biol.*, 44, 591, 1993.
37. Reinbothe, S., Mollenhauer, B., and Reinbothe, C., JIPs and RIPs: the regulation of plant gene expression by jasmonates in response to environmental cues and pathogens, *Plant Cell*, 6, 1197, 1994.
38. Leah, R., Skriver, K., Knudsen, S., Ruud-Hansen, J., Raikhel, N. V., and Mundy, J., Identification of an enhancer/silencer sequence directing the aleurone-specific expression of a barley chitinase gene, *Plant J.*, 4, 579, 1994.

12 Plant Defensins

Alexandre da Silva Conceição, and
Willem F. Broekaert

CONTENTS

12.1 Introduction 247
12.2 Protein Structure 247
12.3 Antimicrobial Activities 248
12.4 Structure-Activity Relationships 251
12.5 Mode of Action 252
12.6 Expression of Plant Defensin Genes 253
12.7 Contribution of Plant Defensins to Host Defense 255
Acknowledgments 256
References 257

12.1 INTRODUCTION

Plant defensins belong to a superfamily of antimicrobial peptides with a cysteine-stabilized $\alpha\beta$ motif that has representatives in vertebrates, invertebrates, and plants. This feature underscores the importance of these defense molecules as central components of a widespread defense strategy of multicellular organisms. In plants, antimicrobial plant defensins can be constitutively expressed during critical developmental stages and in particular cell types. Other plant defensin genes can be induced upon microbial infection. At least in *Arabidopsis*, induction of a plant defensin gene as a response to fungal attack appears to involve the synthesis of jasmonates as signaling molecules. The signal transduction pathway dependent on salicylic acid (SA) and leading to systemic acquired resistance (SAR) is not required for the induction of expression of this plant defensin gene. Similar to defensins from insects and mammals, the action of plant defensins on microorganisms involves changes in ion fluxes across the plasma membrane. However, a direct interaction with phospholipids of microbial membranes does not appear to occur, and the role of a receptor as an anchor for membrane insertion of plant defensins has been postulated.

12.2 PROTEIN STRUCTURE

Plant defensins are a family of small (about 5 kD), usually basic, peptides which are rich in disulfide-linked cysteine residues. The first plant defensins to be isolated

0-8493-0697-3/99/$0.00+$.50

were initially considered as a novel subgroup of thionins, which are also cysteine-rich peptides of about 5 kD in size, and were therefore called γ-thionins.[1,2] However, subsequent work has established that thionins and plant defensins are structurally unrelated.[3-5] To avoid confusion between both families of peptides, the name *plant defensins* was introduced in 1995 by Terras et al.[6] based on the structural and functional similarities with insect defensins (Figure 12.1), and this name is now widely accepted in recent literature (for other reviews, see references 7 and 8).

Plant defensins have been isolated from various monocot and dicot species[1,2,9-17] or identified via the sequencing of cDNA clones.[18-25] When the amino acid sequences of a series of plant defensins and plant defensin-like peptides originating from 10 different plant species are compared, it becomes clear that relatively few residues are conserved in all plant defensins (Figure 12.1). Indeed, the conserved residues are restricted to the eight cysteines and a glycine at position 48 (numbering as in Figure 12.1). These conserved residues are part of the cysteine-stabilized αβ motif characterized by the CXXXC, GXC, and CXC blueprint (with X being any amino acid). This motif occurs in various peptides of animal origin, and all of these peptides have at least two disulfide bridges linking the CXXXC portion embedded in an α-helix to the CXC portion contained within a β-strand (see below).[26]

The 3D structure of the plant defensins γ1-purothionin, γ1-hordothionin, ω-hordothionin, and Rs-AFP1 has been determined.[3-5] All these plant defensins share a similar fold which is dominated by a α-helix packed against an antiparallel triple-stranded β-sheet. For instance, the secondary structure elements of Rs-AFP1, a plant defensin from radish seeds, comprise an *N*-terminal β-strand from lys-2 to pro-7 (β-strand I), a α-helix spanning residues asn-18 to asn-27, a second β-strand from arg-32 to tyr-38 (β-strand II), and a C-terminal β-strand from his-43 and pro-50 (β-strand III). β-strand II and β-strand III are linked by a Type-VI β-turn (Figure 12.2). As predicted by the presence of the cysteine-stabilized αβ motif, β-strand III is connected by two disulfide bridges to the α-helix.[5]

Interestingly, the overall three-dimensional folding pattern of plant defensins corresponds very well to that of drosomycin, a pathogen-inducible antifungal peptide from the fruitfly *Drosophila melanogaster.*[27] Notably, drosomycin and Rs-AFP1 share 38% homologous residues.[28] Despite the fact that there is no significant primary structure homology between plant defensins and scorpion venom neurotoxins, many of these proteins also have adopted a fold with a α-helix and a triple-stranded β-sheet resembling that of plant defensins.[29-31] Finally, the plant defensin fold is also similar to that of the pathogen-inducible insect defensins exhibiting antibacterial activity, except that the latter features a double-stranded β-sheet instead of a triple-stranded one.[32] All proteins mentioned here are members of a superfamily of peptides containing the cysteine-stabilized αβ-motif.

12.3 ANTIMICROBIAL ACTIVITIES

The first plant defensins that were demonstrated to possess antifungal activity were the two plant defensin isoforms Rs-AFP1 and Rs-AFP2 isolated from radish seed.[10] To date, a whole range of plant defensins have been thoroughly analyzed to determine their antimicrobial activity spectrum.[11-13,15,16] Based on the antimicrobial effects

```
AA number              1      10       20       30        40       50       60       70       80       90
                       |   |   |   |   |   |   |*****  |   |  ***   |  ***   |   |   |   |   |   |   |
U18557 Rs-AFP1            QKLCERPSGT--WSGVCGNNNA---CKNQCINLEKAR----HGSCNYVFPAHK-CICYFPC
X97319 Rs-AFP3             KLCERSSGT--WSGVCGNNNA---CKNQCIRLEGAQ----HGSCNYVFPAHK-CICYFPC
       Dm-AFP1             ELCEKASKT--WSGNCGNTGH---CDNQCKSWEGAA----HGACHVRNGKH-MCFCYFNC
P20230 g-thionin           RICRRRSAG--FKGPCVSNKN---CAQVCMQ-EGWG----GGNCDG--PLRR-CKCMRRC
       o-thionin           RICTGKSQHHHF--PCISDKS---CARNCVS-EH-GAHWTAGYCH----LRR-CTCQREC
P21924 amyl inhibit 2     RVCMGKSAG--FKGLCMRDQN---CAQVCLQ-EGWG----GGNCDGV--MRQ-CKCIRQCW
Z11748 tob FST             RECKTESNT--FPGICITKPP---CRKACIS-EKFTD----GHCSKL--LRR-CLCTKPCVFDEKMIKTGAETLVEEAKTLAA
X53375 sunfl SF18       HGKICEEPSKT--WFGNCKDTDK---CDKRCIDWEGAK----HGACHQREAKHM-CFCYFDCDPQK
X84208 mustard mti-2   DSECLEEYGGDVGFPFCAPRIFPTICYTRC-R-ENKGAK--GGRCIWGEGTNVKCLCDY-CNDSPFDQILRGGI
X52874 rape Bp4A       QKKNCP[illegible]KIPIK-G-SYCAPTI----CLDMCKKQHGT-V----GSCAEKGF----CNCA--CK
X97054 PCP1         QEKKTDNQCNLPLINQSG-T-CAANQ----CQAACVKRHKDGV----GKCTTNPDKKMRCICFYLCPR
X75595 drosomycin           DCL--SGRYKG-P-CAVWDNET-CRRVCKE-EGRSS----GHCSPSLK----CWCE-GC
                                                 *****             ***          ***

Z11748                  ALLEEEIMDN
```

FIGURE 12.1 Alignment of plant defensin and plant defensin-like protein sequences. Amino acid sequences of the mature and proprotein domains of the plant defensin (-like) proteins as deduced from cDNAs or genes or as experimentally determined. Available accession numbers are indicated on the left. Gaps introduced for optimal alignment are indicated by hyphens. Cysteines are boxed. Cysteine-stabilized α-helix footprints are indicated by asterisks above and under the sequences. Rs-AFP1 and 3: *Raphanus sativus* antifungal proteins 1 and 3 (Terras et al.[6]); Dm-AFP1: *Dahlia merckii* antifungal protein 1 (Osborn et al.[15]); g-thionin: barley γ-thionin (Méndez et al.[2]); o-thionin: barley ω-thionin (Méndez et al.[47]); amyl inhibit 2: sorghum α-amylase inhibitor 2 (Bloch and Richardson[9]); tob FST: tobacco flower-specific thionin (Gu et al.[21]); sunfl SF18: sunflower cDNA SF18 (Domon et al.[48]); mustard mti2: *Sinapis alba* trypsin inhibitor 2 (Ceci et al.[49]); rape Bp4A: *Brassica napus* pollen-coating protein (Albani et al.[50]); PCP1: B. oleracea pollen-coating protein (Stanchev et al.[45]) and drosomycin: *Drosophila* antifungal protein (Fehlbaum et al.[28]).

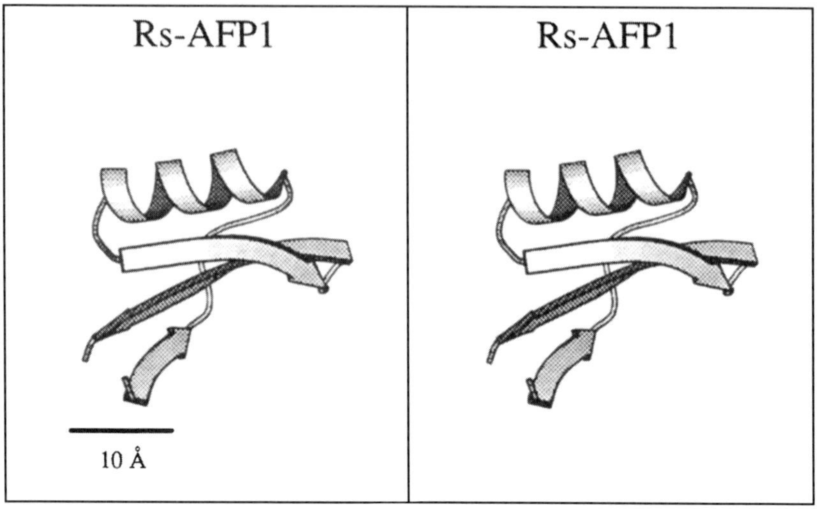

FIGURE 12.2 Stereoview of the 3D structure of the plant defensin Rs-AFP1 from radish seeds. (From Fant, F., Vranken, W., Broekaert, W. F., and Borremans, F., Determination of the three-dimensional solution structure of *Raphanus sativus* antifungal protein 1 by NMR, *J. Mol. Biol.*, 279, 257, 1998. With permission.)

observed on fungi, at least two groups of plant defensins can be distinguished. The "morphogenic" plant defensins cause reduced hyphal elongation with a concomitant increase in hyphal branching, whereas the "nonmorphogenic" plant defensins only slow down hyphal elongation but do not induce marked morphological distortions. The antifungal activity of plant defensins, whether morphogenic or not, is reduced by increasing the ionic strength of the fungal growth assay medium. This antagonism was found to be due to the cations, with divalent cations being at least one order of magnitude more potent than monovalent cations.[10-11] Ca^{2+}, Mg^{2+}, and Ba^{2+} are equally effective at inhibiting the antifungal activity of plant defensins (W.F. Broekaert, unpublished results).[11] The antagonistic effect of cations is strongly dependent on the fungus and on the plant defensin type.[15] When comparing different plant defensins for their relative activity on various fungi, marked differences in activity spectrum can be observed, and such differences are even accentuated upon increasing the ionic strength of the medium. For instance, in a potato dextrose broth medium supplemented with 1 m*M* $CaCl_2$ and 50 m*M* KCl, the plant defensin from horse chestnut (Ah-AMP1) is active against *Leptosphaeria maculans* with an IC_{50} value of 6 μg/ml, but it is basically inactive against *Fusarium culmorum.* In contrast, a plant defensin from *Heuchera sanguinea* (Hs-AFP1) is inactive against the former fungus but inhibits the latter with an IC_{50} value of 3 μg/ml.[15] Some plant defensins have also been reported to exert antibacterial activity such as a *Clitoria ternatea* plant defensin (Ct-AMP1), which is active against *Bacillus subtilis,*[15] and a potato tuber plant defensin, which inhibits *Ralstonia* (*Pseudomonas*) *solanacearum* and *Clavibacter michiganensis* (Moreno et al.[12]). None of the plant defensins has yet been found to have detrimental effects on cultured human cells or plant cells.[10]

12.4 STRUCTURE–ACTIVITY RELATIONSHIPS

As a first approach to identify protein domains which are important for the antifungal activity of plant defensins, De Samblanx et al.[33] synthesized a series of overlapping 15-mer peptides based on the Rs-AFP2 sequence. Rs-AFP2 is the most potent plant defensin from radish and is similar to Rs-AFP1, from which it differs by only two amino acid residues. Like Rs-AFP1, Rs-AFP2 is expressed in seeds of radish.[6,10] The analysis of the antimicrobial activities of each peptide showed that the region from cys-27 to cys-47 is very important for the Rs-AFP2 toxicity against fungi. Therefore, according to the 3D-structure model of Rs-AFP1, a homologous plant defensin,[5] the active Rs-AFP2-derived peptides map to the region extending from β-strand II to β-strand III. The important role of the interconnecting Type-VI β-turn was further demonstrated by the antifungal activity exerted by a cyclized synthetic peptide corresponding to this turn.[33]

Our laboratory has performed a mutational analysis of Rs-AFP2 for the precise identification of amino acid residues which are responsible for the antifungal activity of this plant defensin.[34] The Rs-AFP2 variants were generated by PCR site-directed mutagenesis and expressed in yeast. A couple of Rs-AFP2 variants, which had antifungal activities significantly different from the wild-type protein, had their circular dichroism spectra analyzed to show that no major alterations of backbone secondary structure elements had occurred.

By modification of particular amino acid residues to introduce sequence similarities to SIα2, a plant defensin from sorghum seeds which appears to lack antifungal activity,[15] De Samblanx et al.[34] produced a first series of Rs-AFP2 variants. In addition, another group of variants was produced to replace certain Rs-AFP2 residues by the basic amino acid arginine.[34] The rationale for these substitutions was the correlation between a reduction in antifungal activity and a decrease in net positive charges while comparing Rs-AFP1 and Rs-AFP2.[10] The mutational analysis resulted in the identification of two subsites important for the antifungal activity of Rs-AFP2 against the fungus *Fusarium culmorum*: (i) the protruding domain consisting of the Type-VI β-turn and the first part of the β-strand III (e.g., tyr-38, phe-40, pro-41, ala-42, lys-44, and ile-46); (ii) residues in the loop connecting β-strand I and the α-helix (e.g., thr-10 and ser-12) and contiguous residues on the α-helix and the last part of β-strand III (e.g., leu-28 and phe-49). In another study,[33] the important role of tyr-38 has been further confirmed by the decreased antifungal activity of a cyclized synthetic peptide containing an amino acid substitution for the tyrosine residue. The results from De Samblanx et al.[33,34] showed that the Rs-AFP2 active regions form an elongated patch along the same side of the molecule. Furthermore, the active subsites comprise sequence regions which strongly differ from those of plant defensins closely related to SIα2 (e.g., γ1-purothionin and γ1-hordothionins) that have no antifungal activity.

The contribution of basic amino acids to the antifungal strength of plant defensins has also been demonstrated by point mutagenesis. The substitution of the neutral residues gly-9 and val-39 by arginines (G9R and V39R variants, respectively) increased the toxicity of Rs-AFP2 to *F. culmorum,* while a neutral glutamine at lys-

44 reduced it.[34] Introduction of arginines at other positions did not result in enhanced activity to *F. culmorum,* and sometimes it even decreased the activity.

A more detailed analysis of Rs-AFP2(V39R) and Rs-AFP2 was made using a set of several phytopathogenic fungi.[34] The difference between the antifungal activity of the Rs-AFP2 variant and that of Rs-AFP2 was found to be dependent on the fungal species. On three fungi (*F. culmorum, Nectria haematococca,* and *Verticillium dahliae*), Rs-AFP2(V39R) was more active than Rs-AFP2. On the other hand, Rs-AFP2(V39R) either had antifungal activity similar to that of Rs-AFP2 or was less potent against other fungi. Therefore, these changes in the range of fungal species which are affected by Rs-AFP2 upon site-directed mutagenesis confirmed the existence of highly specific recognition mechanisms between plant defensins and their fungal target sites.

Interestingly, changes in the amount of salts in the growth medium had a great effect on the antifungal activity of both Rs-AFP2 variants and the wild-type protein.[34] In general, point mutations that resulted in loss of antifungal activity were easier to identify on a medium supplemented with $CaCl_2$ and KCl than in a low ionic strength medium. Furthermore, the differences in potency between Rs-AFP2(V39R) and Rs-AFP2 against several fungi were always more pronounced in media with added salts than in low ionic strength conditions. In this particular case, De Samblanx et al.[34] suggested that the introduction of an extra charged amino acid residue may reinforce ionic interactions. Therefore, the ionic strength effect of the medium indicates that not only are stereospecific interactions between particular amino acids of plant defensins and target sites on fungal hyphae important, but that electrostatic interactions also contribute to the antifungal effect.

12.5 MODE OF ACTION

The mechanisms through which plant defensins inhibit fungal growth are just starting to be investigated. A recent study by Thevissen et al.[35] showed that plant defensins induce a rapid Ca^{2+} uptake and K^+ efflux across fungal membranes, with the concomitant alkalinization of the medium. These results were obtained by treatment of *Neurospora crassa* hyphae with either Rs-AFP2 or Dm-AMP1, a morphogenic and a nonmorphogenic plant defensin, respectively. Rs-AFP2 was also shown to stimulate Ca^{2+} uptake by *Fusarium culmorum.*[34] Interestingly, an arginine substitution variant of Rs-AFP2 with enhanced antifungal activity, Rs-AFP2(V39R), caused an even higher increase of Ca^{2+} influx. On the other hand, a variant virtually devoid of antifungal activity, Rs-AFP2(Y38G), did not stimulate Ca^{2+} uptake.[34]

Rs-AFP2 and Dm-AMP1, in contrast to a thionin from barley seeds (α-hordothionin), did not form ion-permeable pores in artificial phospholipid membranes.[35] Consistent with these observations, a plant defensin from potato tubers did not cause a significant permeabilization of artificial phospholipid vesicles for entrapped fluorescent dye molecules.[36] Therefore, since membrane permeabilization through direct protein–lipid interactions does not appear to be the primary cause of fungal growth inhibition, the inhibitory effects of plant defensins might involve a receptor-mediated mechanism.

To further analyze the binding properties of plant defensins, Thevissen et al.[37] radiolabeled Hs-AFP1, a plant defensin isolated from *Heuchera sanguinea* and known to have morphogenic effects on fungal growth similar to those of Rs-AFPs. High-affinity binding sites with a K_d of about 30 nM were identified on both the intact hyphae and membrane microsomes of *Neurospora crassa.* This observation constitutes the first circumstantial evidence for the participation of a membrane receptor. Furthermore, binding of Hs-AFP1 was reduced by structurally related plant defensins as the possible result of a competition for a common binding site. Interestingly, the Rs-AFP2(Y38G) variant with reduced antifungal activity that was unable to induce ion fluxes in fungal hyphae,[34] was also unable to compete for the Hs-AFP1 binding sites.[37] These results suggest a causal link between receptor binding, ion fluxes, and antifungal activity of plant defensins.

Many aspects of the mode of action of plant defensins remain to be investigated. The nature of the putative membrane receptor is not known at present. It is also unclear how receptor binding and generation of ion fluxes are interrelated. One outstanding possibility is that the membrane receptor allows plant defensins to insert in the membrane and to form ion-permeable pores. Another possibility is that binding of the plant defensins to their receptors activates endogenous signal transduction components, which in turn affect the activity of endogenous ion channels or ion transporters. Finally, the actual cause of growth inhibition needs to be further explored. As controlled Ca^{2+}-influx is essential for polar elongation of tip-growing cells,[38] it can be envisaged that increased Ca^{2+}-influx caused by plant defensins somehow interferes with growth-sustaining processes in fungal hyphae.

12.6 EXPRESSION OF PLANT DEFENSIN GENES

Plant defensins have now been isolated from over 20 different plant species, and most probably occur in most, if not all, plant species. Particular plant defensins have been found to be expressed in nearly every plant organ, including leaves,[6,13] roots,[25] tubers,[12] floral organs,[12,13,21-24] fruits,[16,20] and seeds.[1,2,9-11,14,15] In *Arabidopsis*, five different plant defensin genes (*PDF1.1, PDF1.2, PDF2.1, PDF2.2,* and *PDF2.3*) have been identified via random cDNA sequencing. Examination of their expression has revealed that they all have a distinct organ-specific expression pattern. In healthy *Arabidopsis* plants, *PDF1.1* is expressed exclusively in seeds and seed-bearing siliques,[17] *PDF2.1* in roots, siliques, and seeds, *PDF2.2* in all organs except stems and seeds,[38a] and *PDF2.3* in all organs except roots.[39] Hence, at least one member of the plant defensin gene family is constitutively expressed in every *Arabidopsis* organ. The gene *PDF1.2,* on the other hand, is not (or only weakly) expressed constitutively but is strongly induced in leaves upon attack by fungal pathogens (see below).[17]

Localization of constitutively expressed plant defensins in their tissue of origin by immunological or *in situ* hybridization techniques has, in many of the cases reported so far, revealed that they accumulate preferentially in peripheral cell layers. In radish seeds, the highest concentrations of plant defensins occur in the outer cell wall lining the epidermis of cotyledons, hypocotyl, and endosperm.[6] Unopened tobacco flower buds transcribe a plant defensin gene in the epidermis of the adaxial surface of petals and in the peripheral cell layers of the style, the ovary, the stamen filaments, and anthers.[21] Likewise, in potato tubers, highest transcription of a plant defensin gene is in the epidermis and in leaf primordia.[12] In healthy sugar beet leaves, on the other hand, plant defensins were located predominantly in the xylem, in stomatal cells, and cell walls lining the substomatal cavities.[13] The preferential location of plant defensins in peripheral cell layers is consistent with a role in protection of the organ against microbial challenge. Furthermore, their occurrence in stomatal cells and cells surrounding the substomatal cavity also fits in with a role in defense, since stomata are well-known entry points for fungi.

In at least four different plant species, namely pea,[20] tobacco,[21] radish,[6] and *Arabidopsis*,[17] plant defensins have been shown to be induced upon fungal infection of vegetative tissues. In contrast, no overall increase in plant defensin content was observed in sugar beet leaves infected with the fungal pathogen *Cercospora beticola.* However, in this case, infection triggered the appearance of extracellular globular bodies containing plant defensins in the leaf mesophyll cells surrounding the infection zone, whereas such bodies were apparently absent from the corresponding tissue in healthy leaves.[13] At least in radish and in *Arabidopsis,* the induction by fungal pathogens is systemic in the sense that it can be detected not only in infected leaves but also in uninfected leaves.[6,17] Some constitutively expressed plant defensin genes have also been reported to be downregulated upon fungal infection. This is the case for the plant defensin gene *Pth-St1* in potato leaves infected with the bacterium *Pseudomonas syringae*,[12] the gene *SPI1* in Norway spruce roots infected with the fungus *Pythium dimorphum,*[25] and *PDF2.2* in Arabidopsis leaves infected with the fungus *Alternaria brassicicola.*[38a] Downregulation of these constitutively expressed genes may be the result of secretion by the pathogen of suppressor molecules, although this aspect has not yet been explored in the case of plant defensin genes.

To study the signal transduction pathway which controls the expression of plant defensins upon pathogen attack, the induction of *PDF1.2* from *Arabidopsis* has been used as a model.[17] Gene expression analysis after both chemical treatments of wild-type plants and fungal infections of Arabidopsis mutants and transgenic lines indicated that induction of *PDF1.2* is neither mediated by SA nor dependent on other signaling components of the SAR pathway in which SA plays a central role. On the other hand, the exogenous application of either methyl jasmonate (MeJA) or ethylene strongly induced the expression of *PDF1.2.* Epple et al.[39] similarly showed that *PDF1.2,* but not *PDF2.3,* is induced upon treatments with MeJA, silver nitrate, and by inoculation with plant pathogenic fungi. Measurements of jasmonic acid (JA) showed that the endogenous levels of this MeJA analog increased simultaneously to the systemic accumulation of plant defensins in the leaves of Arabidopsis plants which had been infected with *A. brassicicola.*[17] Interestingly, leaves from the *Arabidopsis acd2* mutant have mRNA levels of *PDF1.2* and jasmonic acid contents much higher than those detected in wild-type plants.[17] The *Arabidopsis acd2* mutant was isolated in a genetic screening for plants which spontaneously develop necrotic lesions in their leaves.[40] As previously reported by other authors, Penninckx et al.[17] also observed that the *acd2* mutant accumulates high levels of the PR-1 transcript. Although the increases in gene expression and JA levels were observed throughout the plant, necrotic rather than asymptomatic leaves displayed the most significant changes.

According to Epple et al.,[39] ethylene released by hydrolysis of ethephon does not induce the expression of *PDF1.2.* However, in the work of Penninckx et al.,[17] ethylene, paraquat, and rose bengal all acted as inducers of this same plant defensin. Experimental differences, due for instance to ethylene concentrations, growth conditions, and *Arabidopsis* ecotypes could explain the contradiction between the results of Epple et al.[39] and Penninckx et al.[17] The use of *Arabidopsis* mutants which are insensitive either to MeJA (*coi1* plants),[41] or ethylene (*ein2* and *etr1* plants)[42,43] confirmed that induction of *PDF1.2* involves a pathway in which jasmonate and ethylene act as signaling components. Recent work in our laboratory has shown that the jasmonate and ethylene response pathways act in parallel and not in sequence for the activation of *PDF1.2*.[43a]

The induction patterns of *PDF1.2* and the *Arabidopsis* thionin named *THI2.1*[39,44] are very similar. These observations suggest that *PDF1.2* and *THI2.1* may be active compounds of the same plant defense pathway which, in contrast to the induction of PR-1 proteins, is independent of signaling events mediated by salicylate.

12.7 CONTRIBUTION OF PLANT DEFENSINS TO HOST DEFENSE

It has been proposed that seed-expressed plant defensins from radish play an important role in the protection of seeds or seedlings against invasion by soil-borne fungi.[6] This hypothesis is based on the observation that radish seeds germinating on a medium supporting the growth of a fungal colony cause a growth inhibition halo which can be mimicked by application on the medium of drops containing as little as 1 μg of the purified radish plant defensins Rs-AFP1 or Rs-AFP2. Seeds that were kept dormant by external application of abscisic acid did not produce the inhibition halo unless their seed coats were mechanically perforated. Analysis of the imbibition solution of seeds with a mechanically incised seed coat, revealed that plant defensins accounted for 30% of released proteins, although Rs-AFPs are minor proteins in the seed (0.5% of total seed proteins). The amount of plant defensins released from a single seed was estimated to be at least 1 μg, which is the amount of peptide required to mimic the inhibition halo formed around a germinating seed.[6] All of these experiments indicate that plant defensins are released from radish seeds when the seed coat is perforated (either by the radicle of the germinating embryo under natural conditions or artificially with the aid of a scalpel), and moreover, that the released amounts are sufficient to create a zone around the seeds in which fungal growth is suppressed. Hence, these findings strongly suggest that plant defensins play a role in the protection of seedling tissues during the early stage of emergence and thus may contribute to the enhancement of seedling survival rates. The simple fact that chemical fungicides are commonly used for coating crop seeds to increase seedling stand illustrates that soil-borne or seed-borne fungi form a considerable threat to germinating seeds.

Indirect evidence also suggests that plant defensins contribute to the resistance of vegetative tissues to fungal attack. As mentioned above, the jasmonate-insensitive *coi1* and ethylene-insensitive *ein2 Arabidopsis* mutants are impaired in their ability to induce the plant defensin gene *PDF1.2* upon fungal attack.[17] Interestingly, both the *coi1* and *ein2* mutants were found to be highly susceptible to infection by the soft rot fungus *Botrytis cinerea,* whereas mutants affected by the salicylate-dependent defense pathway were as resistant as wild-type plants.[44a] However, as other defense-related genes, including a chitinase and a PR-4-type gene, are induced coordinately with *PDF1.2* along the same pathway, the reduced resistance of *coi1* and *ein2* mutants to *B. cinerea* cannot be attributed solely to *PDF1.2.*

To investigate more directly the ability of plant defensins to contribute to disease resistance in plants, tobacco plants were transformed with a chimeric gene construct consisting of the coding region of the radish Rs-AFP2 plant defensin cDNA driven by a constitutive promoter. T2-generation tobacco plants expressing Rs-AFP2 at a level of 0.2% of total leaf protein showed a sevenfold reduction in lesion size upon infection with the foliar fungal pathogen *Alternaria longipes* relative to untransformed plants.[6] Hence, constitutive high-level expression of at least some plant defensin genes can confer enhanced resistance to particular microbial diseases.

Although there is evidence that at least some types of plant defensins play a role in the protection of plant tissues from microbial attack, it is highly unlikely that such a role can be coined to all plant defensin types. Indeed, the very high sequence divergence within this peptide family suggests that plant defensins can exert various biological activities, some of which remain to be identified. Some plant defensins, such as SIα2 from sorghum seeds, apparently lack antimicrobial activity. However, SIα2 has been demonstrated to inhibit α-amylases from insects[9] and may therefore interfere with digestive processes in phytofagous insects and thus act as antifeedants against these animals. Interestingly, plant defensins with proven antimicrobial properties, such as Rs-AFP2, Dm-AMP1, or Hs-AFP1, all lacked α-amylase inhibitor activity.[15] Particular plant defensin types may also play a role in developmental processes. Genes from *Brassica oleracea* encoding peptides with the cysteine-stabilized β-motif have been found to be expressed exclusively in microspores,[45] and the peptides themselves were found to be present in the pollen coating and to interact with the stigmatically expressed SLG glycoprotein.[46] Hence, in this case a specific role in pollen–pistil interactions appears to be more likely.

ACKNOWLEDGMENTS

The authors acknowledge financial support by grants from the Commission of the European Community (AIR2-CT94-1356) and from the Fonds voor Wetenschappelijk Onderzoek-Vlaanderen (G.0218.97). AdSC is a recipient of a postdoctoral fellowship from the Research Council of the Katholieke Universiteit Leuven. The authors wish to thank Dr. Franky Fant for the illustration of the 3D-structure of Rs-AFP1.

REFERENCES

1. Colilla, F. J., Rocher, A., and Méndez, E., γ-purothionins: amino acid sequence of two polypeptides of a new family of thionins from wheat endosperm, *FEBS Lett.*, 270, 191, 1990.
2. Méndez, E., Moreno, A., Collila, F., Pelaez, F., Llamas, G. G., Méndez, R., Soriano, F., Salinas, M., and de Haro, C., Primary structure and inhibition of protein synthesis in eukaryotic cell-free system of a novel thionin, γ-hordothionin, from barley endosperm, *Eur. J. Biochem.*, 194, 533, 1990.
3. Bruix, M., Jiménez, M. A., Santoro, J., González, C., Colilla, F. J., Méndez, E., and Rico, M., Solution structure of γ1-H and γ1-P thionins from barley and wheat endosperm determined by ^{1}H-NMR: a structural motif common to toxic arthropod proteins, *Biochemistry*, 132, 715, 1993.
4. Bruix, M., González, C., Santoro, J., Soriano, F., Rocher A., Méndez, E., and Rico, M., ^{1}H-NMR studies on the structure of a new thionin from barley endosperm, *Biopolymers*, 36, 751, 1995.
5. Fant, F., Vranken, W., Broekaert, W. F., and Borremans, F., Determination of the three-dimensional solution structure of *Raphanus sativus* antifungal protein 1 by 1H NMR, *J. Mol. Biol.*, 279, 257, 1998.
6. Terras, F. R. G., Eggermont, K., Kovaleva, V., Raikhel, N. V., Osborn, R. W., Kester A., Rees, S. B., Torrekens, S., Van Leuven, F., Vanderleyden, J., Cammue, B. P. A., and Broekaert, W. F., Small cysteine-rich antifungal proteins from radish (*Raphanus sativus* L.): their role in host defense, *Plant Cell,* 7, 573, 1995.
7. Broekaert, W. F., Terras, F. R. G., Cammue, B. P. A., and Osborn, R. W., Plant defensins: novel antimicrobial peptides as components of the host defense system, *Plant Physiol.*, 108, 1353, 1995.
8. Broekaert, W. F., Cammue, B. P. A., De Bolle, M. F. C., Thevissen, K., De Samblanx, G. W., and Osborn, R. W., Antimicrobial peptides from plants, *Crit. Rev. Plant Sci.*, 16, 297, 1997.
9. Bloch, C. and Richardson, M., A new family of small (5 kDa) protein inhibitors of insect α-amylases from seeds of sorghum (*Sorghum bicolor* (L) Moench) have sequence homologies wheat γ-purothionins, *FEBS Lett.*, 279, 101, 1991.
10. Terras, F. R. G., Schoofs, H. M. E., De Bolle, M. F. C., Van Leuven, F., Rees, S. B., Vanderleyden, J., Cammue, B. P. A., and Broekaert, W. F., Analysis of two novel classes of antifungal proteins from radish (*Raphanus sativus* L.) seeds, *J. Biol. Chem.*, 267, 15301, 1992.
11. Terras, F. R. G., Torrekens, S., Van Leuven, F., Osborn, R. W., Vanderleyden, J., Cammue, B. P. A., and Broekaert, W. F., A new family of basic cysteine-rich plant antifungal proteins from Brassicaceae species, *FEBS Lett.*, 316, 233, 1993.
12. Moreno, M., Segura, A., and García-Olmedo, F., Pseudothionin, a potato peptide active against potato pathogens, *Eur. J. Biochem.*, 223, 135, 1994.
13. Kragh, K. M., Nielsen, J. E., Nielsen, K. K., Dreboldt, S., and Mikkelsen, J. D., Characterization and localization of new antifungal cysteine-rich proteins from *Beta vulgaris, Mol. Plant-Microbe Interact.*, 8, 424, 1995.
14. Nitti, G., Orrù, S., Bloch, C., Jr., Morhy, L., Marino, G., and Pucci, P., Amino acid sequence and disulphide-bridge pattern of three γ-thionins from *Sorghum bicolor, Eur. J. Biochem.*, 228, 250, 1995.

15. Osborn, R. W., De Samblanx, G. W., Thevissen, K., Goderis, I., Torrekens, S., Van Leuven, F. Attenborough, S., Rees, S. B., and Broekaert, W. F., Isolation and characterization of plant defensins from seeds of Asteraceae, Fabaceae, Hippocastanaceae and Saxifragaceae, *FEBS Lett.,* 368, 257, 1995.
16. Meyer, B., Houlné, G., Pozueta-Romero, J., Schantz, M. L., and Schantz, R., Fruit-specific expression of a defensin-type gene family in bell pepper. Upregulation during ripening and upon wounding, *Plant Physiol.,* 112, 615, 1996.
17. Penninckx, I. A. M. A., Eggermont, K., Terras, F. R. G., Thomma, B. P. H. J., De Samblanx, G. W., Buchala, A., Métraux, J. P., Manners, J. M., and Broekaert, W. F., Pathogen-induced systemic activation of a plant defensin gene in *Arabidopsis* follows a salicylic acid-independent pathway, *Plant Cell,* 8, 2309, 1996.
18. Stiekema, W. L., Heidekamp, F., Dirkse, W. G., Van Beckum, J., De Haan, P. Ten Bosch, C., and Louwerse, J. D., Molecular cloning and analysis of four potato tuber mRNAs, *Plant Mol. Biol.,* 11, 255, 1988.
19. Ishibashi, N., Yamauchi, D., and Minamikawa, T., Stored mRNA in cotyledons of *Vigna unguiculata* seeds: nucleotide sequence of cloned cDNA for a stored mRNA and induction of its synthesis by precocious germination, *Plant Mol. Biol.,* 15, 59, 1990.
20. Chiang, C. C. and Hadwiger, L. A., The *Fusarium solani*-induced expression of a pea gene family encoding high cysteine content proteins, *Mol. Plant-Microbe Interact.,* 4, 324, 1991.
21. Gu, Q., Kamata, E. E., Morse, M. J., Wu, H. M., and Cheung, A. Y., A flower-specific cDNA encoding a novel thionin in tobacco, *Mol. Gen. Genet.,* 234, 89, 1992.
22. Karunanandaa, B., Singh, A., and Kao, T. H., Characterization of a predominantly pistil-expressed gene encoding a γ-thionin-like protein of *Petunia inflata, Plant Mol. Biol.,* 26, 459, 1994.
23. Milligan, S. B. and Gasser, C. S., Nature and regulation of pistil-expressed genes in tomato, *Plant Mol. Biol.,* 28, 691, 1995.
24. Brandstädter, J., Roßlach, C., and Theres, K., Expression of genes for a defensin and a proteinase inhibitor in specific areas of the shoot apex and the developing flower in tomato, *Mol. Gen. Genet.,* 252, 146, 1996.
25. Sharma, P. and Lönneborg, A., Isolation and characterization of a cDNA encoding a plant defensin-like protein from roots of Norway spruce, *Plant Mol. Biol.,* 31, 707, 1996.
26. Cornet, B., Bonmatin, J.-M., Hétru, C., Hoffmann, J. A., Ptak, M., and Vovelle, F., Refined three-dimensional solution structure of insect defensin A, *Structure,* 3, 435, 1995.
27. Landon, C., Sodano, P., Hétru, C., Hoffman, C., and Ptak, M., Solution structure of drosomycin, the first inducible antifungal protein from insects, *Protein Sci.,* 6, 1878, 1997.
28. Fehlbaum, P., Bulet, P., Michaut, L., Lageux, M., Broekaert, W. F., Hétru, C., and Hoffmann, J. A., Insect immunity. Septic injury of *Drosophila* induces the synthesis of a potent antifungal peptide with sequence homology to plant antifungal proteins, *J. Biol. Chem.,* 269, 33159, 1994.
29. Darbon, H., Weber, C., and Braun, W., Two-dimensional ^{1}H nuclear magnetic resonance study of AaH IT, an anti-insect toxin from the scorpion *Androctonus australis* Hector. Sequential resonance assignments and folding of the polypeptide chain, *Biochemistry,* 30, 1836, 1991.

30. Fontecilla-Camps, J. C., Habersetzer-Rochat, C., and Rochat, H., Orthorombic crystals and three-dimensional structure of the potent toxin II from the scorpion *Androctonus australis* Hector, *Proc. Natl. Acad. Sci. U.S.A.,* 85, 7443, 1988.
31. Landon, C., Cornet, B., Bonmatin, J. M., Kopeyan, C., Rochat, H., Vovelle, F., and Ptak, M., ^{1}H-NMR-derived secondary structure and the overall fold of the potent anti-mammal and anti-insect toxin III from the scorpion *Leiurus quinquestriatus, Eur. J. Biochem.,* 236, 395, 1996.
32. Bonmatin, J.-M., Bonnat, J.-L., Gallet, X., Vovelle, F., Ptak, M., Reichhart, J. M., Hoffmann, J. A., Keppi E., Legrain, M., and Achstetter, T., Two-dimensional ^{1}H-NMR study of recombinant insect defensin A in water: resonance assignments, secondary structure and global folding, *J. Biomol. NMR,* 2, 235, 1992.
33. De Samblanx, G. W., Fernandez del Carmen, A., Sijtsma, L., Plasman, H. H., Schaaper, W. M., Posthuma, G. A., Fant, F., Meloen, R. H., Broekaert, W. F., and van Amerongen, A., Antifungal activity of synthetic 15-mer peptides based on the Rs-AFP2 (*Raphanus sativus* antifungal protein 2) sequence, *Peptide Res.,* 9, 262, 1996.
34. De Samblanx, G. W., Goderis, I. J., Thevissen, K., Raemaeker, S. R., Fant, F., Borremans, F., Acland, D., Osborn, R. W., Patel, S., and Broekaert, W. F., Mutational analysis of a plant defensins from radish (*Raphanus sativum*) reveals two adjacent sites important for antifungal activity, *J. Biol. Chem.,* 272, 1171, 1997.
35. Thevissen, K., Ghazi, A., De Samblanx, G. W., Brownlee, C., Osborn, R. W., and Broekaert, W. F., Fungal membrane responses induced by plant defensins and thionins, *J. Biol. Chem.,* 271, 15018, 1996.
36. Caaveiro, J. M. M., Molina, A., Gonzales-Manas, J. M., Rodriguez-Palenzuela, P., García-Olmedo, F. and Goni, F. M., Differential effects of five types of antipathogenic plant peptides on model membranes, *FEBS Lett.,* 410, 338, 1997.
37. Thevissen, K., Osborn, R. W., Acland, D. P., and Broekaert, W. F., Specific, high affinity binding sites for an antifungal plant defensin on *Neurospora crassa* hyphae and microsomal membranes, *J. Biol. Chem.,* 272, 32176, 1997.
38. Cai, G., Moscatelli, A., and Cresti, M., Cystoskeletal organization of pollen tube growth, *Trends Plant Sci.,* 2, 86, 1997.
38a. Thomma, B. P. H. J. and Broekaert, W. F., Tissue-specific expression of plant defensin genes PDF2.1 and PDF2.2 in *Arabidopsis thaliana, Plant Physiol. Biochem.,* 36, 533, 1998.
39. Epple, P., Apel, K., and Bohlmann, H., ESTs reveal a multigene family for plant defensins in *Arabidopsis thaliana, FEBS Lett.,* 400, 168, 1997.
40. Greenberg, J. T., Guo, A., Klessig, D. F., and Ausubel, F. M., Programmed cell death in plants: a pathogen triggered response activated coordinately with multiple defense functions, *Cell,* 77, 551, 1994.
41. Feys, B. J. F., Benedetti, C. E., Penfold, C. N., and Turner, J. G., Arabidopsis mutants selected for resistance to the phytotoxin coronatine are male sterile, insensitive to methyl jasmonate, and resistant to a bacterial pathogen, *Plant Cell,* 6, 751, 1994.
42. Guzmán, P. and Ecker, J. R., Exploiting the triple response of Arabidopsis to identify ethylene-related mutants, *Plant Cell,* 2, 513, 1990.
43. Chang, C., Kwok, S. F., and Bleecker, A. B., Arabidopsis ethylene-response gene *etr1*: similarity of product to two-component regulators, *Science*, 262, 539, 1993.
43a. Penninckx, I. A. M. A., Thomma, B. P. H. J., Buchala, A., Métraux, J.-P., Broekaert, W. F., Concomitant activation of jasmonate and ethylene responses is required for induction of a plant defensin gene in Arabidopsis, *Plant Cell,* 10, 2130, 1998.

44. Epple, P., Apel, K., and Bohlmann, H., An *Arabidopsis thaliana* thionin gene is inducible via a signal transduction pathway different from that for pathogenesis-related proteins, *Plant Physiol.,* 109, 813, 1995.
44a. Thomma, B. P. H. J., Eggermont, K., Penninckx, I. A. M. A., Mauch-Mani, B., Vogelsang, R., Cammue, B. P. A., and Broekaert, W. F., Separate jasmonate-dependent and salicylate-dependent defense response pathways in Arabidopsis are essential for resistance to distinct microbial pathogens, *Proc. Natl. Acad. Sci. U.S.A.,* 95, 15107, 1998.
45. Stanchev, B. S., Doughty, J., Scutt, C. P., Dickinson, H., and Croy, R. R. D., Cloning of *PCP1*, a member of a family of pollen coat protein (PCP) genes from *Brassica oleracea* encoding novel cysteine-rich proteins involved in pollen-stigma interactions, *Plant J.,* 10, 303, 1996.
46. Doughty, J., Hedderson, F., McCubbin, A., and Dickinson, H., Interaction between a coating-borne peptide of the *Brassica* pollen grain and S (incompatibility)-locus-linked stigmatic glycoproteins, *Proc. Natl. Acad. Sci. U.S.A.,* 89, 467, 1993.
47. Méndez, E., Rocher, A., Calero, M., Girbéz, T., Citores, L., and Soriano, F., Primary structure of ω-hordothionin, a member of a novel family of thionins from barley endosperm, and its inhibition of protein synthesis in eukaryotic and prokaryotic cell-free systems, *Eur. J. Biochem.,* 239, 67, 1996.
48. Domon, C., Evrard, J.-L., Herdenberger, F., Pillay, D. T. N., and Steinmetz, A., Nucleotide sequence of two anther-specific cDNAs from sunflower (*Helianthus annuus* L.), *Plant Mol. Biol.,* 15, 643, 1990.
49. Ceci, L. R., Spoto, N., de Virgilio, M., and Gallerani, R., The gene coding for the mustard trypsin inhibitor is discontinuous and wound-inducible, *FEBS Lett.,* 364, 179, 1995.
50. Albani, D., Robert, L. S., Donaldson, P. A., Altosaar, I., Arnison, P. G., and Fabijanski, S. F., Characterization of a pollen-specific gene family from *Brassica napus* which is activated during early microspore development, *Plant Mol. Biol.,* 15, 605, 1990.

13 Expression and Function of PR-Protein Genes in Transgenic Plants

Karabi Datta, Subbaratnam Muthukrishnan, and Swapan K. Datta

CONTENTS

13.1 Introduction 261
13.2 Over-Expression of PR Proteins in Transgenic Plants as a Defense Strategy 262
13.2.1 PR Proteins as Antifungal Proteins 263
13.2.2 PR Proteins as Insecticidal Proteins 266
13.3 Expression of PR Protein Genes in Rice 266
13.4 Co-Suppression of PR Genes in Transgenic Plants 268
13.5 Conclusion 271
Acknowledgments 272
References 272

13.1 INTRODUCTION

Over millions of years, plants, fungi, and insects have evolved together and have developed attack and defense mechanisms that have ensured their mutual survival. While plant resistance is the predominant outcome in most plant–pathogen interactions, virulent strains of pathogens can cause considerable damage to plant tissues in a given environment. In order to survive a continuous onslaught by pathogens (bacteria, fungi, and viruses) and pests (insects, nematodes, etc.), plants deploy an assortment of defensive responses soon after infection or exposure to abiotic stress. These responses involve activation of host defense genes that were either inactive or expressed at basal levels previously. The major defense response involves the biosynthesis and accumulation of pathogenesis-related proteins (PR proteins). Both plants and pathogens possess certain sets of resistance and avirulence genes that determine the outcome of the plant–pathogen interaction in what is known as the gene for gene hypothesis. These genes represent upstream defense responses, but PR proteins are expressed in both resistant and susceptible interactions and in situations where a necrotrophic pathogen invades a plant host, usually after some

0-8493-0697-3/99/$0.00+$.50

delay following the perception of the pathogen. Thus PR proteins represent downstream generalized defense responses. The induction of PR proteins has been often interpreted as an attempt by the plant to prevent or limit the spread of the pathogen. In most cases, an assortment of PR proteins belonging to diverse subclasses, rather than a single member of a single family of PR proteins, is induced, indicating that such a combinatorial defense may have been selected during coevolution of the plants and pathogens. In some cases, there is a good correlation between a rapid and high-level expression of one or more PR proteins and the resistance reaction of the host plant. Even though it has not been demonstrated that loss of expression of a specific or multiple PR proteins results in greater susceptibility to disease, over-expression of genes in plants encoding certain PR proteins has been shown to provide protection against specific pathogens. These findings have led to the hope that a genetic engineering strategy involving constitutive, high-level expression of combinations of PR proteins with different modes of action against target organisms may provide broad-spectrum, durable resistance to a variety of diseases and pests. Hence, an understanding of the nature, distribution, and function of such proteins becomes very important in management of modern biotechnology-based crop improvement. In this chapter, we will attempt to summarize some of the data obtained with transgenic crop plants in which PR proteins are expressed at high levels and the effect of such over-expression on the progression of the disease or pest survival.

13.2 OVER-EXPRESSION OF PR PROTEINS IN TRANSGENIC PLANTS AS A DEFENSE STRATEGY

Plant protection is a major challenge to agriculture worldwide. The application of molecular genetic approaches in the study of plant–pathogen interactions has resulted in the relatively rapid accumulation of information concerning mechanisms underlying virulence in pathogens and resistance in plants. The most desirable and effective strategy for disease resistance in plants has been the incorporation of disease-resistant genes into commercially acceptable cultivars. There have been experimental successes with the transformation of plants with various foreign genes to effect disease resistance. For example, the genes *Pto* and *Xa21* confer resistance to fungal and bacterial pathogens on tomato and rice, respectively. However, the resistance conferred by these genes is specific to a particular host and in some cases to a limited assortment of pathovars. Further, complete resistance to a specific fungus may provide an opportunity for other fungi to emerge as new threats to the cultivated crop. Because plants are exposed to a broad and unpredictable assortment of pathogens in nature, it may be more advantageous to deploy an assortment of general defenses as plants have learned to do over evolutionary time. In this strategy, one would ensure that PR proteins would be present in host plants at levels needed for effective resistance (even if the resistance achieved is partial) before the attack by pathogens and pests. Further, the PR proteins would be located in regions where they are most effective against the invading pathogens (for example, the extracellular space). The transgene-encoded PR proteins are deployed in addition to, and not in place of, the inducible defenses of the host plant, including the endogenous PR proteins. It is to be borne in mind, however, that this strategy will exact the metabolic price of diverting some of the plant's resources into the synthesis of these defense proteins as insurance against a pathogen attack that may or may not ensue. The goal is to stabilize rather than enhance yield. In areas where diseases are endemic, this strategy may result in enhanced yield per unit land area compared to plots where control plants are grown.

In the following, we will first review the literature dealing with the expression of transgenic crop plants, mostly involving dicotyledonous species such as tobacco, canola, tomato, and potato. In the second half, we will focus on the utilization of the same strategy in monocot plants. Because of our familiarity with published and ongoing efforts with transgenic rice plants and the limited amount of published data on other transgenic monocot plants, our review is weighted heavily toward transgenic rice plants. Work is in progress in several laboratories with transgenic maize, wheat, and sorghum that intend to exploit PR proteins to enhance their resistance to diseases and pest attack.

13.2.1 PR-Proteins as Antifungal Proteins

13.2.1.1 Chitinases and β-1,3 Glucanases

The most attractive initial candidates for manipulation of the single gene defense mechanisms approach are genes encoding chitinases or β-1, 3-glucanases because these two enzymes hydrolyze chitin and β-1-3-glucans which are structural components of the cell walls of several fungi. In the first report of success with this approach, the expression of a bean vacuolar chitinase gene under the control of the constitutive 35S promoter from cauliflower mosaic virus in tobacco and *Brassica napus* was shown to result in decreased formation of symptoms by the necrotrophic pathogen, *Rhizoctonia solani.*[1] Significant reduction in fungal growth and delay in disease development were observed. Similarly, a fungal chitinase gene from *Rhizopus oligosporus* has been shown to operate as an antifungal system in transgenic tobacco.[2] The constitutive expression of vacuolar tobacco chitinases increased the resistance of *Nicotiana sylvestris* to the root pathogen, *Rhizoctonia solani,* but did not reduce symbiotic potential of the plants.[3] Transgenic cucumber harboring the rice chitinase genes exhibited enhanced resistance against gray mold, *Botrytis cinerea.*[4] However, enhanced plant resistance to *Cercospora nicotianae* was not observed with over-expression of a tobacco chitinase in tobacco.[5] Furthermore, transgenic tobacco plants accumulating acidic class III chitinases from sugar beet did not show an increase in resistance against *Cercospora nicotianae.*[6] Thus while some chitinases are effective as defense mechanisms in some hosts, they are not universally effective against all species of chitin-containing fungi. It is likely that the nature of the specific chitinase deployed, its cellular location, the accessibility of this chitinase to the cell wall-associated chitin (and possibly differences in the chitin substrate), and the mode of infection of the plant host by the pathogen may have contributed to the negative outcome.

While it is clear that it is possible in several cases to alter the expression of chitinase transgenes to generate plants with increased resistance to the pathogen, it is not clear whether constitutively expressed chitinase alone is responsible for the reduction of disease symptoms as observed in the case of tobacco and canola. The presence of increased chitinase levels prior to infection initiation may have triggered the host to mount a more rapid defensive response involving other host defenses.[1,7] Introduction of bacterial chitinase gene from *Serratia marcescens* in transgenic tobacco cells showed up to an eightfold increase in the amount of chitinase protein in the plants and conferred resistance to *R. solani*.[8] The *S. marcescens* chitinase A gene in transgenic tobacco plants with CaMV-35S promoter resulted in increased resistance to *Alternaria longipes* by reducing necrotic lesions and chlorosis during infections. This increase in resistance, however, appears to diminish during maturity.[9] Chitinase protein accumulated to about 0.25% of the total soluble leaf protein in transgenic tobacco generated using promoters from petunia plants.[10] At least 50% of the chitinase protein produced in transgenic tobacco cells with the *S. marcescens* chitinase A gene was of the same molecular weight as *S. marcescens* chitinase.[11] Using immunocytochemical methods, Benhamou et al.[12] observed that hyphal tips of *R. solani* in root tissues of transgenic plants appeared damaged and suffered increased vacuolization and cell lysis compared with the active fungi found in control plants. Fungal colonization was restricted to the cortex in the transgenic plants, whereas fungal colonization occurred extensively in all root tissues in the wild type.

Over-expression of an elicitor-releasing β-1,3-glucanase from soybean in transgenic tobacco plants was shown to result in enhanced resistance to *Alternaria alternata* and the oomycete, *Phytophthora parasitica* var *nicotianae*.[13] Furthermore, β-1,3-glucanases and chitinases have been shown to act synergistically against fungi in *in vitro*[14] and *in vivo* assays. Transgenic tomato plants expressing only a chitinase transgene or a β-1,3-glucanase transgene were susceptible to *Fusarium oxysporum*, but plants expressing both genes had significantly higher resistance than the plants expressing only chitinase or β-1,3-glucanase.[15] Similarly, Jach et al.[16] demonstrated that tobacco plants expressing a barley β-1,3-glucanase and a chitinase gene had a greatly enhanced resistance compared to plants expressing only one of these two enzymes. Tobacco plants expressing a gene encoding rice basic chitinase and others expressing an alfalfa acidic glucanase were developed separately. Crossing both transgenic lines generated hybrid plants. Evaluation of these hybrids, which were heterozygous for each transgene, and the homozygous selfed progeny, showed that the combined effect of two transgenes in conferring protection against *Cercospora nicotianae* was greater than that afforded by the single transgenes. These results suggest that the combined expression of PR2 (β-1,3-glucanase) and PR3 (chitinase) gives effective protection against fungal infection.[17]

13.2.1.2 Ribosome-Inactivating Proteins (RIPs)

Expression of the barley ribosome-inactivating protein cDNA under the control of a wound-inducible promoter in transgenic tobacco plants confers protection against the soil-borne pathogen *R. solani* as judged by height differences between control and transgenic plants grown in infected soil.[18] High-level expression of the transferred genes was detected in transgenic plants when cDNAs encoding three proteins from barley (*Hordeum vulgare*), a class II chitinase (CHI), a class II β-1,3-glucanase (GLU), and a Type-I ribosome inactivating protein (RIP) were expressed in tobacco plants.[19] The performance of tobacco plants co-expressing the barley transgene GLU/CHI or CHI/RIP in a *Rhizoctonia solani* infection assay showed significantly enhanced protection against fungal attack. These data indicated synergistic protective interaction of the co-expressed antifungal proteins as antifungal defenses.

13.2.1.3 Thaumatin-like Protein

Transgenic tobacco and potato plants constitutively over-expressing a full-length osmotin gene accumulated osmotin to a level of about 2% of the total cellular proteins.[20] While the transgenic potato plants expressing the tobacco osmotin gene showed a delay in the development of disease symptoms, there was no effect of over-expression of the tobacco osmotin gene in transgenic tobacco plants, leading to the speculation that the heterologous nature of the osmotin was responsible for the effectiveness of the tobacco osmotin gene in potato against the potato pathogen. When transgenic potato plants over-expressing a potato osmotin gene were tested against the potato late blight pathogen, transgenic potato plants showed high levels of potato OLP and delayed development of disease symptom when inoculated with *P. infestans.* These results do not support a coevolution hypothesis, but do support the interaction between a specific osmotin and the particular pathogen. Tobacco and potato plants over-expressing full-length osmotin gene accumulated osmotin mostly in an intracellular compartment, probably the vacuole. In contrast, in plants over-expressing a truncated form of the osmotin gene with a deletion of region encoding the C-terminal 20 amino acids, osmotin was totally secreted into the extracellular matrix. Transgenic potato plants over-expressing the truncated osmotin protein also showed resistance to *Phytophthora infestans.*[21] Even though there was some improvement in reduction of disease symptoms soon after infection, it could not be established whether the extracellular expression resulted in significantly greater resistance than the transgenic plants in which osmotin accumulated in vacuoles.

Zhu et al.[22] reported transgenic potato plants expressing sense or antisense RNAs for an osmotin-like protein (OLP) from potato. Sense transgenic potato plants showed high levels of potato OLP and delayed development of disease symptoms when inoculated with *P. infestans.* Antisense transformants showed accumulation of sense OLP mRNA, induced by fungal infection, but no corresponding protein. To study the mechanism of OLP gene expression as induced by ethylene, Sato et al.[23] cloned the gene from *Nicotiana sylvestris* and ancestors of *N. tabacum.* Sequence analysis showed that it has two AGCCGCC sequences in its promoter region which are conserved in most basic PR protein genes. Transgenic tobacco plants with wild and mutated OLP promoter: *Gus* fusion genes showed that mutated AGCCGCC sequences clearly inhibited the *gus* expression induced by ethylene. An EREBP2 protein, one of the proteins binding to the AGCCGCC sequence of tobacco β-1,3-glucanase gene, is also shown to bind to the AGCCGCC sequence(s) of the *OLP* gene, which suggests that ethylene-induced expression of OLP is regulated by a transacting factor common to basic PR proteins.

13.2.1.4 Other PR Proteins

Constitutive expression of the *PR-1* gene has no apparent effect on symptoms development in transgenic tobacco inoculated with tobacco mosaic virus,[24,25] but recent data indicate increased tolerance to infection by two oomycete pathogens, *Perenospora tabacina* and *Phytophthora parasitica* var. *nicotianae*[26] in transgenic tobacco expressing PR1a.

Tobacco plants were transformed with a chimeric gene construct consisting of the coding region of the radish Rs-AFP2 plant defensin cDNA driven by a constitutive promoter. The T2 plants in expressing Rs-AFP2 promoter showed a sevenfold reduction in lesion size upon infection with the fungal pathogen *Alternaria longipes* compared to untransformed plants.[27]

13.2.2 PR Proteins as Insecticidal Proteins

A bean class I endochitinase gene was used to transform potato plant to enhance resistance to the tomato moth, *Lacanobia oleracea.*[28] The bean chitinase accumulated up to 2% of the total soluble protein, but it did not provide any protective effect against tomato moth larvae. Transgenic rice plants expressing high levels of a rice chitinase gene also failed to have any inhibitory effect on the growth or mortality of stem borer (our unpublished data). On the other hand, Gatehouse et al.[29] observed a small but significant decrease in the growth of the peach-potato aphid, *Myzus persicae,* on transgenic potato plants expressing genes encoding bean chitinase and a snow drop lectin.

Hilder et al.[30] showed that transgenic tobacco plants expressing a cowpea trypsin inhibitor gene were more resistant to attack by larvae of the lepidopteran insect, *Heliothis virescens,* than untransformed control plants. Johnson et al.[31] introduced the genes coding for tomato proteinase inhibitor I (chymotrypsin inhibitory), tomato and potato inhibitor II (trypsin and chymotrypsin inhibitory) to tobacco. Plants expressing high levels of inhibitor II protein showed severe growth retardation when fed to larvae of the tobacco hornworm *Manduca sexta,* while a reduced induction of inhibitor I and II was noted in transgenic tomato plants expressing prosystemin gene (encoding a systemic signal for wound-responsive genes) in antisense orientation. Prosystemin is the precursor protein of systemin, an 18-amino acid polypeptide that activates protease inhibitor genes in tomato leaves. Growth rates of *Manduca sexta* larvae feeding on the antisense plants were three times higher than larvae feeding on untransformed control plants.[32] Expression of insect-encoded Pis in transgenic plants has been elaborately discussed in Heitz, Chapter 6.

13.3 EXPRESSION OF PR PROTEIN GENES IN RICE

Rice is the world's most important food crop. Almost two billion people, i.e., one third of the world's population, depend primarily on rice for basic food. Rice fields cover more than 360 million acres of land around the world and yield 560 million tons of grain per year. Every year there are significant yield losses due to biotic and abiotic stresses, making plant protection against these stresses the major challenge to agriculture. One of the most devastating biotic stresses of rice is sheath blight disease, caused by the fungus *Rhizoctonia solani.* It is prevalent in all rice-growing countries, especially in Asia. The genetic sources for high levels of resistance have not been identified in cultivated rice nor in its wild relatives in spite of an extensive effort.[33] The development of genetic engineering techniques for rice opens up possibilities for testing the effects of expression of agronomically important genes, including PR genes with antifungal activity, as strategies to overcome sheath blight disease of rice.[34,35]

Expression of a PR-protein gene in rice was first reported by Lin et al.[36] in 1995. A rice class I chitinase gene which is normally expressed in seeds, Chi-11,[37] under the control of the cauliflower mosaic virus (CaMV) 35S promoter, which confers constitutive expression, was introduced to the indica rice cultivar, Chinsurah Boro II by PEG (polyethyleneglycol) mediated protoplast transformation system. Several independent transgenic lines were regenerated, with chitinase expression ranging from 2- to 19-fold over controls. Western blot analyses of transgenic plants and their progeny revealed the presence of two proteins with apparent molecular weights of 30 and 35 kDa that reacted with the chitinase antibody, while no chitinase bands were detectable in uninfected control plants. Progeny from the transgenic chitinase-positive plants expressing chitinase at high and low levels were tested for their resistance to the sheath blight pathogen, *Rhizoctonia solani.* Disease ratings were monitored for the progression of the disease as described by Ou.[38] The number and size of lesions in transgenic plants were smaller compared to control. The lesions had spread to the upper half of the control plants, while in the transgenic plants, the lesions were confined to the lower half of the sheaths and covered an area smaller than those in the controls. The degree of resistance displayed by the transgenic plants to the *R. solani* correlated with the level of chitinase expression.

Regulation, expression, and function of a new basic rice chitinase gene in transgenic rice was described by Xu et al.[39] A new basic chitinase gene, designated RC24, was isolated from a rice genomic library, and the encoded RC24 chitinase expressed in *Escherichia coli* exhibited chitinase activity and strongly inhibited bacterial growth. Because the RC24 promoter contained several putative stress-related elements, regulation of RC24 expression by fungal elicitor and wounding stimuli was examined. Gene-specific primer extension analysis showed that the RC24 transcript accumulation in leaf tissues was also induced by wounding. The regulation of the *RC24* gene by wounding stimulus was further confirmed by analysis of GUS activities in tissues of transgenic rice plants containing the RC24 promoter/*GUS* gene fusion. These results indicate that the RC24 promoter is very responsive to fungal elicitor and wounding stimuli.

Recently a pathogen-inducible chitinase cDNA clone, RC7, that encoded a class I chitinase with a size of 35 kDa molecular weight was isolated from *R. solani*-infected IR58 rice plants.[40] Its deduced amino acid sequence was found to be significantly different from the other rice chitinases used in other studies with transgenic rice plants. We observed constitutive expression of this isolated chitinase gene in the transgenic plants obtained from IR72 and IR64 (both derived from biolistic gene delivery system) and from Chinsurah boro II (from protoplast transformation system), and the plants showed an enhanced resistance to sheath blight (Datta et al. in preparation). However, independent fertile lines had variable protein expression. Expression and segregation of chitinase transgene was noted in the following generations. Two different homozygous lines (T_2 generation) were obtained from each transgenic progeny of the two cultivars, IR64 and IR72. Homozygous progeny of these chitinase transgenic plants obtained from both IR72 and IR64 were assayed for resistance to the fungal pathogen *R. solani.* Bioassay data showed variation of fungal protection, and the protective effect of the chitinase transgene on the transgenic homozygous lines was shown to be about 50% both in lesion size and number.

More recently, we have expressed yet another PR-protein gene in transgenic rice in our laboratory. A 1.1 kb DNA fragment containing the coding region of a thaumatin-like protein (TLP), a member of the PR-5 group of PR proteins from rice, was introduced into different indica rice cultivars. We have described the over-expression of 23 kDa rice TLP in genetically engineered rice under the control of the CaMV 35S promoter.[41] Western blot analyses of protein extracts of transgenic plants showed variation in their protein expression levels. Although the accumulation level of this protein varied among individual transformants, several primary transgenic plants and their progeny showed a considerable amount of protein accumulation (levels approaching 0.5% of total protein). Several lines of TLP-transformants of the T1 generation with high-level constitutive expression of rice TLP were subjected to a bioassay for resistance to the fungal pathogen *R. solani.* The plants expressing high levels of TLPs had significantly less infection (rated on the basis of lesion number, area size, and localization) compared to nontransgenic control. This indicates that over-expression of a rice TLP can result in protection against the spread of the disease. Molecular analysis of TLP rice plants is shown in Figure 13.1. We are now in the process of developing transgenic rice plants with multiple PR-protein genes in an attempt to engineer durable resistance against fungal attack in agreement with multigene strategies for crop breeding.[42]

Several foreign protease inhibitor genes have also been introduced into rice. The constitutive expression of the cowpea trypsin inhibitor gene[43] or the wound-inducible expression of potato proteinase inhibitor II gene[44] in rice confers resistance to stem borer, the major lepidoptera rice pest. Tissue-specific promoters have been shown to be very active in rice for Bt gene expression,[45] which can eventually be designed to maximize PR protein expression in transgenic rice. Irie et al.[46] introduced a corn cystatin gene in the rice genome, and accumulation of up to 2% of total soluble seeds protein was observed and exhibited a wider inhibition spectrum against insect gut proteinases than the endogenous oryzacystatin. Table 13.1 summarizes the selected list of work done on PR proteins in transgenic plants.

13.4 CO-SUPPRESSION OF PR GENES IN TRANSGENIC PLANTS

It has been reported that transgenes introduced into plants can interact with homologous host genes leading to decreased expression of both genes.[47-49] The report of

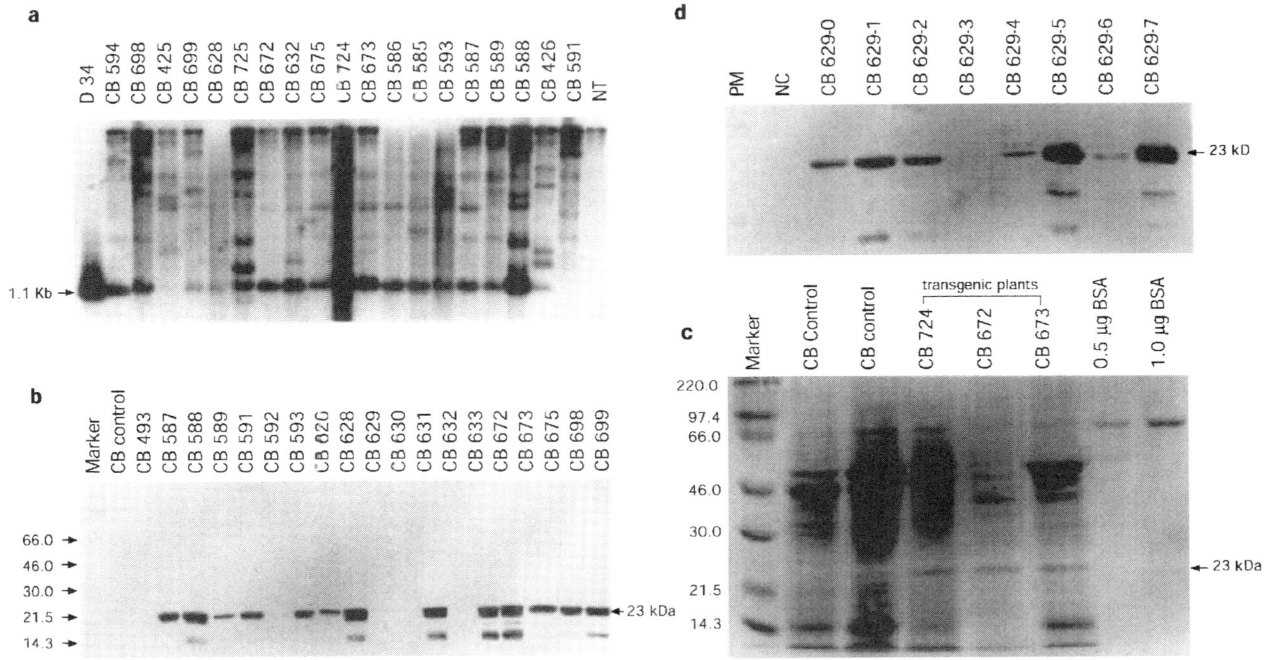

FIGURE 13.1 (**a**) Southern blot analysis of several independently transformed plants of CBII with *tlp-D34* gene in T0. Most plants except CB425 and CB591 showed the expected 1.1 Kb fragment indicative of the *tlp-D34* gene digested by BamHI, HindIII. NT nontransformed rice DNA, D34 plasmid DNA used for transformation (**b**) Western blot analysis showing TLP-D34 protein level in T0 plants. Molecular weight of markers given in kilodaltons (kDa) at the extreme left. The TLP-D34 protein of transgenic CBII plants showed 23 kDa molecular weight protein. CB493, CB592, CB630, CB631, CB633 did not show any protein expression. Fifty micrograms of leaf extract protein from freeze-dried samples was analyzed by 10% SDS-PAGE (**c**) Coomassie blue-stained SDS-PAGE of total proteins from rice plants (T0) transformed with *tlp-D34* gene. Three transgenic plants, CB724, CB672, and CB673, showed the expression of 23 kDa protein. Molecular weight of marker given in kilodalton (kDa) at the extreme left. Bovine serum albumin (BSA) used as standard protein in known concentration. (**d**) Western blot analysis showing expression of TLP-D34 protein (23 kDa) in T0 plants and its T1 progenies (CB629). PM = protein marker of known molecular weight, NC = negative control; CB629-0 is the T0 plant and CB629-1 to CB629-7 are the progenies of CB629-0. (From Datta, K., Velazhahan, R., Oliva. N., Ona, I., Mew, T., Khush, G.S., Muthukrishnan, S., and Datta S. K., *Theor. Appl. Genet.* (in press), 1999.)

TABLE 13.1
PR Protein Genes Introduced in Transgenic Plants

PR Genes	Transgenic crop	References
Chitinase; β-1, 3-glucanase	Tobacco, rape	1, 14, 15, 16, 17, 53
(Bacterial) chitinase	Tobacco	54
Acidic and basic chitinase	Tobacco	55
Chitinase	Tobacco	3, 5, 11
Chitinase (Bean chitinase promoter)	Tobacco	56
Class II chitinase, β-1, 3-glucanase, type I	Tobacco	19
Ribosome-inactivating protein		
35-S bean chitinase	Canola	12
Ribosome-inactivating protein (barley)	Tobacco	18
β-1, 3-glucanase	Tobacco, alfalfa	13, 57, 58, 59
Chitinase (rice)	Cucumber	4
PR 1a	Tobacco	26, 60, 61
Class I chitinase	Tobacco	50
Phenyl alanine ammonia lyase (PAL)	Tobacco	62
Rice chitinase promoter	Tobacco	63
Class II chitinase (peanut)	Tobacco	64
RIP b-32 (maize)	Tobacco	65
Wun I promoter (wound inducible)	Tobacco	66
H_2O_2-generating glucose oxidase	Potato	67
Chitinase (Chi1)	Tobacco	2
PRB-1 promoter	Tobacco	68
PR-1a promoter	Tobacco	69
PR-1 and the class II isoforms of PR-2 and PR-3	Tobacco	70
Osmotin (PR5)	Tobacco, potato	21
Pokeweed antiviral protein (PAP)	Tobacco	71
Osmotin-like protein (OLP) PR5	Tobacco	23
β-1, 3-glucanase	Alfalfa	72
1, 3-1, 4 β-glucanase	Barley	73
PR-1a, 1b, 1c	Tobacco	74
PR-1a	Tobacco, carrot	75
PR-1 pseudogenes (W38/1 and W 38/3)	Tobacco (not active after stable transformation only transient expression with GUS reporter gene)	76
Chitinase A (Chi A)	Tobacco	8
Plant defensin (Rs-AFP2)	Tobacco	27
Chitinase	Rice	36
Basic chitinase (RC24)	Rice, rose	39, 77
Proteinase inhibitor II	Rice	44
Cystatin gene	Rice	46
Cowpea trypsin inhibitor gene	Rice	43
Thaumatin-like PR-5 gene	Rice	41

Hart et al.[50] addresses the problem of stability and developmental regulation and inhibitory interaction between homologous tobacco class I chitinase gene and transgenes in *Nicotiana sylvestris.* A vacuolar chimeric class I chitinase gene from tobacco with CaMV promoter was introduced into *Nicotiana sylvestris,* and the transgenic homozygous T_2 plants accumulated high levels of chitinase. However, some of the transgenic plants accumulated levels of chitinase lower than nontransformed plants. The heritability of silent phenotypes was examined, and the report indicated that silencing can be regulated both developmentally and by environmental factors. Gene-specific measurement of chitinase and chitinase mRNA indicated that the silent phenotype results from stable but potentially reversible states of gene expression that are not meiotically transmitted and from co-suppression of both host and transgene expression. This silent state was not correlated with cytosine methylation of the transgene.

A similar type of suppression was observed in the case of β-1,3-glucanase transgene expression in homozygous plants. A basic β-1,3-glucanase isoform, GN1, accumulates in the vacuoles of the transgenic plants. Progeny of the transgenic plants in the hemizygous condition contained high levels of gn1 mRNA and 14fold higher activity than untransformed plants. Constitutive high-level expression of GN1 protein did not cause any apparent phenotypic abnormality. However, the expression of GN1 was completely suppressed in the homozygous progeny. This suppression occurs at a post-transcriptional level and is developmentally controlled. By generating haploid plants, it was shown that this silencing is correlated with the transgene dose in the plant genome.[51]

In our studies with transgenic rice, all T3 progeny of a homozygous parent transformed with a rice chitinase gene, Chi-11, under control of CaMV 35S promoter expressed the gene constitutively at 3 weeks after germination. However, about 20% of these progeny did not show any detectable chitinase 8 weeks after germination, indicating silencing of the transgene. A similar percentage of transgene silencing was also observed among the progeny of a heterozygous parent. There was co-silencing of the selectable marker gene whose expression was also driven by CaMV 35S promoter. Northern blot analysis and nuclear run-on transcription studies showed that the metabolic block in the silent plants was transcriptional.[52]

13.5 CONCLUSION

Transgenic cereals expressing PR proteins, especially rice plants, are now widely available in several laboratories. Efforts are under way to identify the combinations of PR proteins that are most effective against a specific pathogen or broad range of fungi. A better understanding of all of the PR proteins, their mode of action, and their cellular targets would undoubtedly lead to practical uses by manipulation of their expression and cellular or extracellular targeting. A desirable avenue of research will involve the use of inducible and/or tissue-specific promoters that will allow expression of transgenic PR proteins only when needed and only in desired tissues. Because there is an increasing demand for food crops (quantitatively and qualitatively) which need to be produced on less land and with minimum application of chemical input, genetic engineering offers excellent opportunities to complement conventional breeding for further improvement of crops. The availability of a large assortment of well-characterized stress-related PR genes (for biotic or abiotic stress), further understanding of stress signals and transduction mechanisms, and identification of additional defense genes will provide attractive opportunities for enhancing crop protection. Breeding for resistance in transgenic plants with multigene combination of PR protein genes (gene pyramiding) may be a good strategy to achieve longer-lasting resistance by reducing the probability of the breakdown of crops' resistance to pathogens, akin to the recent conventional breeding programs that aim to select for more durable resistance using "multigene" combinations.[42]

ACKNOWLEDGMENTS

We thank BMZ (Germany) and the Rockefeller Foundation (U.S.A.) for the financial support of our work on transgenic rice. Thanks are due to L. Torrizo, E. Abrigo, N. Oliva, M. F. Alam, J. Tu, N. Baisakh, and M. Viray for their help and contributions. We also thank T. Mew, G.S. Khush, and G. Wenzel for their suggestions and support.

REFERENCES

1. Broglie, K., Chet, I., Holliday, M., Cressman, R., Biddle, P., Knowlton, S., Mauvais, C. J., and Broglie, R., Transgenic plants with enhanced resistance to the fungal pathogen *Rhizoctonia solani, Science,* 254, 1194, 1991.
2. Terakawa, T., Takaya, N., Horiuchi, H., Koike, M., and Takagi, M., A fungal chitinase gene from *Rhizopus oligosporus* confers antifungal activity to transgenic tobacco, *Plant Cell Reports,* 16, 439, 1997.
3. Vierheilig, H., Alt, M., Neuhaus, J-M., Boller, T., and Wiemken, A., Colonization of transgenic *Nicotiana sylvestris* plants, expressing different forms of *Nicotiana tabacum* chitinase, by the root pathogen *Rhizoctonia solani* and by the mycorrhizal symbiont *Glomus mosseae, Mol. Plant-Microbe Interact.,* 6, 261, 1992.
4. Tabei, Y., Kitade, S., Nishizawa, Y., Kikuchi, N., Kayano, T., Hibi, T., and Akutsu, K., Transgenic cucumber plants harboring a rice chitinase gene exhibit enhanced resistance to gray mold (*Botrytis cinerea*), *Plant Cell Reports,* 17, 159, 1998.
5. Neuhaus, J. M., Ahl-Goy, P., Hinz, U., Flores, S., and Meins, F., Jr., High level expression of a tobacco chitinase gene in *Nicotiana sylvestris.* Susceptibility of transgenic tobacco plants to *Cercospora nicotianae, Plant Mol. Biol.,* 16, 141, 1991.
6. Nielsen, K. K., Mikkelsen, J. D., Kragh, K. M., and Bojsen, K., An acidic class III chitinase in sugar beet induction by transgenic tobacco plants, *Mol. Plant–Microbe Interact.,* 6, 495, 1993.
7. Broglie, R. and Broglie, K., Chitinase gene expression in transgenic plants: a molecular approach to understanding plant defense responses, *Phil. Trans. R. Soc. Lond.,* 342, 265, 1993.
8. Jach, G., Logemann, S., Wolf, G., Oppenheim, A., Chet, I., Schell, J., and Logemann, J., Expression of a bacterial chitinase leads to improved resistance of transgenic tobacco plants against fungal infection, *Biopractice*, 1, 1, 1992.

9. Suslow, T. V., Matsubara, D., Jones, J., Lee, R., and Dunsmuir, P., Effect of expression of bacterial chitinase on tobacco susceptibility to leaf brown spot, *Phytopathology*, 78, 1556, 1988.
10. Jones, J. D. G., Dean, C., Gidoni, D., Gilbert, D., Bond-Nutter, D., Lee, R., Bedbrook, J., and Dunsmuir, P., Expression of bacterial chitinase protein in tobacco leaves using two photosynthetic gene promoters, *Mol. Gen. Genet.*, 212, 536, 1988.
11. Taylor, J., Jones, J. D. G., Sandler, S., Mueller, G. M., Bedbrook, J., and Dunsmuir, P., Optimizing the expression of chimeric genes in plant cells, *Mol. Gen. Genet.*, 210, 572, 1987.
12. Benhamou, N., Broglie, K., Broglie, R., and Chet, I., Antifungal effect of bean endochitinase on *Rhizoctonia solani* ultrastructural changes and cytochemical aspects of chitin breakdown, *Can. J. Microbiol.*, 39, 318, 1993.
13. Yoshikawa, M., Tsuda, M., and Takeuchi, Y., Resistance to fungal diseases in transgenic tobacco plants expressing the phytoalexin elicitor-releasing factor, β-1,3-endoglucanase, from soybean, *Naturwiss.*, 80, 417, 1993.
14. Mauch, F., Mauch-Mani, B., and Boller, T., Antifungal hydrolases in pea tissue II. Inhibition of fungal growth by combinations of chitinase and β-1,3-glucanase, *Plant Physiol.*, 88, 936, 1988.
15. Jongedijk, E., Tigelaar, H., van Roekel, J. S. C., Bres-Vloemans, S. A., Dekker, I., van den Elzen, P. J. M., Cornelissen, B. J. C., and Melchers, L. S., Synergetic activity of chitinases and β-1,3-glucanases enhances fungal resistance in transgenic tomato plants, *Euphytica*, 85, 173, 1995.
16. Jach, G., Gornhardt, B., Mundy, J., Logemann, J., Pinsdorf, E., Leah, R., Schell, J., and Mass, C., Enhanced quantitative resistance against fungal disease by combinatorial expression of different barley antifungal proteins in transgenic tobacco, *Plant J.*, 8, 97, 1995.
17. Zhu, Q., Maher, E. A., Masoud, S., Dixon, R. A., and Lamb, C. J., Enhanced protection against fungal attack by constitutive co-expression of chitinase and glucanase genes in transgenic tobacco, *Bio/Technology*, 12, 807, 1994.
18. Logemann, J., Jach, G., Tommerup, H., Mundy, J., and Schell, J., Expression of a barley ribosome-inactivating protein leads to increased fungal protection in transgenic tobacco plants, *Bio/Technology*, 10, 305, 1992.
19. Jach, G., Gornhardt, B., Mundy, J., Logemann, J., Pinsdorf, E., Leah, R., Schell, J., and Maas, C., Enhanced quantitative resistance against fungal disease by combinatorial expression of different barley antifungal proteins in transgenic tobacco, *Plant J.*, 8, 97, 1995.
20. Liu, D., Raghothama, K. G., Hasegawa, P. M., and Bressan, R. A., Osmotin overexpression in potato delays development of disease symptoms, *Proc. Natl. Acad. Sci. U.S.A.*, 91, 1888, 1994.
21. Liu, D., Rhodes, D., D-Urzo, M. P., Xu, Y., Narasimhan, M. L., Hasegawa, P. M., Bressan, R. A., and Abad, L., *In vivo* and *in vitro* activity of truncated osmotin that is secreted into the extracellular matrix, *Plant Science*, 121, 123, 1996.
22. Zhu, B., Chen, T. H. H., and Li, P. H., Analysis of late-blight disease resistance and freezing tolerance in transgenic potato plants expressing sense and antisense genes for an osmotin-like protein, *Planta*, 198, 70, 1996.
23. Sato, F., Kitajima, S., Koyama, T., and Yamada, Y., Ethylene-induced gene expression of osmotin-like protein, a neutral isoform of tobacco (*Nicotiana sp.*) PR-5, is mediated by the AGCCGCC cis-sequence, *Plant Cell Physiol.*, 37, 249, 1996.

24. Cutt, J. R., Harpster, M. H., Dixon, D. C., and Carr, J. P. D., Disease response to tobacco mosaic virus in transgenic tobacco plants that constitutively express the pathogenesis related *PR-1b* gene, *Virology*, 173, 89, 1989.
25. Linthorst, H. J. M., Meuwissen, R. L. J., Kauffman, S., and Bol, J. F., Constitutive expression of pathogenesis-related proteins PR-1, GPR and PR-S in tobacco has no effect on virus infection, *Plant Cell*, 1, 285, 1989.
26. Alexander, D., Goodman, R. M., Gut-Rella, M., Glascock, C., Weymann, K., Friedrich, L., Maddox, D., Ahl-Goy, P., Luntz, T., Ward, E., and Ryals, J., Increased tolerance to two oomycete pathogens in transgenic tobacco expressing pathogenesis-related protein 1a, *Proc. Natl. Acad. Sci. U.S.A.*, 90, 7327, 1993.
27. Terras, F. R. G., Eggermont, K., Kovaleva, V., Raikhel, N. V., Osborn, R. W., Kester, A., Rees, S. B., Torrekens, S., Leuven, F. V., Vanderleyden, J., Commue, B. P. A., and Broekaert, W. F., Small cysteine-rich antifungal proteins from radish: their role in host defense, *Plant Cell*, 7, 573, 1995.
28. Gatehouse, A. M. R., Davison, G. M., Newell, C. A., Merryweather, A., Hamilton, W. D. O., Burgess, E. P. J., Gilbert, R. J. C., and Gatehouse, J. A., Transgenic potato plants with enhanced resistance to the tomato moth, *Lacanobia oleracea*: growth room trials, *Mol. Breeding*, 3, 49, 1997.
29. Gatehouse, A. M. R., Down, R. E., Powell, K. S., Sauvion, N., Rahbe, Y., Newell, C. A., Merryweather, A., Hamilton, W. D. O., and Gatehouse, J. A., Transgenic potato plants with enhanced resistance to the peach-potato aphid *Myzus persicae, Entomol. Exp. Appl.*, 79, 295, 1996.
30. Hilder, V. A., Gatehouse, A. M. R., Sheerman, S. E., Barker, R. F., and Boulter, D., A novel mechanism of insect resistance engineered into tobacco, *Nature*, 330, 160, 1987.
31. Johnson, R., Narvaez, J., An, G., and Ryan, C. A., Expression of proteinase inhibitors I and II in transgenic tobacco plants: effects on natural defense against *Manduca sexta* larvae, *Proc. Nat. Acad. Sci. U.S.A.*, 86, 9871, 1989.
32. Orozco-Cardenas, M., McGurl, B., and Ryan, C. A., Expression of an antisense prosystemin gene in tomato plants reduces resistance toward *Manduca sexta* larvae, *Proc. Natl. Acad. Sci. U.S.A.*, 90, 8273, 1993.
33. Bonman, J. M., Khush, G. S., and Nelson, R. J., Breeding rice for resistance to pests, *Annu. Rev. Phytopathol.*, 30, 507, 1992.
34. Datta, S. K., Peterhans, A., Datta, K., and Potrykus, I., Genetically engineered fertile indica-rice recovered from protoplasts, *Bio/Technology*, 8, 736, 1990.
35. Datta, S. K., Torrizo, L. B., Tu, J., Oliva, N. P., and Datta, K., Production and molecular evaluation of transgenic rice plants, *IRRI Discussion Paper*, 21, International Rice Research Institute, Philippines, 1997.
36. Lin, W., Anuratha, C. S., Datta, K., Potrykus, I., Muthukrishnan, S., and Datta, S. K., Genetic engineering of rice for resistance to sheath blight, *Bio/Technology*, 13, 686, 1995.
37. Huang, J., Wen, L., Segle, M., Tran, H., Thin, T. H., Naylor, H. M., Muthukrishnan, S., and Reeck, G. R., Nucleotide sequence of a rice genomic clone that encodes a class I endochitinase, *Plant Mol. Biol.*, 16, 479, 1991.
38. Ou, S. H., *Rice Diseases*, Second ed., p. 280. Commonwealth Mycological Institute Publication, Kent, Surrey, UK., 1985.
39. Xu, Y., Zhu, Q., Panbangred, W., Shirasu, K., and Lamb, C., Regulation expression and function of a new basic chitinase gene in rice (*Oryza sativa* L.), *Plant Mol. Biol*, 30, 387-401, 1996.

40. Anuratha, C. S., Zen, K. C., Cole, K. C., Mew, T., and Muthukhrishnan, S., Induction of chitinases and β-1, 3-glucanases in *Rhizoctonia solani*-infected rice plants: isolation of an infection-related chitinase cDNA clone, *Physiol. Plant.,* 97, 39, 1996.
41. Datta, K., Velazhahan, R., Oliva, N., Ona, I., Mew, T., Khush, G. S., Muthukrishnan, S., and Datta S. K., Over expression of cloned rice thaumatin-like protein (PR-5) gene in transgenic rice plants enhances environmental friendly resistance to *Rhizoctonia solani* causing sheath blight disease, *Theor. Appl. Genet.* (in press), 1999.
42. Wenzel, G., Strategies in unconventional breeding for disease resistance, *Annu. Rev. Phytopathol.,* 23, 149, 1985.
43. Xu, D. P., Xue, Q. Z., and McElroy, D., Constitutive expression of a cowpea trypsin inhibitor gene, *CpTi,* in transgenic rice plants confers resistance to two major rice insect pests, *Mol. Breed.,* 2, 167, 1996.
44. Duan, X., Li, X., Xue, Q., Abo-El-Saad, M., Xu, D., and Wu, R., Transgenic rice plants harboring an introduced potato proteinase inhibitor II gene are insect resistant, *Nat. Biotechnol.,* 14, 494, 1996.
45. Datta, K., Vasquez, A., Tu, J., Torrizo, L., Alam, M. F., Oliva, N., Abrigo, E., Khush, G. S., and Datta, S. K., Constitutive and tissue-specific differential expression of the *cryIA(b)* gene in transgenic rice plants conferring resistance to rice insect pest, *Theor. Appl. Genet.,* 7, 20, 1998.
46. Irie, K., Hosoyama, H., Takeuchi, T., and Iwabuchi, K., Transgenic rice established to express corn cystatin exhibit strong inhibitory activity against insect gut proteinases, *Plant Mol. Biol.,* 30, 149, 1996.
47. Napoli, C., Lemieux, C., and Jorgensen, R., Introduction of a chimeric chalcone synthase gene in Petunia results in reversible co-suppression of homologous genes in trans, *Plant Cell,* 2, 279, 1990.
48. Van der Krol, A. R., Mur, L. A., Beld, M., Mol, J. N. M., and Stuitje, A. R., Flavonoid genes in Petunia: addition of a limited number of gene copies may lead to a suppression of gene expression, *Plant Cell,* 2, 291, 1990.
49. Smith, C. J. S., Watson, C. F., Bird, C. R., Ray, J., Schuch, W., and Grierson, D., Expression of a truncated tomato polygalacturonase gene inhibits expression of the endogenous gene in transgenic plants, *Mol. Gen. Genet.,* 224, 477, 1990.
50. Hart, C. M., Fischer, B., Neuhaus, J. M., and Meins, F., Jr., Regulated inactivation of homologous gene expression in transgenic *Nicotiana sylvestris* plants containing a defense-related tobacco chitinase gene, *Mol. Gen. Genet.,* 235, 179, 1992.
51. De Carvalho, F., Gheysen, G., Kushnir, S., Van Montagu, M., Inze, D., and Castresana, C., Suppression of β-1, 3-glucanase transgene expression in homozygous plants, *EMBO J.,* 11, 2595, 1992.
52. Chareonpornwattana, S., Thara, K. V., Wang, L., Datta, S. K., Panbangred, W., and Muthukrishnan, S., Inheritance, expression, and silencing of a chitinase transgene in rice, *Theor. Appl. Genet.,* (in press), 1999.
53. Sela-Buurlage, M. B., Ponstein, A. S., Bres-Vloemans, S. A., Melchers, L. S., van den Eltzen, P. J. M., and Cornelissen, B. J. C., Only specific tobacco (*Nicotiana tabacum*) chitinases and β-1, 3-glucanases exhibit antifungal activity, *Plant Physiol.,* 101, 857, 1993.
54. Jones, J. D. G., Expression of bacterial chitinase genes encoding two chitinase enzymes from *Serratia marcescens, EMBO J.,* 5, 467, 1988.

55. Linthorst, H. J. M., Van Loon, L. C., Van Rossum, C. M. A., Mayer, A., Bol, J. F., Van Roekel, J., Meulenhoff, E. J. S., and Cornelissen, B. J. C., Analysis of acidic and basic chitinases from tobacco and petunia and their constitutive expression in transgenic tobacco, *Mol. Plant–Microbe Interact.,* 3, 252, 1990.
56. Roby, D., Broglie, K., Cressman, R., Biddle, P., Chet, I., and Broglie, R., Activation of bean chitinase promoter in transgenic tobacco plants by phytopathogenic fungi, *Plant Cell,* 2, 999, 1990.
57. Yoshikawa, M., Tsuda, M., and Takeuchi, Y., Resistance to fungal diseases in transgenic tobacco plants expressing phytoalexin elicitor-releasing factor, β-1,3-endoglucanase from soybean, *Naturwissenschaften,* 80, 417, 1993.
58. Leubner-Metzger, G., Frundt, C., Vogeli-Lange, R., and Meins, F., Jr., Class I β-1,3-glucanases in the endosperm of tobacco during germination, *Plant Physiol.,* 109, 751, 1995.
59. Beffa, R. S., Hofer, R. M., Thomas, M., and Meins, F. Jr., Decreased susceptibility to viral disease of β-1, 3-glucanase-deficient plants regenerated by antisense transformation, *Plant Cell,* 8, 1001, 1996.
60. Ohshima, M., Itoh, H., Matsuoka, M., Murakami, T., and Ohashi, Y., Analysis of stress-induced or salicylic acid-induced expression of the pathogenesis-related la protein gene in transgenic tobacco, *Plant Cell,* 2, 92, 1990.
61. van de Rhee, M. D., van Kan, J. A. L., Gonzales-Jaen, M. T., and Bol, J. F., Analysis of regulatory elements involved in the induction of two tobacco genes by salicylate treatment and virus infection, *Plant Cell,* 2, 357, 1990.
62. Elkind, Y., Edwards, R., Mavandad, M., Hedrick, S., Ribak, O., Dixon, R. A., and Lamb, C. J., Abnormal plant development and down regulation of phenyl-propanol biosynthesis in transgenic tobacco containing a heterologous phenylalanine ammonia lyase gene, *Proc. Natl. Acad. Sci. U.S.A.,* 87, 9057, 1990.
63. Zhu, Q., Doemer, P. W., and Lamb, C. J., Stress induction and developmental regulation of a rice chitinase promoter in transgenic tobacco, *Plant J.,* 3, 203, 1993.
64. Kellmann, J. W., Kleinow, T., Engelhardt, K., and Philipp, C., Characterization of two class II chitinase genes from peanut and expression studies in transgenic tobacco plants, *Plant Mol. Biol.,* 30, 351, 1996.
65. Maddaloni, M., Forlani, F., Balmas, V., Donini, G., Stasse, L., Corazza, L., and Motto, M., Tolerance to the fungal pathogen *Rhizoctonia solani* AG4 of transgenic tobacco expressing the maize ribosome inactivating protein B-32, *Transgenic Res.,* 6, 393, 1997.
66. Seibertz, B., Logemann, J., Willmitzer, L., and Schell, J., Cis-analysis of the wound-inducible promoter wun1 in transgenic tobacco plants and histochemical localization of its expression, *Plant Cell,* 1, 961, 1989.
67. Wu, G., Shart, B. J., Lawrence, E. B., Lovine, E. B., Fitzsimmons, K. C., and Shah, D. M., Disease resistance conferred by expression of a gene encoding H_2O_2-generating glucose oxidase in transgenic potato plants, *Plant Cell,* 7, 1357, 1995.
68. Sessa, G., Meller, Y., and Fluhr, R., A GCC element and a G-box motif participate in ethylene-induced expression of the *PRB-1b* gene, *Plant Molec. Biol.,* 28, 145, 1995.
69. Chang, Y. C., Pfitzner, U. M., and Pfitzner, A. J. P., Fusion of the inducible promoter of the *PR-1*a gene to the activator transposase gene can transactive excision of a non-autonomous transposable element by external and by internal stimuli, *Plant Sci.,* 106, 141, 1995.
70. Beffa, R., Szell, M., Meuwly, P., Pay, A., Vogeli-Lange, R., Metraux, J. P., Neuhaus, G., Meins, F., Jr., and Nagy, F., Cholera toxin elevates pathogen resistance and induces pathogenesis-related gene expression in tobacco, *EMBO J.,* 14, 5753, 1995.

71. Smirnov, S., Shulaev, V., and Tumer, N. E., Expression of pokeweed antiviral protein in transgenic plants induces virus resistance in grafted wild-type plants independently of salicylic acid accumulation and pathogenesis-related protein synthesis, *Plant Physiol.,* 114, 1113, 1997.
72. Masoud, S. A., Zhu, Q., Lamb, C., and Dixon, R. A., Constitutive expression of an inducible β-1,3-glucanase in alfalfa reduced disease severity caused by the oomycete pathogen *Phytophthora megasperma* f. sp *medicaginis*, but does not reduce severity of chitin-containing fungi, *Transgenic Res.,* 5, 313, 1996.
73. Jensen, L. G., Olsen, O., Kops, O., Wolf, N., Thomsen, K. K., and von Wettstein, D., Transgenic barley expressing a protein-engineered, thermostable β-1,3, β-1,4-glucanase during germination, *Proc. Natl. Acad. Sci, U.S.A.,* 93, 3487, 1996.
74. Dixon, D. C., Cutt, J. R., and Klessig, D. F., Differential targeting of the tobacco PR-1 pathogenesis-related proteins to the extracellular space and vacuoles of crystal idioblasts, *EMBO J.,* 10, 1317, 1991.
75. Beilmann, A., Pfitzner, A. J. P., Goodman, H. M., and Pfitzner, U. M., Functional analysis of the pathogenesis-related 1a protein gene minimal promoter region. Comparison of reporter gene expression in transient and in stable transfections, *Eur. J. Biochem.,* 196, 415, 1991.
76. Pfitzner, A. J. P., Beilmann, A., Goodman, H. M., and Pfitzner, U. M., Molecular analysis of two PR-1 pseudogenes from tobacco, *Plant Mol. Biol. Int. J. Fund. Res. Genet. Eng.,* 16, 129, 1991.
77. Marchant, R., Davey, M. R., Lucas, J. A., Lamb, C. J., Dixon, R. A., and Power, J. B., Expression of a chitinase transgene in rose (*Rosa hybrida* L.) reduces development of blackspot disease (*Diplocarpon rosae* Wolf), *Mol. Breed.,* 4, 187, 1998.

Index

A

ABA, *see* Abscisic acid
Abscisic acid (ABA)
  ß-1,3-glucanases, 55, 56
  hypersensitive response and plant stress, 10
  proteinase inhibitor induction, 145
  thaumatin-like protein induction, 117–118
  thionin induction, 219
ACC, *see* 1-Aminocyclopropane-1-carboxylic acid
Access channel, 175
acd2 mutant, 254–255
Acidic domain, 217–218, *see also* Thionin
Acidity/basicity
  PR proteins, 7–9
  PR-1 proteins, 22–24
  PR-2 proteins, 50, 51, 52
  thaumatin-like proteins, 109
Acquired resistance, *see* Systemic acquired resistance
Action mode, 252–253
Adenosine triphosphate (ATP), 212–213
Adenylate cyclase, 214
Aegilops squarrosa, 208, 220
Agglutination, 164
AGPs, *see* Arabinogalactan proteins
Alfalfa, 58, 140
Allergy, 97
Allosamidin, 89–90
αβ motif, 247, 248
Alternaria spp., 254, 256, 264, 266
Amino acid
  peroxidase characterization, 174, 175, 180, 181, 182
  sequences of plant defensins, 248, 249
1-Aminocyclopropane-1-carboxylic acid (ACC), 30
3-Amino-1,2,4-triazole (3AT), 28
Amphipathicity, 213, 214, *see also* Thionin
α-Amylase, 97, 115, 215
Ankyrin repeats, 200, 201
Anthers, 52–53, *see also* ß-1,3-Glucanases
Antibacterial activity, *see* Antimicrobial activity; Bacteria
Anti-feedant proteins, 115, 256
Antifreeze proteins, 97, 115, 121
Antifungal activity, *see also* Antimicrobial activity; Fungi
  chitinases, 95–96, 263–264
  ß-1,3-glucanases, 57
    chitinases and, 263–264
  plant defensins, 250
  thaumatin-like proteins, 114–115
  transgenic plants, 263–266
Antimicrobial activity, *see also* Bacteria; Fungi; Virus
  plant defensins, 248, 250, 256
  thaumatin-like protein, 116
  thionin, 211, 215, 218
Antioxidants, 28
AP24 gene, 122
APX, *see* Ascorbate peroxidase
Arabidopsis spp.
  chitinases, 86, 87, 94, 95
  ß-1,3-glucanases, 62, 63, 64, 65
  plant defensins, 253, 254
  PR-1 gene induction/expression, 26–27, 30
  PR gene expression, 197, 200, 201
  thaumatin-like protein, 116, 118
  thionin, 211, 215, 220–222
Arabinogalactan proteins (AGPs), 97, 164
Arachidonic acid, 120
Arginine substitutions, 251–252
as-1 element, 35
Ascorbate peroxidase (APX), 27
Aspartyl proteinase, 134, 136, 138, 140, *see also* Proteinase inhibitors
Aspergillus giganteus, 236
3AT, *see* 3-Amino-1,2,4-triazole
AtERF1 gene, 199
ati gene, 140
ATP, *see* Adenosine triphosphate
Auxin, 29, 60–61, 143
Auxin 1-naphthaleneacetic acid (NAA), 29
Avocado, 97
avr genes, *see* Pathogen virulence genes

B

Bacillus thuringiensis, 143
Bacteria
  hydroxyproline-rich glycoproteins, 164

proteinase inhibitors, 137–138, 139
PR protein induction in plants, 3
thaumatin-like protein, 116
thionin, 212, 218
Bacterial speck disease, 197
Barley, *see Hordeum vulgare*
Basic-leucine-zipper (bZip) transcription factor, 35, 64
Bean, 96, 118
Benzothiadiazole (BTH), 222
6-Benzylaminopurine, 29
BGL32, 240–241
Biochemical marker, 158–159
Botrytis cinerea, 138, 240, 256, 263
Bowman–Birk inhibitor, *see also* Proteinase inhibitors
digestive proteinases and activity, 141
fungi induction, 136
occurrence and potential role in plants, 134, 135
wounding induction, 140
Brassica spp., 162, 163, 263
Brassicaceae, 211, *see also* Thionin
BTH, *see* Benzothiadiazole
bZip, *see* Basic-leucine-zipper transcription factor

C

Cabbage, 138
Cactus proteins, 201
CAD, *see* Cinnamyl alcohol dehydrogenase
Caenorhabditis elegans, 139
Calcium, 213, 221, 252, 253
Calcium chloride, 252
Callose, 7, 52, 53, 59–60
CaMV, *see* Cauliflower mosaic virus
Candida albicans, 114
Cardiotoxin, 213, 214, *see also* Thionin
Carrot, 97, 158, 162–163
Caster bean, 236
Catalase, 27, 28
Catalytic domain, 236–237, 238, *see also* Ribosome-inactivating proteins
Catalytic mechanisms/specificity
family 18 chitinases, 89–91
family 19 chitinases, 91–92
PR-4 chitinases, 92
Caterpillars, 143
Cations, 250, 252, *see also* Defensins (plant)
Cauliflower mosaic virus (CaMV), 53, 54, 63, 264
CBD, *see* Chitin-binding domain
CCPI, *see* Cowpea cystatin
cDNA
classification of PR proteins, 6, 9
PR-1 proteins, 23, 24
thaumatin-like proteins, 119
thionin genes, 220
Cell layers, peripheral, 253–254, *see also* Defensins (plant)
Cell walls
hydroxyproline-rich glycoproteins, 159, 163–164
peroxidase, 172
thionin, 217, 219
Cellulase, 120
Cercospora spp.
ß-1,3-glucanases, 59
pathogenesis-related proteins induction in tobacco, 9
plant defensins, 254
thaumatin-like proteins, 114
transgenic plants, 263, 264
Cereals, 208, *see also* Individual entries
Cf resistance gene, 197
Chi-11 transgene, 271, *see also* Chitinase transgenes
Chia genes, 6, 7, 94–95
Chib genes, 6, 7, 94
Chic genes, 6, 7, 94
Chid genes, 6, 7, 95
Chimeric constructs, 242, 243, 256
Chitin, 196
Chitin-binding domain (CBD)
antifungal activity of chitinases, 95–96
chimeric constructs of ribosome-inactivating proteins, 242, 243
PR-3 family structure, 80, 82, 83, 84
role, 93
solanaceous lectins relation, 165
Chitin synthase, 138
Chitinase 26, 245–241
Chitinase transgenes, 263–264, 266, 267, 271
Chitinases
catalytic mechanisms and specificities, 88–84
functions, 95–98
gene structure and regulation, 94–95
HYP-containing glycoproteins, 165
protein structure, 79–88, 89
PR protein classification, 6–7, 9
reproduction role, 12
Rhizobium–legume interactions, 11–12
Saccharomyces cerevisiae growth, 11

Chitosan, 144
2-Chloroethylphosphonic acid, *see* Ethephon
Chn family, 57, 61
Cholera toxin (CTX), 65
Chymotrypsin, 136, 138, 142–143, *see also* Proteinase inhibitors
Chymotrypsin inhibitor transgene, 266
Cinnamyl alcohol dehydrogenase (CAD), 185
Circadian regulation, 218
cis-Acting elements, *see also trans*-Acting elements
ß-1,3-glucanase expression/ regulation, 63–64
peroxidase gene regulation, 184–185
PR-1 promoter regulation, 32–34
transcriptional regulation of PR protein gene expression, 198
Cladosporium fulvum, 7, 197
Class I chitinases, 80, *see also* Chitinases
Class II chitinases, 80, 82–83, *see also* Chitinases
Class IV chitinases, 83, *see also* Chitinases
Class V chitinases, 83, *see also* Chitinases
Class VI chitinases, 83, *see also* Chitinases
Class VII chitinases, 83–84, *see also* Chitinases
Classification
PR proteins, 5–9
ribosome-inactivating proteins, 236–237, 238
thionin, 211
coi1 mutant, 256
Coix lacrima-jobi, 97
Coleoptera, 140–141, *see also* Individual entries
Collectotrichum spp.
ß-1,3-glucanases, 58
hydroxyproline-rich glycoproteins, 159, 161, 162
proteinase inhibitors, 136, 138
Colonization, fungi, 138, *see also* Proteinase inhibitors
Colorado potato beetles, 139, 141, 142
Comoviruses, 137
Conspecific translation, 239, 240
Corn rootworm larvae, 141
Cosuppression, transgenic plants, 268, 271, *see also* Transgenic plants
Cowpea cystatin (CCPI), 139
Cowpea trypsin inhibitor gene, 269, *see also* Trypsin inhibitor
Cowpea weevil, 141
cpr genes, 200, 201
cpTI gene, 143
Crambe abyssinica, 211
Crambin, 210, 211, 214, 217, *see also* Thionin
Crop plants, 222, *see also* Individual entries
Cross-linking, 163–164, 172, *see also* Hydroxyproline-rich glycoproteins
cryA gene, 143
Crystal idioblasts, 23
CTX, *see* Cholera toxin
Cucumber
chitinase, 84, 85, 94, 95
chitinase transgene expression, 263
peroxidase, 173
Cultured cells, 212
Cycloheximide, 26
Cydia pomonella, 141
Cystatin, 141, 142, 269, *see also* Cysteine proteinase
Cysteine proteinase, *see also* Proteinase inhibitors
antimicrobial activity, 138
digestive proteinases from insects, 141
mechanical wounding and induction in plants, 140
occurrence and potential role, 134, 135–136
orzyacystatin and activity, 137
Cysteine residues, 211, 217
Cystine nooses, 92
Cytokinins, 29, 60–61

D

2,4D, *see* 2,4 Dichlorophenoxyacetic acid
def1 mutants, 145
Defense reactions, 10–11, 163–164, 207, 255–256
Defensins (plant)
antimicrobial activity, 248, 250
gene expression, 253–255
host defense and contribution, 255–256
inclusion in PR protein family, 6
mode of action, 252–253
protein structure, 247–248, 249
structure–activity relationships, 251–252
Denaturation, 109
Developmental processes, 11, 31–32, 256
2,6-Dichloroisonicotinic acid, 118
2,4-Dichlorophenoxyacetic acid (2,4D), 29
Dicots, 208, 210

Digestive proteinases, 140–143, *see also* Proteinase inhibitors
Dm-AMP1, 252, *see also* Defensins (plant)
Double mutant analysis, 201
Drechslera graminea, 219, 220
Drosomycin, 248
Drosophila spp., 201. 248
Drought stress, 117, *see also* Stress

E

E2 gene, 119
E22 gene, 119
Eicosanoids, 163
ein2 mutant, 256
EIN3 protein, 199
EIX, *see* ß-1,4-Endoxylanase
Elastase, 138, 142–143
Elicitors
 ß-1,3-glucanases, 57–58
 hydroxyproline-rich glycoproteins, 162, 163
 PR protein gene expression, 196
 role of PR proteins, 11
 thaumatin-like protein, 109, 118
Embryogenesis, 55, 97
ß-1,3-Endoglucanases, 6–7, 9, 11
ß-1,4-Endoxylanase (EIX), 30
Endocytosis, fluid-phase, 242
Endoplasmic reticulum (ER), 217
Endosperm
 ß-1,3-glucanases, 54–56
 ricin, 236
 thionin, 208, 217
Endo-ß-1,4-glucosaminidases, *see* Chitinases
Enzymes, 52, 138, 215
Epidermal cells, 218
ER, *see* Endoplasmic reticulum
ERE, *see* Ethylene-responsive element
EREBPs, *see* Ethylene-responsive element binding proteins
Erwinia spp., 65, 137
Erysiphe graminis, 116, 184, 219
Ethephon, 30, 61, 199
Ethylene
 gene expression of PR proteins, 197, 198, 199, 201
 ß-1,3-glucanase expression, 55, 61
 hydroxyproline-rich glycoproteins induction, 162, 163
 hypersensitive response and PR proteins, 10
 induction of PR proteins, 9
 osmotin induction, 117–118, 120, 265
 plant defensins gene expression, 255
 PR-1 proteins, 29–31, 34, 36
 proteinase inhibitor induction, 145, 146
 thaumatin-like protein induction, 118
Ethylene-responsive element (ERE), 63, 198
Ethylene-responsive element binding protein (EREBP), 36, 64–65
Ethylene-responsive element binding protein (EREBP) gene, 198
Evolution, PR-1 proteins, 24
Exocortis virus, 146
Extensin,172, *see also* Hydroxyproline-rich glycoprotein
Extracellular space, 23, 24, 50
Extracellular targeting, 122–123

F

Family 18, 84, 89–91
Family 19, 91–92, 94–95
Floral tissues, 12, 31–32, *see also* Individual entries
Freezing tolerance, 121, *see also* Antifreeze proteins
Fungi, *see also* Individual entries
 ß-1,3-glucanases, 57–59
 peroxidase, 172
 plant defensins, 254, 255
 PR protein induction, 3, 9
 PR-1 proteins, 25
 proteinase inhibitors, 137–138
 thaumatin-like protein, 116
 thionin, 212, 213, 218
Fusarium spp.
 ß-1,3-glucanases, 55, 57, 58
 gene expression of PR proteins, 200
 proteinase inhibitor activity, 138
 ribosome-inactivating proteins, 240, 241
 Rs-AFP2 activity, 251, 252
 thionin, 215, 221
 tobacco and PR proteins, 9
 transgenic plants, 264

G

GA, *see* Gibberellins
G-box element, 184–185
GCC box, 198–199

GE, *see* ß-Glucan elicitor
GE-binding protein (GEBP), 196
GEBP, *see* ß-Glucan elicitor binding protein
Gene expression
hydroxyproline-rich glycoproteins, 161–163
peroxidases, 176–179, 183–185
plant defensins, 253–255
PR-1 proteins, 25–32
thaumatin-like proteins, 116–117
thionin, 219–220
Genes, *see also* Individual entries
peroxidase organization, 181, 183
suppression/overexpression and potential role, 185–186
thaumatin-like proteins, 119–120
Genetic engineering, 222
Germ tube elongation, 138
Gibberellins (GA), 55, 143
GIP-1, *see* ß-1,3-Glucanase inhibitor protein
Globodera pallida, 139
ß-Glucan elicitor (GE), 196
ß-Glucan elicitor binding protein (GEBP), 57–58
Glucan endo-1,3-ß-glucanases, *see* PR-2 family
Glucanase, classification, 6–7, 9
ß-1,3-Glucanase inhibitor protein (GIP-1), 58
ß-1,3-Glucanase transgene, 263–264, 271
ß-1,3-Glucanases, *see also* PR-2 family
Glucans, 196
ß-Glucuronidase (GUS) reporter-gene
ß-1,3-glucanase expression, 55, 62, 65
hydroxyproline-rich glycoproteins, 163
PR-1 proteins, 27, 30, 32–34
thaumatin-like proteins, 119
GN1 transgene, 271
G-proteins, *see* GTP binding proteins
GT-1 factors, 34–35
GTP binding proteins (G-proteins), 65
GUS, *see* ß-Glucuronidase reporter-gene

H

Heat shock, 173
Heat-shock proteins (HSP), 10
Heat shock regulatory element (HSE), 33
Heliothis spp., 25, 141, 266
Heme-binding domains, 174–175
Hemiptera, 141, *see also* Individual entries
Hemolysis, 212, 213
Herpes simplex virus, 137
Heterodera spp., 139
Heuchera sanguinea, 253
Hevamine, 89–91
Hevea brasiliensis, 84, 86, 87
Hevein, 87, 96, 242, 243
Histidine residues, 174, 175, 182
Histochemical analysis, 27
Hordeum vulgare
peroxidase, 184
PR-4 proteins, 88, 89
proteinase inhibitors, 136, 138–139
ribosome-inactivating protein, 236, 239
thionin, 208, 215, 218–220
Hordiothionin, 215, 217, 218, *see also* Thionin
Hormones, 29–31
H/PRP, *see* Hyp/proline-rich proteins
HR, *see* Hypersensitive response
HRGP, *see* Hydroxyproline-rich glycoproteins
Hs-AFP1, 253, *see also* Defensins (plant)
HSE, *see* Heat shock regulatory element
HSP, *see* Heat-shock proteins
Hydrolysis, 89, *see also* Chitinases
Hydroxyproline-rich glycoproteins (HRGP)
accumulation in diseased plants, 159–161
defense role, 163–164
gene expression and regulation, 161–163
other HYP-containing glycoproteins, 164–165
plant–microbe interactions, 172
structural features, 158–159
Hyp/proline-rich proteins (H/PRP), 165
Hypersensitive response (HR)
ß-1,3-glucanases, 62, 63, 57
hydroxyproline-rich glycoproteins, 164
PR proteins
function, 11
plants, 1, 2, 3
Hyphae, 114, 250

I

IAA, *see* Indoleacetic acid
IκB protein, 201
IDT, *see* Isodityrosine bridges
Incompatibility reactions, 164
Indoleacetic acid (IAA), 118
Inflorescences, 32, *see also* Floral tissues

Inhibition halo, 255
Insect defensins, 248
Insects, *see also* Individual entries
  pathogenesis-related proteins, 6
  PR-1 proteins, 25
  proteinase inhibitors, 139–143
  thionin toxicity, 212
  transgenic plants, 266
Intercellular spaces, 6, 164
Ionic strength, 250, 252, *see also* Defensins (plant)
Irradiation, 31
Isodityrosine (IDT) bridges, 164
Isoforms
  ß-1,3-glucanases, 50
  pathogenesis-related proteins, 5–9
  PR-1 proteins, 22–23
  thionin, 208–211

J

JA, *see* Jasmonic acid
Jasmonates, 197, 201, 247
Jasmonic acid (JA)
  ethylene-induced *PR-1* gene expression, 30–31
  hydroxyproline-rich glycoproteins induction, 163
  plant defensin gene expression, 254, 255
  proteinase inhibitor induction, 144, 145, 146
  thionin induction, 218
JIP60 protein, 236, 239
Job's Tears, *see Coix lacrima-jobi*

K

Killer toxin, 84
Kimura protein distance correction, 175
Kinetin, 29
Kluyveromyces lactis, 84
Kunitz-type inhibitors, 134, 135, 138, 140, *see also* Proteinase inhibitors

L

Lacanobia oleracea, 266
Larvae, hatching, 139
Lemmings, 143
Lepidoptera, 140–141, *see also* Individual entries
Light responsiveness, 218
Lignification, 172
Lily, 53
Linolenic acid, 144, 145
Lipid peroxidases, 27
Loblolly pine, *see Pinus taeda*
Lth locus, 219
Lysis, 114

M

Maize
  gene expression in PR proteins, 200
  proteinase inhibitors, 136, 140
  ribosome-inactivating proteins, 236
  thaumatin-like proteins, 115
Manduca sexta, 25, 142, 143
Medicago truncatula, 184
MeJa, *see* Methyl ester methyl jasmonate
Melon
  hydroxyproline-rich glycoproteins, 158, 159, 162, 163
  proteinase inhibitors induction, 136
Membrane receptor, 252–253
Membranes, permeability, 114, 212
Messenger ribonucleic acid (mRNA), 6, 26
Metal ions, 213
Metallocarboxypeptidase, 140
Metallo-proteinase, 134, 136, 137, *see also* Proteinase inhibitors
Metals, heavy, 218
Methyl ester methyl jasmonate (MeJa), 30–31
Methyl jasmonate
  plant defensin gene expression, 254
  proteinase inhibitor induction, 142, 144, 145, 146
  ribosome-inactivating proteins induction, 239
  thaumatin-like protein induction, 118
  thionin induction, 215, 218, 221
Microbial signals, 145–146
Microsporogenesis, 53
Mistletoe, 208
Monocots, 208, 209
MPI gene, 146
mRNAS, *see* Messenger ribonucleic acid
Multigene family, 219–220, *see also* Thionin
Mybl protein, 29, 35
Myzus persicae, 266

N

NAA, *see* Auxin 1-naphthaleneacetic acid
Narbonins, 87
Necrosis, 1–3, 4, 21–22
Nectria haematococca, 252
Neighbor joining method, 175
Nematodes, 139, *see also* Individual entries
Neurospora crassa, 114, 252, 253
Nicotiana spp., *see also* Tobacco plants
chitinase transgene expression, 263
ß-1,3-glucanase
deficiency and virus susceptibility, 59
expression regulation, 61, 62, 63–64
structural classes, 50
hypersensitive reaction, 21–22
PR protein induction, 1, 2, 5, 9
PR-1 proteins, 22, 23, 25
serine proteinase isolation, 135
Nikkomycin, 114
Nitric oxide, 197
NLS, *see* Nuclear localization sequence
NMR, *see* Nuclear magnetic resonance
Nodulation factors, 96–97
Norbornadiene, 118, 120
NP24, 108, 114
NPR1 gene, 200, 201
Nuclear localization sequence (NLS), 201
Nuclear magnetic resonance (NMR), 175

O

OBFs, *see ocs* element binding factors
Oc-IΔD86, 139, *see also* Oryzacystatin
ocs element, 35
ocs element binding factors (OBFs), 64
Octadecanoid pathway, 221
OGA, *see* Oligogalacturonides
Okadaic acid, 118
Oligogalacturonides (OGA), 162, 163
Oligosaccharide elicitors, 57–58, *see also* Elicitors
Oligouronides, 144
OLP, *see* Osmotin-like protein
Oryza sativa, 28
Orzyacystatin, 137, 139, 141, 142, *see also* Proteinase inhibitors
Osmotic stress, 108, 117, *see also* Stress
Osmotin
abscisic acid induction, 117–118
accumulation in seedlings, 109
antifungal activity, 114, 115
disease resistance
extracellular targeting, 122
overexpression, 120–121
gene isolation, 119
plant stress induction, 10, 117
signal molecule induction, 118
thaumatin similarity, 108
Osmotin-like protein (OLP), 109, 120, 121
Osmotin-like protein (OLP) transgene, 265
Osmotin transgene, 265
Overexpression
PR genes in transgenic plants, 262–266
thionin, 222

P

PAP, *see* Pokeweed antiviral protein
Papillae, 159, 161, 219
Parsley, 196, 199–200
Pathogen virulence genes (*avr*) genes, 196
Pathogenesis-related protein (PR) genes
genetic studies on expression, 200–201
pathogens activated by different pathways, 197–198
signals and putative receptors that activate expression, 196–197
transcriptional regulation of expression, 198–200
transgenic plants
cosuppression, 268, 271
overexpression, 262–266
rice, 266–268, 269, 270
Pathogenesis-related proteins (PRs)
functions, 9–12
induction in pathological or related situations, 1–5
occurrence and properties, 5–9
PCD, *see* Programmed cell death
PDF1 genes, 201, 253, 254
Pea, 254
Pep25 elicitor, 200, *see also* Elicitors
Perenospora tabacina, 266
Permeability, membranes, 212
Peroxidase
characterization of plant, 173–181, 182
enzyme activity correlation with resistance, 172–173
gene expression, 183–185
lignin-forming and PR protein homologs, 7
organization of plant genes, 181, 183

potential role in pathogen defense in transgenic plants, 185–186
Peroxide, 27–28
Petunia, 53, 264
Phaseolus vulgaris, 161
Phenolic compounds, 172
Phospholipases, 213–214
Phospholipid binding site, 212, 213, 214, 217
Phylogenetic analysis, 24
Phylogenetic clustering, 175–181, 182
Phylogram, 237, 238
Phytoalexin, 158, 196
Phytohormones, 29–30, 35
Phytophthora spp.
function of PR proteins, 11
gene expression in PR proteins, 196
ß-1,3-glucanases, 58–59
hydroxyproline-rich glycoproteins, 162, 164
osmotin protein overexpression, 121
PR-1 proteins, 25
proteinase inhibitors, 136
transgenic plants, 264, 265, 266
PI, *see* Proteinase inhibitors
Pileup, 237, 238
Pinus taeda, 84
PIR proteins, 123–124
Plant peroxidase superfamily, 173–175, *see also* Peroxidase
Plant
–fungal interactions, 240–241
–insect interactions, 139–143
–microbe interactions, 136–139, 165
–viral interactions, 239–240
Plasmodiophora brassicae, 215
Plasmolysis, 115
Pokeweed antiviral protein (PAP), 236, 239–240, *see also* Ribosome-inactivating proteins
Poliovirus, 137
Pollen, 52–53
–pistil interactions, 256
Pollen tube growth, 53, 54
Polycations, 164
Polyphenoloxidase (PPO), 144
Pores, formation, 212
Potassium chloride, 252
Potato
proteinase inhibitors, 135, 139, 140, 141
plant defensin gene expression, 254
transgenic plants, 121, 265
Win protein structure, 87
Potato I inhibitors, 135, *see also* Serine proteinase
Potato II inhibitors, 135, *see also* Serine proteinase
Potato proteinase inhibitor gene, 269
Potato tuber lectin (PTL), 165, *see also* Solanaceous lectins
Potyvirus, 137
Powdery mildew, *see Erysiphe graminis*
PPO, *see* Polyphenoloxidase
PR-1 family
characterization, 22–25
expression/induction, 25–32
promoter analysis, 32–36
PR1 genes, 22–24, 200
PR-1-like proteins, 31
PR-11 family, 86–87, 94, *see also* Chitinases
PR-2 family
biological functions, 52–60
regulation, 60–65
structural classes and nomenclature, 50–52
PR-3 family, *see also* Chitinases
Chia genes structure and regulation, 94–95
protein structure, 79–84
PR-4 chitinases, 92
PR-4 family, *see also* Chitinases
Chid genes structure and regulation, 95
protein structure, 87–88, 89
structure, 88, 89
PR-5 domains, 123
PR-5 family
biological properties, 113–115
cDNAs and genes, 119–120
disease resistance and overexpression in transgenic plants, 120–122
do other defense proteins have PR-5 domains, 123
extracellular targeting, 122
occurrence, 108–109
physicochemical properties, 109–113
regulation of expression, 116–119
signal transudation pathway and osmotin interference, 123–124
PR-6 family
future prospects, 147
occurrence and structure of proteinase inhibitors with potential defensive capabilities, 133–136
proteinases and proteinase inhibitors plant–insect interactions, 139–143

plant–microbe interactions, 136–139
regulation of proteinase inhibitors as inducible plant defense proteins, 143–147
PR-8 family, 84–86, 94, *see also* Chitinases
Preproproteins, 215, 217, *see also* Thionins
PR-like protein (PRLs), 7–8, 50, 52, 53
PRLs, *see* PL-like proteins
PRms genes, 200
Processing, thionin, 215–218
Programmed cell death (PCD), 123–124, 237, 239, *see also* Suicide factor
Promoter analysis, 32–36
Prosystemin, 266
Prosystemin gene, 142
Protease, 97–98, 109
Protein bodies, 217
Protein kinase C, 30
Proteinase inhibitors (PI)
induction in infected plants, 136–137
isolation and function, 132–133
microbial proteinases and, 137–139
occurrence, structure, and defensive capabilities, 133–136
regulation and induction of plant defense proteins, 143–147
Proteolytic enzymes, 138
PR-P69, 146–147
Pseudolysozymes, 91
Pseudomonas spp.
gene expression in PR proteins, 197
ß 1,3-glucanase expression, 62, 64
proteinase inhibitors, 136, 145–146
peroxidase, 173
plant defensin induction, 254
thaumatin-like protein, 116
thionins, 219
hydroxyproline-rich glycoproteins, 164
Pti proteins, 199
PTL, *see* Potato tuber lectin
Pto genes, 199, 262
Puccinia graminis, 116, 172, 219
Purothionin, 208, 212, 214, 215, *see also* Thionin
Pyrenopeziza brassicae, 138
Pyrenophora teres, 219
Pyrularia pubera, 211
Pyrularia thionin, 213–214, *see also* Thionin
Pythium dimorphum, 254

R

Radish
plant defensin induction, 248, 253–254, 255
thionin isolation, 211
RC24 chitinase transgene, 267
RC7 chitinase transgene, 267–268
Reactive oxygen species (ROC), 27, 28, 31, 197
Red flour beetle, 141
Redox reactions, 214
Regulation
ß-1,3-glucanase expression
cis-acting elements, 63–64
developmental and hormonal, 60–61
pathogenesis-related, 62–63
signal transduction and *trans*-acting elements, 64–65
hydroxyproline-rich glycoproteins, 161–163
plant peroxidase genes, 184–185
ribosome-inactivating proteins, 237, 239
thionin, 220
Reporter-gene, *see* ß-Glucuronidase (GUS) reporter-gene
Reproduction
chitinase role, 12
ß-1,3-glucanase role, 52–56
Resistance (R) gene, 196, 197
Retaining mechanism, 89–91, *see also* Chitinases
Rhizobium spp., 162, 163, 173
Rhizoctonia spp.
overexpression of PR-5 proteins, 121–122
ribosome-inactivating proteins, 240, 241
thaumatin-like protein, 116
transgenic plants, 263, 264, 267–268
Rhynchosporium secalis, 219
Ribonucleases, 7
Ribosome-inactivating proteins (RIPs)
engineering chimeric and targeted, 242, 243
expression in transgenic plants and defense, 264–265
perspectives, 242
physiological roles and regulation, 237, 239
plant
–fungal interaction role, 240–241
–viral interaction role, 239–240

structural classification, 236–237, 238
Ribulosebisphosphate carboxylase (Rubisco), 2
Rice
class VII chitinase isolation, 83
ethylene-induced *PR-1* gene expression, 31
pathogen resistance, 172
proteinase inhibitor activity, 142
stress and peroxidase gene expression, 183
thaumatin-like protein induction, 116
transgenic plants, 121–122, 185–186, 266–268
Rice weevil, 141
Ricin, 236, *see also* Ribosome-inactivating proteins
RIP 30, 236, 240–241, 242, 243, *see also* Ribosome-inactivating proteins
RIPs, *see* Ribosome-inactivating proteins
ROC, *see* Reactive oxygen species
Rs-AFP1, 248, 250, 251, *see also* Defensins (plant)
Rs-AFP2, 251–252, 266, *see also* Defensins (plant)
Rubber tree, *see Hevea brasiliensis*
Rubisco, *see* Ribulosebisphosphate carboxylase

S

SA, *see* Salicylic acid
SA-binding protein (SABP2), 28
SABP2, *see* SA-binding protein
Saccharomyces cerevisiae, 11, 84
SAG, *see* Salicylic acid-ß-glucoside
Salicylic acid (SA)
gene expression in PR proteins, 197, 200, 201
ß-1,3-glucanase expression, 62–63
induction of PR protein in plants, 4–5
plant defensin gene expression, 254
PR-1 proteins
induction/expression, 25–29
promoter analysis, 32–35
proteinase inhibitor induction, 146
thaumatin-like protein induction, 118
thionin induction, 218
Salicylic acid-ß-glucoside (SAG), 26, 29
Salts, 115, 252
SAR, *see* Systemic acquired resistance
α-Sarcin, 236, *see also* Ribosome-inactivating proteins
Secondary messengers, 197
Seed germination
ß-1,3-glucanases, 55, 56
plant defensins, 255
thaumatin-like protein, 109
thionin, 217
Senescence, 32, 237, 239
Septoria nodorum, 219, 220
Serine proteinase, *see also* Proteinase inhibitors
antimicrobial activity, 138
digestive proteinases of insects and activity, 140–141
insecticidal activity of *Bacillus thuringiensis*, 143
mechanical wounding and induction, 140
occurrence and potential role in plants, 134–135
Serratia marcescens, 264
Sheath blight disease, *see Rhizoctonia solani*
Signal molecules, 10–11, 24, 118, 162
Signal transduction pathways
ß-1,3-glucanase expression and *trans*-acting elements, 64–65
induction of PR proteins in plants, 4–5
osmotin interference, 123–124
peroxidase gene regulation, 184–185
plant defensin gene expression, 254–255
proteinase inhibitor induction, 145–146
thionin genes, 220
Silent phenotypes, 271
Silver nitrate, 254
S-locus receptor kinase (SRK), 123
Solanaceous lectins, 165
Somatic embryogenesis, 12, *see also* Embryogenesis
Sorghum *see*ds, 256
Soybean, 57–58
Sp41 genes, 53–54
SPC, *see* Subtilisin-related proprotein convertases
Spodoptera spp., 25, 141, 142
Spores, 114, 115, 138
SRK, *see* S-locus receptor kinase
Stagonospora nodorum, 136
Stem borer, 142
Stem rust fungus, *see Puccinia graminis*
Sterility, 53

Stinging nettle, *see Urtica dioica*
Storage proteins, 215
Streptomyces griseus, 79, 82
Stress
hydroxyproline-rich glycoproteins induction, 162
induction of PR proteins, 9, 10
plant peroxidase gene expression/ regulation, 183–185
thaumatin-like protein accumulation, 108, 109
Structure
chitinases
family 18, 84
other related proteins, 87
PR-3 (family 19, genes *Chia*), 79–84
PR-4 family, 87–88, 89
PR-8/class III, 84–86
PR-11, 86–87
ß-1,3-glucanases classes, 50
hydroxyproline-rich glycoproteins, 158–159
plant defensins, 247–248, 249
activity relationships, 251–252
plant peroxidases, 173–175
ribosome-inactivating proteins, 236
thaumatin-like proteins, 109, 113
thionin, 211–212
Stylossanthes humilis, 184
Subtilisin-related proprotein convertases (SPC), 146–147
Sugar beet, 83, 139, 254, 263
Suicide factor, 235, *see also* Programmed cell death
Sweetness receptors, 113
swin1.1 genes, 140
Synergism, 139, 213–214, 240
Systematic acquired resistance (SAR)
ß-1,3-glucanase expression, 62, 63
pathogenesis-related proteins
relation, 4
virus infection, 7, 21
PR-1 gene induction/expression, 28, 30
Systemin, 144–145

T

Tag1 protein, 52
Taste, 113
TGA1a transcription factor, 35
Th1.2 gene, 255
Thaumatin-like protein, 12, 265–266, 268, 269, *see also* PR-5 family
Thermal denaturation, 214
Thi2 gene, 201, 220–222
Thionins
Arabidopsis spp., 220–222
barley leaf, 218–220
distribution, 208–211
function, 214–215
genetic engineering, 222
inclusion in PR protein family, 6
processing and localization, 215–218
PR protein activity, 9
structure, 211–212
synergism and proteinase inhibitor activity, 139
toxicity, 212–214
TIMP genes, 146
TMV, *see* Tobacco mosaic virus
Tobacco
chitinases and antifungal activity, 96
classification of PR proteins, 5
ß-1,3-glucanases and seed germination, 54, 55, 56
induction of PR proteins, 1–3
plant defensin induction, 254
proteinase inhibitors induction, 136
structure and regulation of genes, 94, 95
thaumatin-like proteins, 119
transgenic plants
chitinase antifungal activity, 263, 264
ß-1,3-glucanases, 58–59
ß-1,3-glucanase transgene expression, 264
hordiothionin function, 215
osmotin expression and defense, 265
overexpression of PR-5 proteins, 121
PR-1 and *Rs-AFP2* gene expression, 266
PR-1 proteins, 25, 32–33
proteinase inhibitor activity, 142–143
ribosome-inactivating proteins expression, 265
salicylic acid-induced *PR-1* gene activation, 28
viral infection and PR proteins, 22
Tobacco hornworm, *see Manduca sexta*
Tobacco mosaic virus (TMV)
antiviral activity of pokeweed antiviral protein, 240

ß-1,3-glucanases, 54, 59, 62
induction of PR protein, 1, 2
PR-1 proteins, 26, 32–35
proteinase inhibitors induction, 136
Tobacco ring spot virus, 116
Tomato
abscisic acid-induced accumulation of osmotin, 117
classification of PR proteins, 5, 7
gene expression of PR protein, 197
ß-1,3-glucanases, 58, 64–65
lignification, 172
PR-1 proteins, 25
proteinase inhibitors induction, 135, 136, 139, 144, 145
thaumatin-like protein induction and osmotic stress, 117
transgenic plants, 186, 264
Tomato moth, *see Lacanobia oleracea*
Tomato planta macho viroid (TPMV), 116
Toxicity
peroxidase and resistance correlation, 175
thionin, 208, 211, 212–214
TPMV, *see* Tomato planta macho viroid
trans-Acting elements, 64–65, *see also cis*-Acting elements
Transcription, 198
Transgene silencing, 271
Transgenic plants, *see also* Individual entries
abscisic acid-induced accumulation of osmotin, 117–118
antifungal activity
chitinases, 96
ribosome-inactivating proteins, 240–241
thaumatin-like proteins, 114
cosuppression of PR proteins, 268, 271
overexpression
pathogenesis-related proteins, 262–268
PR-5 proteins, 120–122
pokeweed antiviral protein and antiviral activity, 240
proteinase inhibitor activity and digestive proteinases of insects, 141–143
thionin induction, 221
Trees, 140, *see also* Wounds
Trichoderma spp., 95, 114, 240, 241
Trichosanthin, 236, *see also* Ribosome-inactivating proteins
Trypsin inhibitor, *see also* Proteinase inhibitors
dietary intake and biological cycle in lemmings, 143
insect reproduction, 142–143
plant–microbe interactions, 136, 138, 139
resistance in transgenic plants, 266
thaumatin-like proteins, 115
Tulipa gesneriana, 208
Turnip, 138
Tyrosine residues, 214

U

Ultraviolet (UV) light, 25, 31
Urtica dioica, 83, 96

V

Vacuoles, 24, 50, 217
Vegetative tissues, 256
Verticillium spp., 172, 184, 252
Vicia narbonensis, 87
Virus
ß-1,3-glucanase-deficient plants, 59–60
PR-1 proteins, 25, 29, 31
properties of PR proteins, 6–7
proteinase inhibitors, 137
solanaceous lectin accumulation, 165
thaumatin-like protein induction, 116
Viscotoxin, *see also* Thionin
isolation from mistletoe, 208, 210, 211
processing and localization, 217
thermal denaturation, 214
toxicity, 212
Viscum album, 208, 211
Volicitin, 145

W

wali genes, 140
W-box, 199–200
Western corn rootworm, 141
Wheat, 139, 140, 208
White clover, 173
White fly, 142–143
Win, *see* Wound-induced protein
win3 genes, 140

Wine, 97
Winter rye, 115
wip 1 genes, 140
Wound-induced (Win) protein, 87
Wounds
  hydroxyproline-rich glycoproteins induction, 162
  induction of PR proteins, 4–5
  osmotin induction, 118–119
  peroxidase induction, 173, 184
  PR-1 proteins induction/expression, 26
  proteinase inhibitor induction, 139–140, 143, 144, 145
  RC24 induction, 267
  solanaceous lectin accumulation, 165
  thionin induction, 219
WRKY genes, 200

X

Xa21 genes, 262
Xanthomonas spp.
  lignification, 172
  peroxidase gene regulation, 183–184, 185
  proteinase inhibitor activity, 137–138

Y

Yeasts, 212
Yeast–two hybrid assay, 199
ypr genes, 5, 8

Z

Zea mays, 162
Zeamatin, 113, 114